Immunological Challenges in Organ Transplantation

Georgi Abraham • Vivek Kute
Narayan Prasad • Dolly Daniel
Shruti Tapiawala
Editors

Immunological Challenges in Organ Transplantation

A Comprehensive Case-Based Approach

Springer

Editors
Georgi Abraham
The Madras Medical Mission Hospital
Chennai, Tamil Nadu, India

Narayan Prasad
Department of Nephrology
Sanjay Gandhi Inst. of Medical Sciences
Lucknow, Uttar Pradesh, India

Shruti Tapiawala
Institute of Renal Sciences
Gleneagles Hospitals
Mumbai, Maharashtra, India

Vivek Kute
Institute of Kidney Diseases
IKDRC-ITS
Ahmedabad, Gujarat, India

Dolly Daniel
Transfusion Medicine &
Immunohaematology
Christian Medical College
Vellore, Tamil Nadu, India

ISBN 978-981-95-1532-5   ISBN 978-981-95-1533-2 (eBook)
https://doi.org/10.1007/978-981-95-1533-2

This Springer imprint is published by the registered company Springer Nature Singapore Pte Ltd.
The registered company address is: 152 Beach Road, #21-01/04 Gateway East, Singapore 189721, Singapore

Foreword

I am delighted to write the foreword for this remarkable book, "Immunological Challenges in Organ Transplantation: A Comprehensive Case-Based Approach," edited by Georgi Abraham, Vivek Kute, Narayan Prasad, Dolly Daniel, and Shruti Tapiawala.

Transplantation immunology has come a long way since the pioneering insights of Sir Peter Medawar, who laid the foundation for our understanding of graft rejection and immune tolerance. Despite significant scientific advancements, a significant gap persists between evolving immunological knowledge and its practical application in clinical transplantation. Among the most formidable hurdles is the management of sensitized transplant recipients—a domain fraught with complexity and requiring nuanced, individualized, and tailored approaches.

This volume offers a timely and in-depth exploration of these challenges. Uniquely structured around case-based discussions, it brings together the expertise of accomplished contributors who address intricate topics such as pediatric transplantation, desensitization strategies, pharmacokinetics and pharmacodynamics of immunosuppressive agents, and ABO-incompatible transplantation. The case-based methodology enriches the content, bridging theory with real-world clinical scenarios and offering valuable insights to both seasoned practitioners and trainees.

The evolving landscape of transplant immunology demands continuous learning and adaptation. This book stands out as a comprehensive resource, not only for nephrologists and transplant surgeons but also for immunologists, intensivists, and allied healthcare professionals involved in transplantation. It is a timely contribution that meets the pressing need for accessible, evidence-informed guidance in a rapidly advancing field.

I highly commend the editors and authors for their dedication in producing a scholarly yet practical guide that will undoubtedly serve the transplant community well.

My sincere thanks and gratitude to all those who have contributed to this great book.

President Royal College of Physicians
London, UK
Manchester Royal Infirmary
Manchester, UK
NHS England, North West
Manchester, UK
University of Liverpool, School of Medicine
Liverpool, UK

Mumtaz Patel

Editor's Note

Transplantation is the best treatment for end-stage organ failure. The challenges in transplantation are enormous, especially in developing countries. The practice of immunological workup is not standardized and differs in different regions. The United States, China, and India are the three leading countries in the world in organ transplantation. In addition to procuring both live and deceased organs, transporting them, maintaining the viability, and successfully transplanting these organs require a multidisciplinary approach with the support of a well-experienced structured team.

Peter Medawar's groundbreaking work established the field of transplantation immunology and made organ transplantation possible. He discovered acquired immunological tolerance, demonstrating that the immune system can tolerate foreign tissue, and his research paved the way for successful organ transplants. This revolutionized medical practice and is now a cornerstone of transplantation medicine.

The contributors of this book "Immunological Challenges in Organ Transplantation: A Comprehensive Case-Based Approach" have experience garnered over many years by facing the immunological challenges with modern approach. We, the editors, express our gratitude to patients and colleagues who have been instrumental in our learning process.

We are thankful to Rene George, Renu Singh, Shruthi, Mrs Suhila Tapiawala, and Ranjan Daniel.

Chennai, India — Georgi Abraham
Ahmedabad, India — Vivek Kute
Lucknow, India — Narayan Prasad
Vellore, India — Dolly Daniel
Mumbai, India — Shruti Tapiawala

Contents

About the Editors

Georgi Abraham, MD, MRCP, FRCP is a practicing and academic kidney specialist who has published extensively on kidney disease, hemodialysis, peritoneal dialysis, and transplantation. He is the founder trustee of the Tanker Foundation, a charity for kidney disease patients. He has received several prestigious awards, including the International Distinguished Medal in Nephrology from the NKF, USA; D G Oreopoulos Lifetime Achievement Award for his contribution to peritoneal dialysis from the International Society for Peritoneal Dialysis; Roscoe Robinson Award from the International Society of Nephrology; and PACES Champion Award from the Federation of the Royal Colleges of Physicians of the UK. He has also published extensively on transplantation and immunosuppressive drugs with 360 publications, 34 book chapters, and many orations in India and abroad. He has served as the President of the Indian Society of Organ Transplantation from 2022 to 2023.

Vivek Kute MD, DM Nephrology (Gold Medal), FASN, and FRCP (London), is a Professor and Unit Head of Nephrology and Transplantation at the Institute of Kidney Diseases and Research Center, Dr. H L Trivedi Institute of Transplantation Sciences (IKDRC-ITS), Gujarat University of Transplantation Sciences (GUTS), and B J Medical College, Ahmedabad, India. He is the Treasurer and Former Secretary of the Indian Society of Organ Transplantation, Treasurer of the Transplantation Society, and a Member of the Finance Committee for WTC 2025 in San Francisco, USA, and the TTS 2026 Congress in Australia. IKDRC-ITS has completed 7200 kidney transplants, including 20% from deceased donors and 600 kidney exchanges. Dr. Kute's vision is to reach out to developing countries and enhance global alliances to improve transplant registry outcomes and access to transplantation and to help promote TTS, its mission, and its ethics in Asia and globally. Dr. Kute is ranked as the number one kidney transplant expert in India and among the top ten in Asia according to Expertscape's global expertise rankings. He has 250 peer-reviewed publications.

Narayan Prasad is Professor and Head of the Department of Nephrology and Renal Transplantation at a tertiary care institute, Sanjay Gandhi Postgraduate Institute of Medical Sciences, Lucknow, India. He is a clinician scientist at par excellence with in-depth basic science and clinical research in clinical nephrology,

dialysis, and transplantation. He has extensively researched in pharmacogenomics, MDR gene, permeable glycoprotein in drug resistance in glomerular disease, and transplantation. He has obtained one patent for basic science research. He is responsible for running transplantation immunogenetics laboratory and renal lab at the institute. The department is currently following 4000 kidney transplant recipients and performs 200 kidney transplantation each year with 20% high-risk ABO-incompatible kidney transplantation and highly sensitized HLA-incompatible transplantation. He has published 364 original research and review articles in journals of repute in the field. He has provided academic leadership, formerly as Councillor and Chair of South Asia regional board, of the International Society of Nephrology, and was local organizing co-chair of the World Congress of Nephrology, 2025, held at New Delhi. He is currently a council member of the Asia Pacific Society of Nephrology. He has formerly provided leadership as an Honorary Secretary of the Indian Society of Nephrology, Indian Society of Organ Transplantation, and Peritoneal Dialysis Society of India and is currently the Secretary of the Society of Renal Nutrition and Metabolism. He has been recognized among the 2% scientists of the world in Stanford University Survey conducted for the last 5 years. He is also a member of Global Burden of Disease of Public Health Foundation. He actively participates in the evaluation of research grants for the Indian Council of Medical Research and brainstorming session at the Directorate of Science and Technology (DST) and Directorate of Biotechnology (DBT), Ministry of Science and Technology, Government of India. He has also contributed as a member of oversight committee of Prime Minister National Dialysis Program, National Health Mission, Government of India.

Dolly Daniel, MD, FRCP is a Professor and Head of the Department of Transfusion Medicine and Immunohematology at Christian Medical College, Vellore. A pathologist by qualification, she branched into transplant immunology, histocompatibility, and immunogenetics. The gap in the field of transplant immunology at the time was a driving force that spurred her growth. In charge of the laboratory since 1995, she set up molecular HLA typing after training at the Royal London Hospital. Since then, platforms including NGS and multiple platforms for anti-HLA and non-HLA antibody testing have been introduced, and a very strong clinical laboratory interface has been built. Their publications span disease associations, estimating the likelihood of finding a matched donor, algorithms for swap transplants, and clinical applicability of antibody testing assays. She is guiding a PG thesis on epitope analysis in renal transplant and was the Co-PI of a recently concluded Indo-Japan project, "Haplobanking," creating a bank of iPSCs from individuals with homozygous HLA haplotypes. She coordinates the 1-year postdoctoral fellowship in transplant immunology, histocompatibility, and immunogenetics at CMC Vellore.

Shruti Tapiawala MD in Internal Medicine and DrNB in Nephrology, is a senior consultant in the Department of Nephrology at Gleneagles Hospitals and Sir HN Reliance Foundation Hospital, Mumbai. She is also a laboratory director at ATDI Labs Private Ltd., Mumbai, which is a specialized transplant immunology and histocompatibility laboratory set up at international standards. She is a nephrologist and transplant physician by qualification trained in the management of multi-organ transplants, HLA medicine, and advanced diagnostics in transplant, making her one of her kinds in the field of nephrology to have established a transplant immunology laboratory catering to state-of-the-art technology for evaluation and monitoring of transplant patients providing a composite analysis. She has established SORTER—a one-of-its-kind acceptable mismatch registry to help sensitized patients find HLA-compatible donors since 2018. She has also numerous chapters and peer-reviewed publications to her credit. She is the course coordinator for transplant immunology simplified biannual course designed for clinicians in transplant to understand the application of transplant immunology in day-to-day practice. She is the President of the Indian Society of Nephrology—West Zone chapter—and is a council member of Women in Nephrology and Women in Transplantation.

HLA Structure, Function, and Clinical Implications in Transplantation: A Case-Based and Molecular Perspective

1

Sathya Selvarajan, Jyothipriya Jyothindrakumar, Milly Mathew, and Georgi Abraham

Case Report: Acute Tubular Necrosis and Suspected T-Cell-Mediated Rejection

A 54-year-old non-diabetic male underwent a live-related, ABO-compatible kidney transplant on March 3, 2025. The donor was his son, with one HLA-B mismatch and a negative CDC crossmatch. The transplant was technically challenging due to the donor kidney having three renal arteries and one vein (Fig. 1.1). Laparoscopic left nephrectomy was performed, with a cold ischemia time of 35 min and warm ischemia time of 45 min. All three arteries were anastomosed separately to the external iliac artery. The patient experienced intraoperative hypotension requiring norepinephrine support. Immunosuppression included a single intraoperative dose of thymoglobulin (75 mg) and methylprednisolone (750 mg), followed by maintenance therapy with prednisolone 35 mg daily and mycophenolate sodium 720 mg twice daily.

Postoperatively, the graft showed good perfusion and immediate urine output. On the first postoperative day, fluid intake was 6207 mL, and urine output was 3445 mL. However, serum creatinine remained elevated at 7.1 mg/dL, and urine output declined over the following days, consistent with non-oliguric acute tubular necrosis (ATN). Thrombocytopenia (platelets 40,000–50,000/μL) precluded allograft biopsy, raising suspicion of ITP. Daily Doppler ultrasound confirmed good

S. Selvarajan (✉)
Laboratory Medicine, Clinical Biochemistry, MGM Healthcare, Chennai, Tamil Nadu, India

J. Jyothindrakumar
MGM Muthoot Hospital, Pathanamthitta, Kerala, India

M. Mathew · G. Abraham
The Madras Medical Mission Hospital, Chennai, Tamil Nadu, India

G. Abraham et al. (eds.), *Immunological Challenges in Organ Transplantation*,
https://doi.org/10.1007/978-981-95-1533-2_1

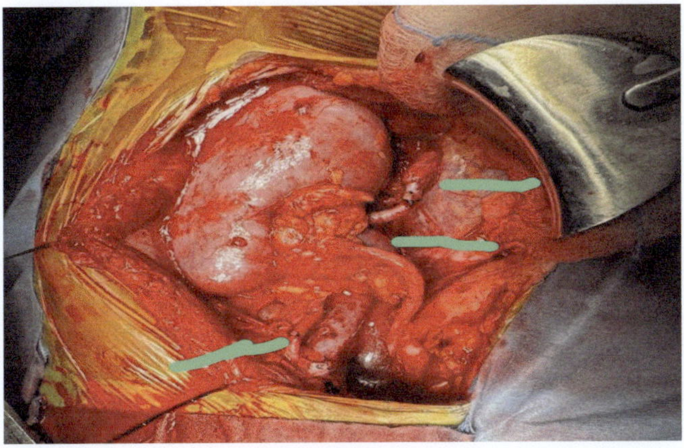

Fig. 1.1 Showing reddish kidney, three renal arteries indicated with green arrows

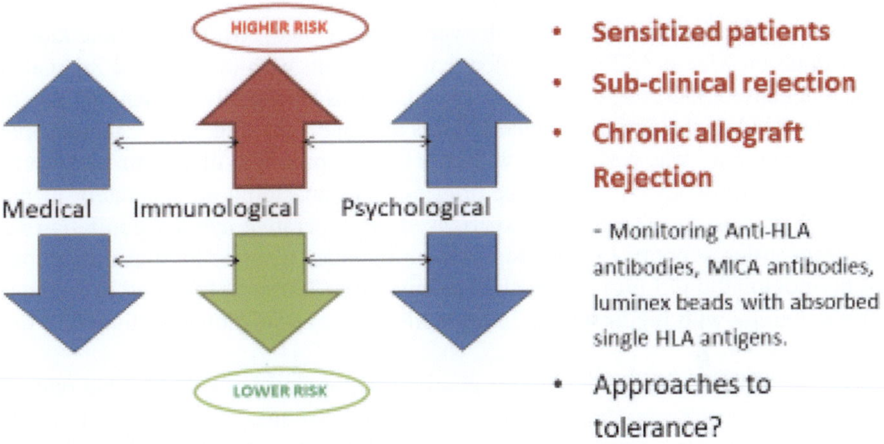

Fig. 1.2 Challenges in transplantation are depicted in Fig. 1.2

flow in all renal arteries and the vein. Despite diuresis (800–1200 mL/day with furosemide), serum creatinine remained between 5 and 6 mg/dL. On postoperative day 7, in the absence of renal function improvement, a provisional diagnosis of T-cell mediated rejection (TCMR) was made. Urine microscopy showed clusters of tubular epithelial cells. In the absence of biopsy, empirical therapy with two doses of intravenous methylprednisolone (500 mg each) was administered. The patient required three hemodialysis sessions. Subsequently, renal function improved steadily, and by day 14, serum creatinine declined to 0.9 mg/dL, indicating recovery from ATN. The patient continues on triple immunosuppressive therapy.

Acute tubular necrosis (ATN) and graft rejection in kidney transplantation share converging molecular pathways (Fig. 1.2), predominantly involving ischemia-reperfusion injury, oxidative stress, and immune-mediated inflammation. ATN is a common

precursor to delayed graft function and is marked by tubular epithelial apoptosis, mitochondrial dysfunction, and elevated pro-inflammatory cytokines such as TNF-α and IL-6. Recent studies also implicate nitric oxide dysregulation and hormonal imbalances—particularly altered ghrelin signaling—in exacerbating renal injury [1].

Graft rejection involves both T-cell and antibody-mediated immune mechanisms leading to endothelial injury and fibrosis. Notably, ATN may enhance alloimmune activation by upregulating class II HLA molecules, especially HLA-DR, along with ICAM-1 expression on renal tubular cells following ischemic insult [2]. This promotes CD4+ T-cell recruitment and cytokine signaling, including IL-2, contributing to rejection risk [3]. Additionally, the immunomodulatory HLA-G is upregulated in ATN and has been associated with acute rejection episodes [4]. At the molecular level, rejection is further driven by increased expression of endothelial adhesion molecules, costimulatory receptors such as CCR7, and enhanced antigen presentation capacity [5].

Understanding these shared biochemical cascades is essential for the development of targeted strategies aimed at reducing kidney injury and improving long-term graft survival.

Human Leukocyte Antigen (HLA): Biochemistry, Function, and Relevance in Nephrology

Introduction

The Human Leukocyte Antigen (HLA) system, encoded on chromosome 6 (Fig. 1.3), plays a pivotal role in immune regulation, particularly in antigen presentation, self/non-self recognition, and immune tolerance. For nephrologists, the HLA system is essential in contexts such as renal transplantation, autoimmune nephropathies, and immune-mediated kidney injury.

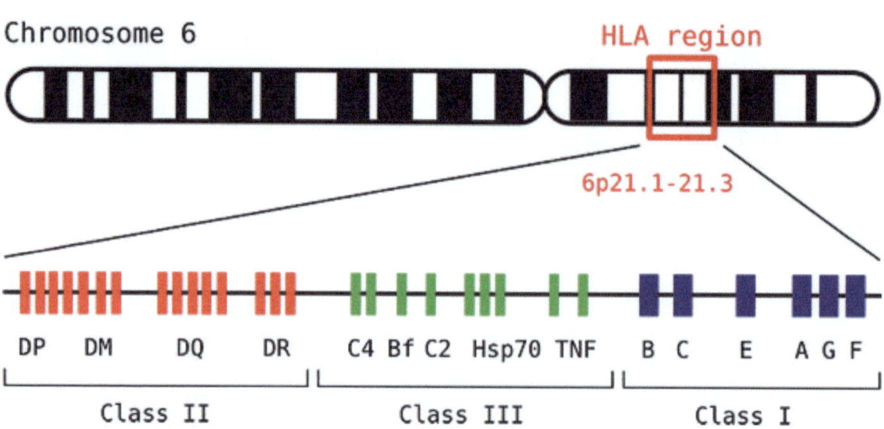

Fig. 1.3 HLA gene complex on Chromosome 6

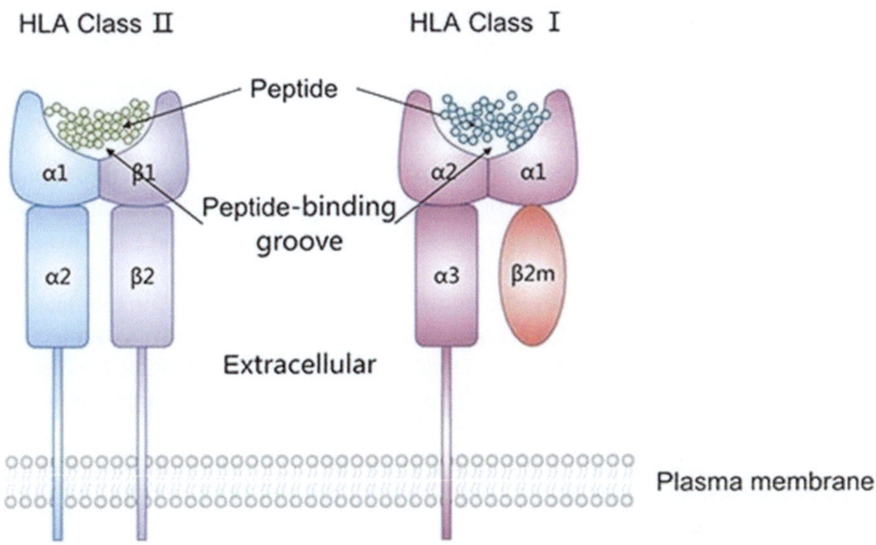

Fig. 1.4 Structure of HLA Class I and II molecules

As of December 2024, the IPD-IMGT/HLA Database has cataloged 38,008 HLA alleles, including 26,610 Class I and 11,398 Class II alleles—a significant increase from 24,093 in 2019—driven by advances in next-generation sequencing (NGS) technologies [6].

HLA Gene Organization and Biochemical Classification

HLA genes are part of the Major Histocompatibility Complex (MHC) and are categorized into:

Class I (HLA-A, -B, -C): Present endogenous antigens to CD8+ T-cells
Class II (HLA-DR, -DQ, -DP): Present exogenous antigens to CD4+ T-helper cells
Class III genes: Encode complement proteins (C2, C4, Bf) and inflammatory mediators like TNF-α [7]

HLA Class I Molecules

Structure and Function

HLA Class I molecules are 57 kDa glycoproteins composed of a heavy α-chain (~45 kDa) and β2-microglobulin (~12 kDa). The peptide-binding groove is formed by α1 and α2 domains, while α3 interacts with β2-microglobulin, conferring structural stability (Fig. 1.4) [8, 9].

These molecules present 8–10 amino acid-long intracellular peptides (from self, viral, or tumor proteins) to CD8+ cytotoxic T-cells, initiating immune responses [10]. They are present on all nucleated cells, excluding mature red blood cells, and can exist in soluble form in plasma or platelets [11].

Antigen Presentation Pathway

Antigen processing begins with the proteasome degrading intracellular proteins into peptide fragments [12]. These peptides are then transported into the endoplasmic reticulum (ER) by Transporter Associated with Antigen Processing (TAP) proteins [13]. Within the ER, peptide loading onto MHC class I molecules is facilitated by chaperones such as calnexin and tapasin. The stable peptide-MHC complexes are subsequently trafficked through the Golgi apparatus to the cell surface [14]. Once presented on the membrane, these complexes are recognized by $CD8^+$ T cells, initiating a cytotoxic immune response [15].

Clinical Relevance

Immune evasion is a common strategy employed by pathogens such as human cytomegalovirus (HCMV), which suppress HLA class I expression to avoid detection by cytotoxic T cells [10]. In oncology, tumors can similarly evade immune surveillance by downregulating HLA class I molecules; therapeutic strategies aimed at restoring their expression have shown promise in enhancing anti-tumor immunity [16]. Furthermore, genetic polymorphisms—such as HLA-C*12—have been associated with increased susceptibility to autoimmune diseases and infectious agents, highlighting the complex role of HLA diversity in immune regulation [17].

HLA Class II Molecules

Structure and Antigen Presentation

HLA Class II molecules are ~63 kDa heterodimers composed of α (~33 kDa) and β (~30 kDa) chains, expressed on APCs such as dendritic cells and B-cells. The α1 and β1 domains form a peptide-binding groove that accommodates 12–25 amino acid peptides (Fig. 1.4) [18].

These molecules bind exogenous peptides in endosomes/lysosomes, facilitated by HLA-DM after removing the invariant chain (Ii). The resulting complex is transported to the cell surface, where CD4+ T-helper cells recognize the antigen and activate adaptive responses [19, 20].

Polymorphisms and Autoimmunity

Polymorphisms in HLA-DRB1 and HLA-DQ alter peptide affinity and are associated with diseases such as rheumatoid arthritis and type 1 diabetes [21–23]. Structural modeling has advanced understanding of HLA-DR antigenic specificity and therapeutic potential [24].

Post-Translational Modifications (PTMs) of HLA

Post-translational modifications—including glycosylation, phosphorylation, and ubiquitination—play critical roles in regulating the stability, trafficking, and immunological functions of HLA molecules [25, 26]. These modifications directly influence how HLAs interact with immune components such as natural killer (NK) cells and modulate immune recognition.

Glycosylation enhances the stability and surface expression of HLA molecules, thereby improving peptide presentation and efficient intracellular transport.

Phosphorylation modulates intracellular signaling cascades, influencing immune cell communication and HLA functional dynamics.

Ubiquitination serves as a regulatory mechanism to tag excess or misfolded HLA molecules for degradation via the proteasome, maintaining protein quality control [27].

Together, these PTMs fine-tune the immune system's ability to detect and respond to both self and non-self antigens.

Natural Killer (NK) Cells and HLA Interactions

Natural killer (NK) cells play a pivotal role in immune surveillance, particularly in identifying and eliminating cells that have downregulated HLA class I molecules—often a sign of viral infection or tumor transformation. NK cell activity is tightly regulated by a balance of signals from activating and inhibitory receptors that interact with specific HLA class I ligands (Table 1.1).

Polymorphisms within HLA-C, a key ligand for KIRs, have profound effects on NK cell behavior. These genetic variations influence immune responses in diverse contexts such as organ transplantation, viral infections, and autoimmune diseases by altering the threshold for NK cell activation and inhibition [28, 29].

MICA/MICB and NK Activation

MHC class I chain-related proteins A and B (MICA, MICB) are stress-inducible ligands expressed on epithelial and immune cells. Unlike classical HLA, they do not bind β2-microglobulin and interact directly with NKG2D receptors, activating NK cells and cytotoxic T-cells [30]. Variants in MICA/MICB are associated with cancer susceptibility and immune escape [31].

Table 1.1 Key NK cell receptors and their HLA Class I ligands

Receptor	Function	HLA ligands
KIRs	Inhibit or activate NK cytotoxicity	HLA-A, HLA-B, HLA-C
CD94/NKG2A	Inhibitory receptor	HLA-E
NKG2D	Activating receptor	MICA, MICB

Table 1.2 Key proteins involved in HLA antigen processing

Protein	Function
Proteasome	Degrades cytosolic proteins into peptide fragments
TAP	Transports peptides from the cytosol into the ER
Calnexin, Tapasin	Chaperones involved in stabilizing and loading peptides onto HLA class I molecules
Invariant chain (ii)	Prevents premature peptide binding to HLA class II in the ER
HLA-DM	Catalyzes peptide exchange on HLA class II, replacing CLIP with antigenic peptides

Key Proteins in HLA Antigen Processing

The processing and presentation of antigens by HLA molecules involve a coordinated network of proteins that ensure proper peptide generation, transport, and loading onto HLA class I and class II molecules. These proteins are essential for efficient immune recognition and response. These key components and their respective functions are summarized in Table 1.2.

These molecular components work in concert to ensure that peptides are correctly processed and presented to T cells, thereby enabling effective immune surveillance by both $CD8^+$ and $CD4^+$ T lymphocytes.

Cross-Reactive Groups (CREGs) in HLA Typing

Cross-reactive groups (CREGs) are clusters of HLA antigens sharing common epitopes and serologic reactivity. Their identification has improved HLA matching in organ transplantation, reducing rejection and improving graft survival [32].

The HLA system is central to immune recognition, transplant compatibility, and autoimmune disease pathogenesis. Its molecular diversity, structural complexity, and immune interactions—especially in kidney transplantation—demand detailed understanding by nephrologists to optimize patient outcomes.

Genetic Organization and Function of HLA

The HLA region is categorized into three major classes:

- Class I (HLA-A, -B, -C) genes encode the α-chain of HLA molecules that present endogenous antigens to CD8+ T cells. Non-classical Class I molecules such as HLA-G play roles in immune tolerance, particularly in pregnancy [33].
- Class II (HLA-DR, -DQ, -DP) genes encode both α- and β-chains and present extracellular antigens to CD4+ T cells. Their polymorphism is central in autoimmune diseases and transplant rejection [34].
- Class III encodes immune-regulatory proteins like complement factors (C2, C4) and TNF, although not directly involved in antigen presentation [35].

HLA Polymorphism and Evolutionary Significance

HLA genes represent the most polymorphic region of the human genome, a characteristic driven by evolutionary pressures to enhance immune adaptability. This diversity is maintained and expanded through several genetic mechanisms:

- Point mutations that subtly alter the peptide-binding groove, thereby modifying antigen presentation capabilities.
- Gene conversion events, which recombine segments of different alleles to generate novel HLA variants.
- Balancing selection, a key evolutionary force that preserves multiple alleles within populations, ensuring broader protection against a rapidly evolving array of pathogens [34, 36].

This extensive polymorphism serves as a double-edged sword: while it bolsters population-level immunity, it also introduces clinical challenges in areas like organ transplantation compatibility and autoimmune disease susceptibility, where specific HLA variants can have pathogenic roles or influence graft rejection risk.

HLA Haplotypes and Linkage Disequilibrium

HLA alleles are inherited as haplotypes due to strong linkage disequilibrium. This genetic arrangement impacts transplantation compatibility and disease associations. For example, specific haplotypes have been found to correlate with higher risk of autoimmune diseases and are used in population-level studies such as the Iranian cohort for transplantation matching [36].

Clinical Applications in Transplantation

Accurate HLA matching remains a cornerstone of successful organ transplantation, as it significantly reduces the risk of graft rejection and enhances long-term survival. The degree to which HLA compatibility influences outcomes varies by organ type:

- Kidney transplantation is highly dependent on matching HLA-A, HLA-B, and HLA-DR alleles, with mismatches being strongly associated with increased rejection risk and reduced graft longevity [37].
- Heart and lung transplants benefit from HLA compatibility as well, although the impact is generally less pronounced than in renal transplants.
- Liver transplantation is relatively tolerant of HLA mismatches, owing to the liver's unique immunological environment and regulatory properties, making HLA matching less critical in this context.

Table 1.3 HLA alleles associated with autoimmune nephropathies

Autoimmune kidney disease	Associated HLA alleles
Lupus nephritis	HLA-DR3, HLA-DR4
IgA nephropathy	HLA-DQ2, HLA-DQ8

Table 1.4 HLA alleles associated with drug-induced toxicity

Drug	Associated HLA allele	Clinical risk
Flucloxacillin	HLA-B*57:01	Nephrotoxicity
Allopurinol	HLA-B*58:01	Severe cutaneous adverse reactions

The advent of next-generation sequencing (NGS) has revolutionized HLA typing by offering high-resolution allele-level analysis. This precision allows for more refined donor-recipient matching, leading to improved clinical outcomes and expanded donor pool utilization [35].

Autoimmune Kidney Diseases and HLA Association

Certain autoimmune kidney diseases are strongly associated with specific HLA alleles (Table 1.3). These associations are believed to result from altered peptide presentation and molecular mimicry, mechanisms by which particular HLA variants facilitate T-cell activation against self-antigens, thereby promoting autoimmunity [36].

HLA and Drug-Induced Nephrotoxicity

Certain HLA alleles are known to predispose individuals to drug-induced adverse reactions, including nephrotoxicity and severe hypersensitivity syndromes. These findings underscore the value of pharmacogenetic screening in clinical nephrology, particularly in ethnically predisposed populations, where identifying high-risk alleles can significantly reduce drug-related complications. The key HLA-drug risk associations are summarized in Table 1.4 [33, 35].

Pharmacogenetic screening can prevent drug hypersensitivity, especially in populations with high-risk alleles [33, 35].

Immune Tolerance and Thymic HLA Expression

Within the thymus, HLA molecules play a central role in shaping the adaptive immune repertoire by directing two critical selection processes:

- Positive selection, which ensures the survival of T-cells capable of recognizing self-MHC molecules.
- Negative selection, which eliminates autoreactive T-cells that could initiate immune responses against self-antigens.

Disruption in these processes—often resulting from HLA polymorphisms—can compromise central tolerance and contribute to the development of autoimmune pathologies, including autoimmune kidney diseases [35].

HLA polymorphism represents a genetic foundation of immune variability. It underlies key clinical phenomena such as:

- Immune diversity in pathogen recognition
- Allograft compatibility in transplantation
- And susceptibility to autoimmune conditions like lupus nephritis or IgA nephropathy

The integration of high-resolution HLA genotyping into routine clinical workflows has significantly enhanced our ability to:

- Clarify disease mechanisms involving HLA-restricted antigen presentation
- Improve organ transplantation outcomes through precise donor-recipient matching
- And anticipate adverse drug reactions via preemptive pharmacogenetic screening

HLA Polymorphism and Evolutionary Significance

The HLA gene complex is the most polymorphic region of the human genome, a feature critical to the adaptability of the immune system. This extraordinary diversity arises from several molecular and evolutionary mechanisms:

- Point mutations that alter the peptide-binding groove, modifying the repertoire of presented antigens.
- Gene conversions, which shuffle genetic segments between alleles, giving rise to novel HLA variants.
- Balancing selection, an evolutionary force that maintains high allelic diversity to enhance population-level resistance to a wide array of pathogens [34, 36].

This rich polymorphism provides a survival advantage by enhancing immune recognition across a diverse pathogen landscape. However, it also introduces clinical complexity in:

- Solid organ transplantation, where HLA mismatches can trigger alloimmune responses, and
- Autoimmune kidney diseases, where certain HLA alleles predispose individuals to immune dysregulation

Thus, understanding HLA diversity is essential not only for its evolutionary implications but also for improving clinical outcomes in nephrology through better donor matching and individualized immunogenetic risk assessment.

HLA Haplotypes and Linkage Disequilibrium

HLA alleles are often inherited as haplotypes—clusters of alleles transmitted together—due to strong linkage disequilibrium (LD) within the major histocompatibility complex (MHC) region. This non-random association of alleles affects both transplant compatibility and disease susceptibility.

Specific HLA haplotypes have been shown to correlate with an increased risk of autoimmune diseases, and are crucial in population-level genetic studies. For instance, in the Iranian transplantation cohort, particular haplotypes are used to optimize donor matching strategies and investigate HLA-disease associations [36].

Clinical Applications in Transplantation

Effective HLA matching plays a pivotal role in improving graft survival rates, particularly in kidney transplantation:

- Kidney transplants exhibit the strongest dependence on compatibility at HLA-A, HLA-B, and HLA-DR loci, where mismatches significantly affect long-term graft function and rejection risk [37].
- Heart and lung transplants benefit from HLA matching to a lesser extent but still show improved outcomes with fewer mismatches.
- Liver transplants are generally more tolerant to HLA disparities due to the liver's intrinsic immune-regulatory properties, making matching less critical.

The advent of next-generation sequencing (NGS) has revolutionized HLA typing, enabling high-resolution allele-level identification. This advancement has led to:

- More precise donor-recipient matching
- Improved understanding of minor histocompatibility antigens, and
- Enhanced outcomes across multiple organ transplant types [35]

Immune Tolerance and Thymic HLA Expression

Within the thymus, HLA molecules are central to establishing central immune tolerance through:

- Positive selection, which promotes survival of T-cells capable of recognizing self-MHC molecules, and
- Negative selection, which eliminates autoreactive T-cells that could target self-tissues

Disruptions in these selection processes—often due to HLA polymorphisms—can impair self-tolerance and contribute to the onset of autoimmune kidney diseases

[35]. This fundamental mechanism underscores the importance of thymic HLA expression in maintaining immune homeostasis.

Conclusion

HLA polymorphism represents a foundational element of immune diversity, influencing not only pathogen defense and transplant compatibility, but also susceptibility to autoimmune kidney diseases. Advances in genotyping technologies, including high-resolution sequencing, have greatly improved our understanding of HLA-associated mechanisms.

Clinically, these insights have translated into:

- Improved transplant outcomes through better HLA matching
- Reduced drug hypersensitivity via pharmacogenetic screening
- And enhanced predictive capabilities for autoimmune risk profiling

As precision medicine continues to evolve, integrating HLA analysis into routine nephrology practice offers a promising avenue for more personalized, immunogenetically informed care.

Transplant Immunology Physiology: Innate Immunity Activation and T Cell Priming Post-Kidney Transplantation

Kidney transplantation triggers a complex immunological cascade involving both innate and adaptive immune responses. The earliest events are initiated by ischemia-reperfusion injury (IRI), which activates the innate immune system and creates a pro-inflammatory environment that primes T cell-mediated adaptive immunity. The subsequent Signal 1 to Signal 4 paradigm of T cell activation forms the foundation for graft survival or rejection.

Innate Immune Activation and Ischemia-Reperfusion Injury (IRI)

IRI is a universal and unavoidable component of kidney transplantation and unfolds in two critical phases:

Ischemia Phase

During cold storage:

- ATP depletion, shift to anaerobic glycolysis, and ROS accumulation.
- Cellular stress in tubular epithelial and endothelial cells increases susceptibility to injury [38].

Reperfusion Phase

Upon revascularization:

- A surge in oxygen generates a burst of reactive oxygen species (ROS), damaging mitochondria.
- Injured cells release damage-associated molecular patterns (DAMPs) such as HMGB1 and heat shock proteins, activating pattern recognition receptors (PRRs) and initiating inflammation [39].

Innate Immune Mechanisms in IRI

Pattern Recognition Receptors (PRRs)

- TLRs (e.g., TLR2, TLR4) on macrophages and endothelial cells detect extracellular DAMPs.
- NLRs (e.g., NLRP3) form inflammasomes in response to intracellular stress.
- RLRs recognize damaged RNA and amplify inflammation [39].

Complement Activation

- Triggered by ischemia via classical and alternative pathways.
- C3b facilitates opsonization; C5b-9 (MAC) causes lysis and inflammation [40].

Myeloid and NK Cell Recruitment

- Macrophages secrete TNF-α, IL-1β, IL-6.
- Neutrophils release ROS and proteases.
- Dendritic cells (DCs) bridge innate and adaptive immunity.
- NK cells enhance graft cytotoxicity.

Platelet-Endothelium Interactions

- Platelet aggregation and endothelial activation promote thromboinflammation, exacerbating perfusion deficits.

Inflammatory Mediators

- ROS induce apoptosis and necrosis.
- Cytokines (TNF-α, IL-1β, IL-6, IL-8) recruit leukocytes.
- Adhesion molecules (ICAM-1, VCAM-1, selectins) promote infiltration.
- NLRP3 inflammasome drives IL-1β and IL-18 release [39].

Clinical Consequences of Innate Activation

- Acute Tubular Necrosis (ATN) leading to Delayed Graft Function (DGF) [41].
- Endothelial dysfunction triggering thrombosis and immune infiltration.
- Adaptive immune priming via activated APCs.
- Chronic inflammation drives Interstitial Fibrosis and Tubular Atrophy (IFTA).

Emerging Therapies for IRI (Table 1.5 Summarizes Key Interventions [42])

Signal 1: Antigen Recognition

Signal 1 initiates T cell activation through recognition of donor HLA molecules:

- HLA Class I (A, B, C): Present peptides to CD8+ T cells.
- HLA Class II (DR, DQ, DP): Present to CD4+ T cells.

Antigen Presentation Pathways

- Direct allorecognition: Recipient T cells recognize intact donor MHC on donor APCs—potent in acute rejection.
- Indirect pathway: Recipient APCs present processed donor peptides—central to chronic rejection.
- Semi-direct pathway: Recipient APCs acquire donor MHC via exosomes [43, 44].

TCR Signaling

- Engagement of the TCR-CD3 complex activates Lck, ZAP-70, and transcription factors like NF-κB, NFAT, and AP-1.
- Signal 1 alone is insufficient for full T cell activation [43].

Table 1.5 Therapies for IRI

Strategy	Example agents
Antioxidants	MitoQ, Bardoxolone (Nrf2 activator)
Complement inhibitors	Zilucoplan
PRR modulators	TLR4 antagonists, NLRP3 inhibitors
Cellular therapies	MSCs, exosomes
Cytokine blockers	Anakinra (IL-1), Tocilizumab (IL-6)
HIF stabilizers	Roxadustat
MicroRNA modulators	miR-21, miR-155
Targeted delivery systems	Nanoparticles

Table 1.6 Key costimulatory pathways

Pathway	Function
CD28-B7	Promotes IL-2 production, proliferation
ICOS-ICOSL	Supports effector and memory T cells
CTLA-4, PD-1	Inhibitory, essential for tolerance

Signal 2: Costimulation

Signal 2 provides a secondary signal critical for T cell fate—activation or tolerance. Key costimulatory pathways are summarized in Table 1.6 [43].

Intracellular Cascades

- PI3K/Akt pathway: Enhances T cell survival via Bcl-2.
- Calcineurin-NFAT pathway: Induces IL-2 transcription.
- Calcium influx through CRAC channels triggers NFAT activation.

Clinical Relevance

- Acute rejection: Dominated by CD28-B7 signaling.
- Chronic rejection: Sustained costimulation drives fibrosis and vasculopathy.
- Tolerance: Loss of CTLA-4/PD-1 signaling leads to rejection.

Therapeutic Modulation

- Belatacept (CTLA-4-Ig): Blocks CD28; approved for kidney transplantation.
- PD-1 agonists: Promote T cell exhaustion and immune tolerance.
- Combination therapy: Tacrolimus + Belatacept enhances graft survival [42].

Signal 3: Cytokine Production and T Cell Differentiation

Signal 3 shapes T cell fate toward inflammatory or regulatory phenotypes.

Key Cytokines
- IL-2: Promotes T cell expansion [43].
- IL-12/IFN-γ: Drive Th1 differentiation.
- IL-4: Facilitates Th2 and humoral immunity.
- IL-17: Promotes fibrosis and chronic rejection.
- IL-10, TGF-β: Induce Tregs and maintain tolerance [45].

T Helper Subsets and Roles *(Summarized in* Table 1.7*)*.

Cytokine Receptors and Therapeutic Targets
- IL-2R (CD25): Blocked by Basiliximab.
- IL-12/23R: Direct Th1/Th17 polarization.
- mTORC1 pathway: Key for proliferation; inhibited by rapamycin to promote Tregs [42].

Signal 3 therapies are summarized in Table 1.8.

Signal 4: Effector Functions and Graft Rejection

Signal 4 represents the immune response's execution phase, culminating in graft injury or tolerance.

Table 1.7 T helper subset and roles

Subset	Function
Th1	Cytotoxicity, acute rejection
Th2	B cell help, antibody-mediated rejection (AMR)
Th17	Fibrosis, chronic rejection
Tregs	Immune regulation, tolerance

Table 1.8 Signal 3 Therapies

Agent	Target
Basiliximab	IL-2Rα
Tocilizumab	IL-6R
Secukinumab	IL-17A
IFN-γ blockade	Acute inflammation
TGF-β modulators	Fibrosis prevention
JAK inhibitors	Downstream cytokine pathways

Table 1.9 Signal 4 therapies

Strategy	Example agents
Cytokine inhibitors	Anti-IFN-γ, anti-IL-17
T/B cell depletion	Alemtuzumab, rituximab
IVIg/ plasmapheresis	DSA clearance
Belatacept	Prevents upstream activation

CD4+ T Cell Responses

- Th1: IFN-γ activates macrophages and cytotoxicity [46].
- Th2: Induces DSA formation → AMR [47].
- Th17: Drives fibrosis and neutrophil recruitment [45].
- Tregs: Suppress inflammation via IL-10, TGF-β.

CD8+ Cytotoxic T Cells

- Recognize donor MHC I and initiate:
- Perforin + Granzyme B pathway.
- Fas-FasL → caspase-mediated apoptosis [46].

Humoral Immunity and AMR

- Tfh and Th2 cells aid in DSA generation.
- DSAs bind to graft endothelium, activate complement (C1q, C4d), leading to microvascular injury [47].

Amplification by Inflammatory Mediators

- T cells secrete CXCL9, CXCL10, CCL2, driving immune recruitment.
- TNF-α and IL-6 exacerbate inflammation [43, 46].

Signal 4 therapies are summarized in Table 1.9.

Clinical Implications

- Biomarkers: IFN-γ, IL-17, and CD25 levels guide immunosuppressive therapy.
- Personalized Medicine: Signal 3/4-targeted agents offer precision immunosuppression.
- Tolerance Strategies: Enhancing Treg expansion may reduce long-term immunosuppression needs [45].

Conclusion

Understanding the immunological sequence—from innate activation during IRI to the multi-signal orchestration of T cell responses—is essential to optimize transplant outcomes. Each signal contributes uniquely to either graft rejection or tolerance, and targeted modulation of Signals 2–4 is the cornerstone of current and emerging therapies in transplant immunology.

References

1. Ibrhim S, Al-Nashar H, Osman M. Ghrelin and nitric oxide in renal ischemia: role in ATN post-transplant. Kidney Int Rep. 2024;9(1):45–53.
2. Zeiler M, Honarpisheh M, Knechtle S. HLA-DR and ICAM-1 expression in ATN post-kidney transplant. Transplantation. 2000;70(2):231–8.
3. Isoniemi H, Taskinen E, Hayry P. Cytokine signaling and CD4+ T-cell involvement in ischemic renal injury. Nephrol Dial Transplant. 1992;7(Suppl 1):98–103.
4. Kantachuvesiri S, Kelly V, Papalois V. HLA-G expression correlates with acute rejection post-ATN. Am J Transplant. 2015;15(6):1635–43.
5. Tamargo RJ, Kant S. Endothelial activation and CCR7-mediated immune pathways in graft rejection. Front Immunol. 2023;14:1155049.
6. IPD-IMGT/HLA database. HLA Alleles Numbers. https://hla.alleles.org/nomenclature/stats.html
7. Garrido F. MHC gene complexity and immune modulation. Springer; 2024.
8. Lu W, et al. Structure and stability of HLA Class I. bioRxiv. 2025.
9. Lei Z, et al. Molecular evolution of HLA domains. Mol Biol Evol. 2024.
10. Sarkar A, et al. Class I HLA in cytotoxic immune response. Preprints. 2024.
11. Wei X, et al. Soluble HLA molecules and immune circulation. Elsevier; 2025.
12. Etxeberria I, et al. HLA peptide-binding groove structures. AACR. 2025.
13. Gaglione P, Birnbaum M. TAP mutations and immune evasion. Nature. 2024.
14. Perišić O, et al. ER chaperones in HLA loading. Front Mol Biosci. 2025.
15. McInerney K, et al. T-cell recognition of peptide-HLA. PNAS. 2024.
16. Singh R, et al. HLA class I re-expression in tumors. AACR. 2025.
17. Yang Y, et al. HLA-C*12 polymorphisms and disease risk. Elsevier; 2025.
18. Hajijafary A, et al. HLA class II and autoimmune diseases. Elsevier; 2025.
19. Wolf E, et al. HLA-DM and antigen exchange regulation. Cell. 2025.
20. Rammensee H, Rötzschke O. HLA glycosylation and immune function. Springer; 2025.
21. Gaglione P, Birnbaum M. HLA-DQ in autoimmune disorders. Nature. 2024.
22. Wei X, et al. HLA-DR micropolymorphisms in therapy. Elsevier; 2025.
23. Lu W, et al. β1-domain polymorphisms in immune modulation. bioRxiv. 2025.
24. Perišić O, et al. Machine learning for HLA-vaccine prediction. Front Mol Biosci. 2025.
25. Kalousková M, et al. HLA PTMs in immune signaling. MDPI; 2024.
26. Sun Z, et al. NK-HLA interactions in pregnancy. ScienceDirect. 2024.
27. Vollmers E, et al. NK recognition via glycosylated HLA. Cells. 2024.
28. Verswyvel T, et al. HLA-C variation and NK modulation. Redox Biol. 2024.
29. Hanna C, et al. CREGs and transplant matching. Mol Immunol. 2024.
30. Petillo A, et al. MICA/MICB in NK and CTL activation. Nature. 2024.
31. Molfetta R, et al. MICA/MICB polymorphisms in cancer. Front Immunol. 2024.
32. Wang H, et al. HLA-DM mutations in autoimmunity. Cancer Lett. 2024.
33. Keating J. HLA-G expression in pregnancy and implications for immune tolerance. Eur J Hum Genet. 2024;32(1):45–53.
34. Keating J. Advances in HLA diversity and its clinical applications. Nature. 2024;621(7962):103–9.
35. Keating J. Pharmacogenomics and HLA-associated drug reactions. Eur J Hum Genet. 2024;32(2):121–7.
36. Hajighasemi M, Shirkavand A. HLA allelic diversity and transplantation compatibility in Iranian populations. J Basic Clin Physiol Pharmacol. 2024;35(1):19–28.
37. Opelz G, Döhler B. Influence of HLA compatibility on kidney graft survival. Transplantation. 2023;107(2):e48–55.
38. Lasorsa F, et al. Ischemia-reperfusion injury in kidney transplantation: mechanisms and potential therapeutic targets. Int J Mol Sci. 2024;25(8):4332. https://doi.org/10.3390/ijms25084332.

39. Jentho E, Weis S. DAMPs and innate immune training. Front Immunol. 2021;12:699563. https://doi.org/10.3389/fimmu.2021.699563.
40. Nauser CL, Farrar CA, Sacks SH. Complement recognition pathways in renal transplantation. J Am Soc Nephrol. 2017;28(9):2571–8. https://doi.org/10.1681/ASN.2017010079.
41. Siedlecki A, et al. Delayed graft function in the kidney transplant. Am J Transplant. 2011;11(11):2279–96. https://doi.org/10.1111/j.1600-6143.2011.03754.x.
42. Solhjou Z, et al. Emerging therapies targeting intra-organ inflammation in transplantation. Am J Transplant. 2015;15(2):305–11. https://doi.org/10.1111/ajt.13073.
43. Tamargo CL, Kant S. Pathophysiology of rejection in kidney transplantation. J Clin Med. 2023;12(12):4130. https://doi.org/10.3390/jcm12124130.
44. Peereboom ETM, et al. T-cell epitopes shared between immunizing HLA and donor HLA associate with graft failure. Front Immunol. 2021;12:784040. https://doi.org/10.3389/fimmu.2021.784040.
45. Huang DL, et al. The immunomodulation role of Th17 and Treg in renal transplantation. Front Immunol. 2023;14:1113560. https://doi.org/10.3389/fimmu.2023.1113560.
46. Ingulli E. Mechanism of cellular rejection in transplantation. Pediatr Nephrol. 2010;25(1):61–74. https://doi.org/10.1007/s00467-008-1020-x.
47. Zhang R. Donor-specific antibodies in kidney transplant recipients. Clin J Am Soc Nephrol. 2018;13(1):182–92. https://doi.org/10.2215/CJN.00700117.

Nutrition and Immunological Challenges in Transplantation

2

N. Vijayashree, Merina Elizabeth Alex, George Kurian, and Georgi Abraham

Case Report

A14-year-old boy with ESKD and hypertension, nutritional assessment done using WHO percentile growth chart, biochemical assessment and dietary assessment. He is a vegetarian and had suboptimal energy–789 kilocalories and 25 g of protein against the requirement of 2500 kilocalories and 45 g of protein. He falls in to the third percentile BMI in the WHO growth chart which showed severe PEW along with hypo albuminemia (serum albumin 2.4 g/dl). He was initially on MHD and switched over to home peritoneal dialysis due to uncontrolled hypertension. He required bilateral embolization of the kidneys for hypertension control. He is being prepared for kidney transplantation. Nutritional assessment showed protein energy wasting and sarcopenia. He urgently required nutritional intervention. Patient and family was educated about the nutrition and oral renal nutritional supplement was initiated by the nutritionist. With constant education and follow up, he gained dry weight and is improving nutritionally as shown in the BCM (Fig. 2.1).

N. Vijayashree (✉)
MGM Healthcare, Chennai, Tamil Nadu, India

M. E. Alex · G. Abraham
The Madras Medical Mission Hospital, Chennai, Tamil Nadu, India

G. Kurian
Matha Amirthanatha Mayi Hospital, Kochi, Kerala, India

G. Abraham et al. (eds.), *Immunological Challenges in Organ Transplantation*,
https://doi.org/10.1007/978-981-95-1533-2_2

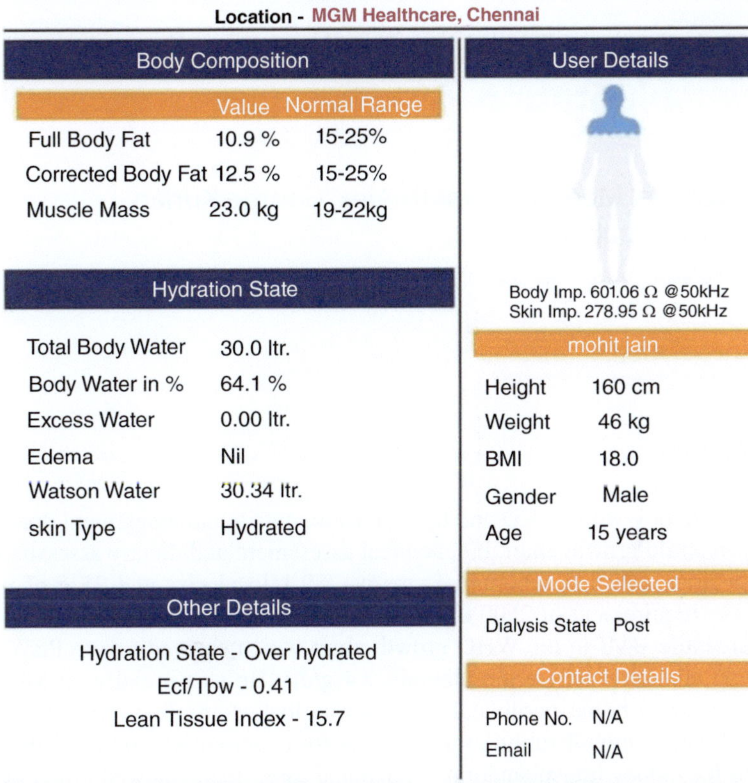

Location - MGM Healthcare, Chennai

Body Composition

	Value	Normal Range
Full Body Fat	10.9 %	15-25%
Corrected Body Fat	12.5 %	15-25%
Muscle Mass	23.0 kg	19-22kg

Hydration State

Total Body Water	30.0 ltr.
Body Water in %	64.1 %
Excess Water	0.00 ltr.
Edema	Nil
Watson Water	30.34 ltr.
skin Type	Hydrated

Other Details

Hydration State - Over hydrated
Ecf/Tbw - 0.41
Lean Tissue Index - 15.7

User Details

Body Imp. 601.06 Ω @50kHz
Skin Imp. 278.95 Ω @50kHz

mohit jain

Height	160 cm
Weight	46 kg
BMI	18.0
Gender	Male
Age	15 years

Mode Selected

Dialysis State Post

Contact Details

Phone No.	N/A
Email	N/A

Fig. 2.1 Body composition analysis indicating improved muscle and fat mass

Introduction

Proper nutrition is crucial for supporting the immune system, as it provides immune cells with the necessary nutrients in sufficient quantities. A wide range of micronutrients, including vitamins and minerals, along with specific macronutrients like amino acids, cholesterol, and fatty acids, have been shown to play key roles in promoting effective immune responses. These nutrients each contribute in distinct ways to the maintenance of optimal immune function [1].

Oral Immune System

The oral cavity's mucosal immune system is continually exposed to various local signals, such as a diverse population of beneficial microbes and their byproducts, repeated tissue stress from chewing, and antigens from food and airborne particles. These factors pose ongoing challenges to the stability of the oral mucosa. Therefore,

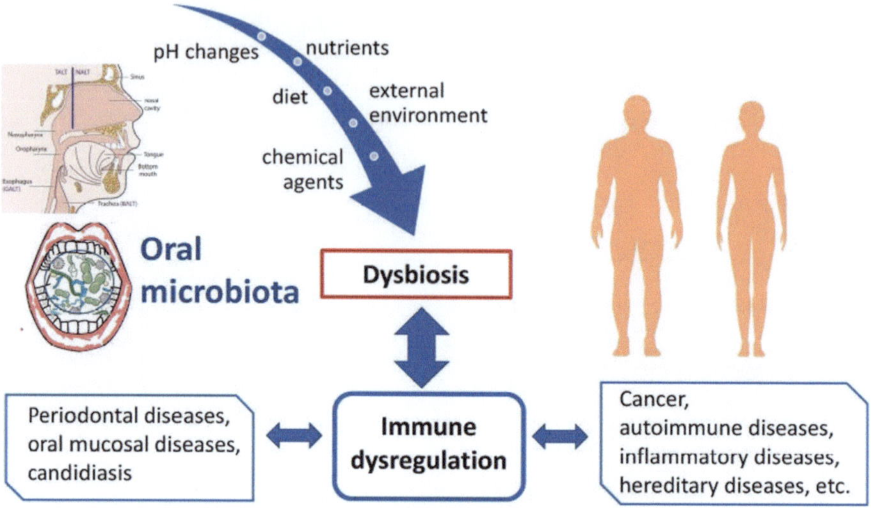

Fig. 2.2 Relation between the immune system and microbiota in the oral cavity

maintaining the health and function of the oral mucosa and its immune components is essential for preserving the body's internal balance. Interaction between epithelial cells and both innate and adaptive immune cells are keys for the swift identification and removal of pathogens at the surface. The immune system and the resident microbiota work in tandem to sustain this equilibrium. However, when the microbial balance is disrupted, known as dysbiosis, it can lead to a breakdown in the local immune regulation (Fig. 2.2) [2].

Immunity and the Gut

The gastrointestinal (GI) tract is a multifaceted organ system made up of various elements, including the intestinal epithelial lining, lamina propria, mucus layer, gut microbes, dietary contents, the enteric nervous system, and a localized immune network. A key part of gut immunity is the gut-associated lymphoid tissue (GALT), along with its draining lymph nodes, which serve as central sites for initiating adaptive immune responses [3]. GALT consists of structures like microfold (M) cells and Peyer's patches (PPs), which are clusters of lymphoid tissue embedded within the mucosal and submucosal layers. M cells are particularly specialized in capturing and transporting antigens that typically cannot cross into the lamina propria [4]. These antigens are then handed off to dendritic cells, which process and present them to T cells—an essential step in triggering immune responses [5]. Located in the small intestine, Peyer's patches contain dense populations of B cell follicles and T cell regions, and they are a primary source of immunoglobulin A (IgA), a crucial antibody for mucosal immunity. Besides Peyer's patches, GALT also includes smaller immune structures such as

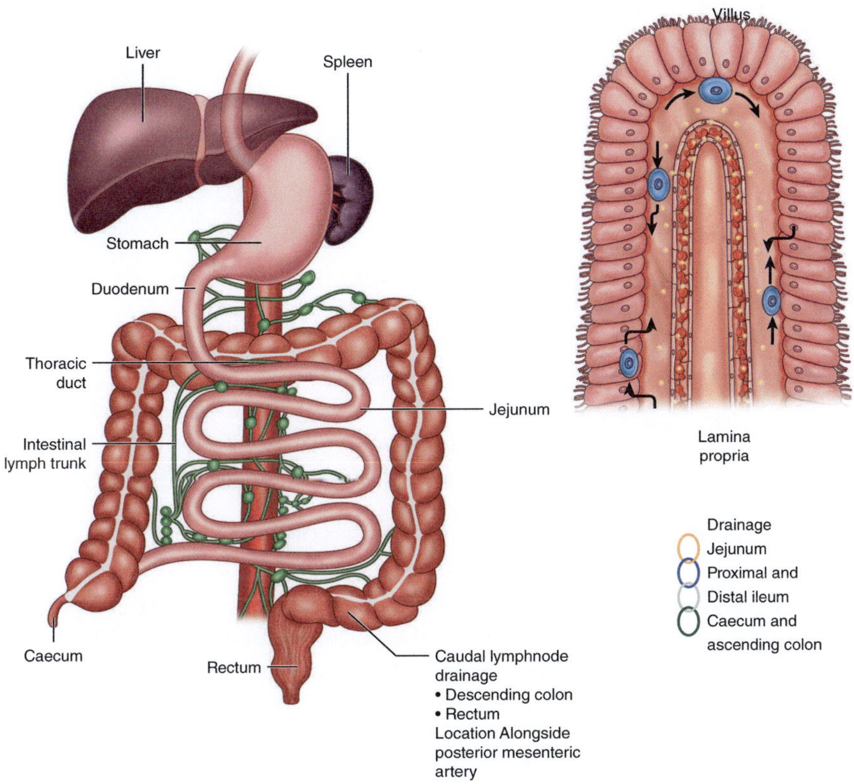

Fig. 2.3 Nutrition and its impact on the immune system

cryptopatches and isolated lymphoid follicles. These areas also house adaptive immune cells, especially B cells, and their structure and cellular makeup can be influenced by aging and by the presence and diversity of gut microbiota [6]. The mesenteric lymph nodes (MLNs) are the largest lymphoid structures associated with the gut and are consistently exposed to immune-stimulating antigens. They facilitate interactions between innate and adaptive immune cells [7, 8]. Notably, different sections of the GI tract drain into specific lymph nodes, indicating that immune responses may differ depending on the anatomical location within the gut (Fig. 2.3).

Relationship Between Nutrition and Immune System

Nutrition plays a vital role in shaping the body's innate immune responses, and any imbalance in this relationship can significantly contribute to disease development. The immune system combats harmful agents using both innate and adaptive immune cells, ultimately producing antibodies that target specific pathogens [8].

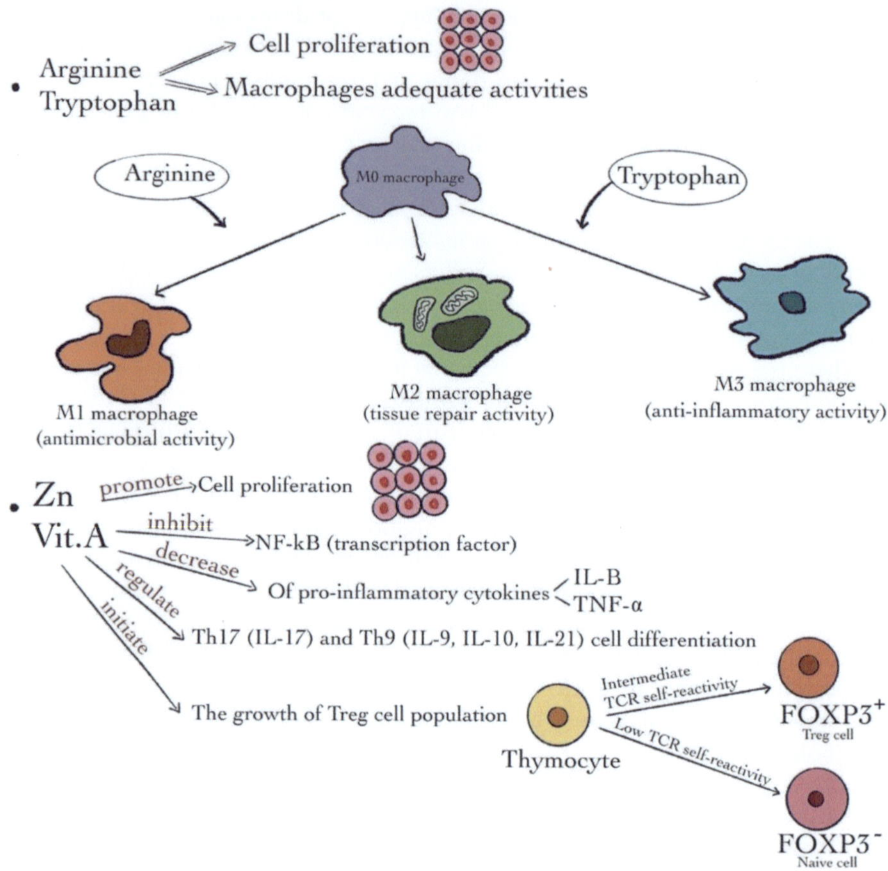

Fig. 2.4 Relationship between nutrition and immune system [8]

Macronutrients like arginine and tryptophan are essential for cell growth and for maintaining the proper function of macrophages. Similarly, micronutrients such as Vitamin A and Zinc help regulate immune activity by encouraging cell proliferation, suppressing the NF-κB signaling pathway, lowering the production of inflammatory cytokines like IL-1β and TNF-α, influencing the development of Th17 and Th9 cells, and promoting the expansion of regulatory T cells (Tregs) (Fig. 2.4).

Nutrients and Immunology in Organ Transplant

Nutrients that are known for its vital immune functions are zinc, selenium, iron, copper and amino acids such as arginine, vitamins A, C, E, D, B-6, folic acid, omega 3 fatty acid, and tryptophan. Nutrients involve in defense system by cell proliferation, cell-mediated immunity, phagocyte functions, and production of complement proteins, secretary immunoglobulin- A, antibodies, and cytokine [10, 11].

Transplant recipients are on immune suppressant drugs and these therapeutic index and large interpatient variabilities in their pharmacokinetic (PK) and pharmaco dynamics (PD) profiles. Cyclosporine and tacrolimus are metabolized in the small intestine by CYP3A4 and CYP3A5 enzymes, and also by P- glycoprotein, which acts as an efflux pump, impacting their oral absorption and bioavailability. Small intestinal diseases can impact their absorption (either increase/decrease the absorption) leading to either toxic blood levels or reduced levels leading to rejection. The discovery of functional single-nucleotide polymorphisms (SNPs) in genes encoding drug-metabolizing enzymes offers significant potential to enhance drug safety and effectiveness. These genetic markers could play a key role in tailoring immunosuppressive therapy to individual patients.

There is a well-established link between the CYP3A5*1/3 polymorphism and how the body processes tacrolimus. However, the effects of CYP3A51/3 and CYP3A41/*1B variants on the metabolism of ciclosporin and sirolimus are still not fully understood. Genetic variations in CYP3A enzymes may contribute to differences in how patients experience drug interactions, and these interactions themselves can influence how these polymorphisms manifest clinically. To address the wide variability in drug responses, researchers are turning to pharmacogenomic models that combine data on drug kinetics, dynamics, and genetic profiles. These advanced models provide a quantitative way to assess how genetic differences impact drug behavior in the body [12]. Ultimately, successful personalized treatment for transplant patients will depend on a comprehensive understanding of how genetic factors affect immune function and drug metabolism.

Immune Enhancing Nutrients in Postrenal Transplant

It was observed in a study that supplementation of immune nutrients (Arginine and omega 3 fatty acids) had a better outcome after transplantation. Daily supplementation (between 30 days and 3 years) of Arginine 9 grams and canola oil/fish frequency of complications associated with kidney transplant recipients (who had steroid- free immune suppression). It was shown to as rejection, infection, new onset diabetes, and cardiovascular disease [13].

Antioxidants and Polyphenols in Transplantation

Other than the macro and micro nutrients, there are also phytochemicals that are responsible for immunity. Food rich in phenoic acids, flavones, flavonols, and isoflavones have strong antioxidant properties. It scavenges and reactive nitrogen species and contributes in reducing the cardiovascular complications in renal transplant recipients. Good sources of antioxidants which impact positively renal transplant

recipients [14]. Patients consuming turmeric curry and ginger, known antioxidants, did not affect the tacrolimus plasma concentrations [15]. Certain fruits such as grape fruit and pomegranate could interact with oral immunosuppressive agents causing interactions when ingested together and hence should be avoided [16].

Pro and Prebiotics in Postrenal Transplant

Growing researches have shown supplementing symbiotic, prebiotics, and probiotics would help to modify and reduce dysbiosis or leaky gut [18]. However use of over aged cheese or unhygienic fermented foods could cause food born infections and hence fermented foods should be taken with caution.

Food Safety in Transplantation

Maintaining food safety is crucial for everyone, but it is especially important for people who have received organ transplants. Proper handling of food helps prevent food borne illnesses, which can cause serious health problems and increase the risk of complications in immune suppressed transplant recipients. Therefore, practicing safe food habits should be part of daily life.

In addition to taking immunosuppressive medications exactly as directed, following food safety measures is vital to protect your health Fig. 2.5 [18].

Some important food safety tips include:

- Always wash your hands before preparing or eating food.
- Thoroughly wash all fruits and vegetables.
- Make sure your drinking water is safe by filtering it or boiling it for at least 1 min.
- Avoid cross-contamination by using different cutting boards for various types of food.
- Cook foods to their recommended safe internal temperatures.
- Steer clear of unpasteurized foods and drinks.
- Properly store food by refrigerating or freezing it promptly (Figs. 2.5 and 2.6).

Handwash, surface cleaning, and washing of fruits and vegetables in running water and using clean kitchen utensils including cutlery, crockeries, cutting boards, and countertops should be ensured. Avoid cross contamination which can spread harmful bacteria. Separate storage, cutting boards, cooking utensil for vegetarian and nonvegetarian foods to be used to avoid cross contamination. All meat and seafood should be stored at the temperature of ≤18 °C and other fresh foods to be stored in refrigerator temperature of ≤4. Foods need to be cooked thoroughly and served hot at temperatures of at least 65 °C (149 °F). Bacteria multiply most rapidly

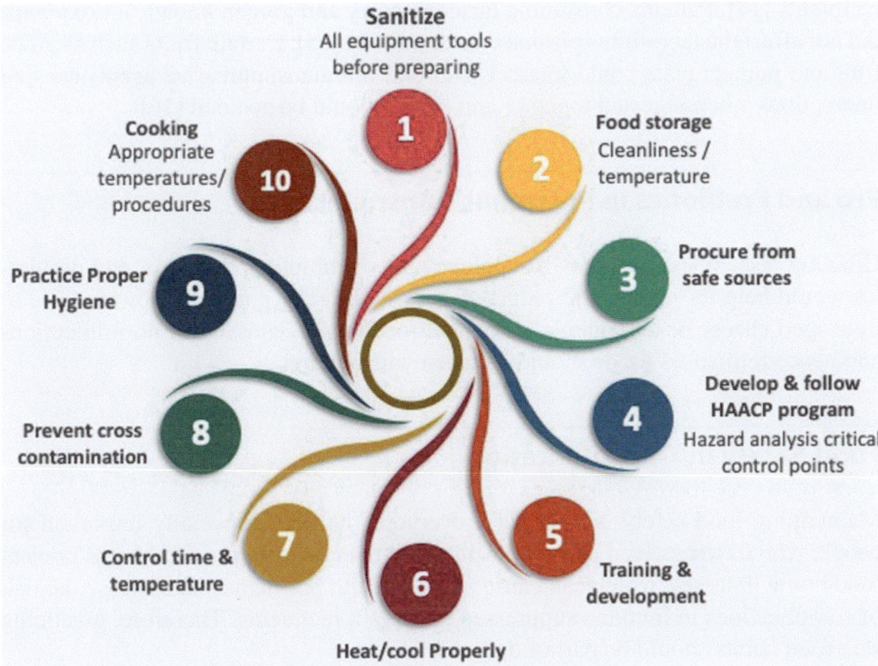

Fig. 2.5 Ten principles of safe food handling and hygiene [18]

between 4 °C (40 °F) and 60 °C (140 °F), which is known as the "danger zone Fig. 2.6." Properly chilling food is one of the best ways to reduce the risk of foodborne illnesses by slowing bacterial growth. Additionally, cooking food completely helps eliminate harmful bacteria.

Maintaining a balanced diet is crucial for overall health and is an important factor when evaluating transplant candidates. Nutrition plays a significant role throughout the entire transplant process, affecting everything from preparation before surgery to long-term well-being after transplantation. Since immunosuppressive drugs are taken orally, their effectiveness depends on their interaction with the gastrointestinal immune system and how well they are absorbed.

Good nutrition also boosts the immune system, helping the body better manage surgery and the medications used to prevent organ rejection. Eating a diet rich in vitamins, minerals, and antioxidants strengthens immune function. Including fruits and vegetables high in vitamin C and antioxidants like beta-carotene supports optimal immune health.

Fig. 2.6 Food storage and service temperature hygiene

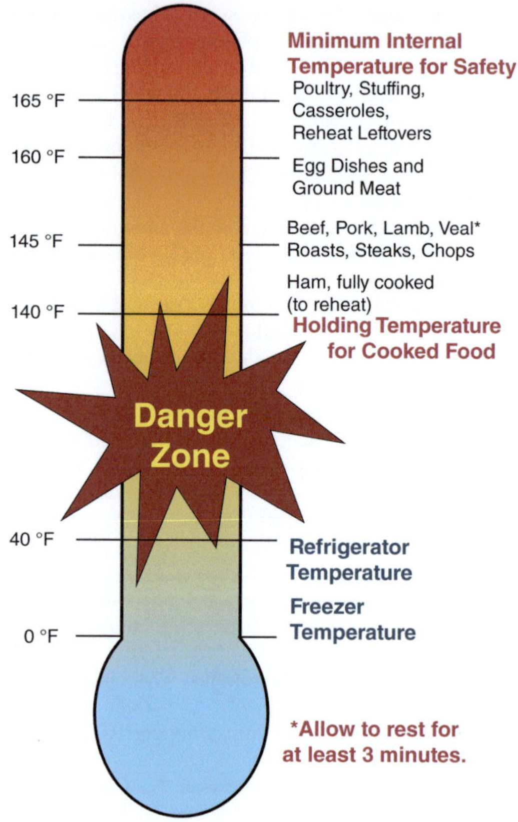

References

1. Gombart A, Pierre A, Maggini S. A review of micronutrients and the immune system–working in harmony to reduce the risk of infection. Nutrients. 2020;12:236. https://doi.org/10.3390/nu12010236.
2. Mowat AM, Agace WW. Regional specialization within the intestinal immune system. Nat Rev Immunol. 2014;14:667–85. https://doi.org/10.1038/nri3738.
3. Ohno H. Intestinal M cells. J Biochem. 2016;159:151–60. https://doi.org/10.1093/jb/mvv121.
4. Reboldi A, Cyster JG. Peyer's patches: organizing B-cell responses at the intestinal frontier. Immunol Rev. 2016;271:230–45. https://doi.org/10.1111/imr.12400.
5. Randall TD, Mebius RE. The development and function of mucosal lymphoid tissues: a balancing act with micro-organisms. Mucosal Immunol. 2014;7:455–66. https://doi.org/10.1038/mi.2014.11.
6. HoustonS A, Cerovic V, Thomson C, Brewer J, Mowat AM, Milling S. The lymph nodes draining the small intestine and colon are anatomically separate and immunologically distinct. Mucosal Immunol. 2016;9:468–78. https://doi.org/10.1038/mi.2015.77.
7. Sommer F, Backhed F. The gut microbiota–masters of host development and physiology. Nat Rev Microbiol. 2013;11:227–38. https://doi.org/10.1038/nrmicro2974.

8. Sevda senel, An Overview of Physical, microbiological and immune barriers of oral mucosa,department of pharmaceutical technology, Hacettepe University, Ankara 06100, Turkey Int. J. Mol. Sci. 2021;22(15):7821. https://doi.org/10.3390/ijms22157821.
9. Maggini S, Pierre A, Calder P. Immune function and micronutrient requirements change over the life course. Nutrients 2018;10:1531. https://doi.org/10.3390/nu10101531.
10. Munteanu C, Schwartz B. The relationship between nutrition and the immune system. Front Nutr. 2022;9:1082500. https://doi.org/10.3389/fnut.2022.1082500. PMID: 36570149; PMCID: PMC9772031.
11. Chandra RK. Nutrition and the immune system: an introduction. The american journal of clinical nutrition. 1997;66(2):460S–463S. ISSN 0002-9165, https://doi.org/10.1093/ajcn/66.2.460S.
12. Fritz A, Busch D, Lapczuk J, Ostrowski M, Drozdzik, Oswald S: Expression of clinically relevant drug-metabolizing enzymes along the human intestine and their correlation to drug transporters and nuclear receptors: An intra-subject analysis. 2018. https://doi.org/10.1111/bcpt.13137.
13. Alexander JW. Immunonutrition in kidney and liver transplant recipients. ExpClin Transplant. 2014;12Suppl 1:34–7. https://doi.org/10.6002/ect.25liver.l24. PMID: 24635789.
14. Bustos NI, Sotomayor CG, Pol RA, Navis GJ, Bakker SJL. Polyphenols and Novel Insights Into Post-kidney Transplant Complications and Cardiovascular Disease: A Narrative Review. Front Cardiovasc Med. 2021;8:751036. https://doi.org/10.3389/fcvm.2021.751036. PMID: 34869655; PMCID: PMC8635050.
15. Boissiere C, Francois E, Vabret E, Le Daré B, Bacle A. Spice-drug interactions: a case report on the use of turmeric, curry and ginger in a renal transplant patient on tacrolimus. Eur J Hosp Pharm. 2023;31(1):68–69. https://doi.org/10.1136/ejhpharm-2023-003871. PMID: 37586787; PMCID: PMC10800271.
16. Tolou-Ghamari Z. Tacrolimus, Cytochrome P450, interactions with food variables in organ transplant recipients; a current and comprehensive review. Curr Drug Metab. 2025;25(8):547–553. https://doi.org/10.2174/0113892002328742241210102522. PMID: 39757635.
17. Cooper TE, Scholes-Robertson N, Craig JC, Hawley CM, Howell M, Johnson DW, Teixeira-Pinto A, Tong A, Wong G. Synbiotics, prebiotics and probiotics for solid organ transplant recipients. Cochrane Database Syst Rev. 2021;2021(6):CD014804. https://doi.org/10.1002/14651858.CD014804. PMCID: PMC8207245.
18. Food safety and standard authority of India, Food safety and standard Act 2011.

Why We Require Immunological Workup: A Patient-Centered-Case Based Approach

3

Priya Haridas Anupama, Priyanka Govindan, Pritika Shrivastava, Dipankar Bhoumik, and Georgi Abraham

Case

A 25-year-old male, born to non-consanguineous parents, has end-stage kidney disease for which he is on maintenance hemodialysis for the past 2 years. He has received over ten packed red blood cells transfusions in the past. He is currently receiving injectable erythropoietin stimulating agents for the anemia.

The patient is undergoing evaluation for an ABO compatible (ABOc) kidney transplant with a 25-year-old male paternal cousin as the potential donor.

Immunological Workup

- **CDC Crossmatch:** Negative
- **Flow cytometry Cross match for T and B lymphocytes**

P. H. Anupama (✉)
Department of Nephrology, SRIHER, Chennai, Tamil Nadu, India

P. Govindan
Fort Collins, CO, USA

P. Shrivastava
Henry Ford Hospital, Detroit, MI, USA
e-mail: pshriva1@hfhs.org

D. Bhoumik
AIIMS, New Delhi, Delhi, India
e-mail: dmbhowmik@aiims.edu

G. Abraham
The Madras Medical Mission Hospital, Chennai, Tamil Nadu, India

G. Abraham et al. (eds.), *Immunological Challenges in Organ Transplantation*,
https://doi.org/10.1007/978-981-95-1533-2_3

Antibody against	MCS	Interpretation	References
T-cells (CD3)	<45	Negative	T-cells MCS < 45 negative T-cells MCS 45–60 borderline T-cells MCS > 60 positive
B-cells (CD19)	194	Positive	B- cells MCS < 86 negative B-cells MCS 86–116 borderline B-cells MCS > 116 positive

Note: MCS is median channel shift (MCN of test-MCN of control)

- **Class I Panel Reactive Antibody (PRA) Percentage:** <4%
- **Class II Panel Reactive Antibody (PRA) Percentage:** 27%
- **Donor-Specific IgG Antibodies (DSA) to HLA Class I and II:** Negative

Antibodies detected against HLA Class I antigen SAB result (Luminex technology) with MFI > 1000 are shown in Table 3.1.

Antibodies detected against HLA Class II antigen SAB result (Luminex technology) with MFI is >1000 are shown in Table 3.2.

Recipient and donor HLA class I and class II is depicted in Table 3.3:

Case immune work up summary is shown in Table 3.4:

Introduction

Kidney transplantation remains the preferred treatment for patients with end-stage kidney disease, offering improved survival and quality of life. However, immunological compatibility between the donor and recipient plays a crucial role in transplant success. Human leukocyte antigen (HLA) mismatches and prior sensitization to HLA can significantly impact the risk of rejection and long-term graft survival. This case report explores the immunological challenges associated with a 25-year-old male recipient with stage V chronic kidney disease undergoing evaluation for an ABO-compatible living-related kidney transplantation. By examining the donor-recipient HLA mismatch and the recipient's sensitization history due to prior blood transfusions, this case highlights key considerations in assessing immunological risk and optimizing transplant outcomes.

Antibodies play a vital role in kidney transplantation, presenting challenges like antibody-mediated rejection (ABMR) while also enabling improved transplant outcomes through enhanced detection, prevention, and treatment strategies.

Table 3.1 Antibodies detected against HLA class I antigen SAB results with MFI > 1000

Allele specificity	MFI
B*37:01	1766.10

Table 3.2 Antibodies detected against HLA Class II antigen with MFI is >1000

Allele specificity	MFI
DQB1*02:01*, *DQA1*05:01	3344.71
DPB1*01:01, DPA1*02:01	2989.66
DRB4*01:01	1430.65
DPB1*05:01, DPA1*02:02	1404.06
DQB1*02:01*, *DQA1*03:01	1205.40
DRB4*01:03	1132.00
DQB1*02:01*, *DQA1*04:01	1047.47

Should We Consider This Donor?

This patient received multiple blood transfusions, leading to sensitization. His Panel Reactive Antibody (PRA) was <4% and 27% for Class I and II, respectively. He has multiple HLA Antibodies but none are donor specific. This makes it possible to pursue the transplant despite the B cell crossmatch positive. Flow cross match was repeated and found to be negative for B and T cell prior to the tranplantation. Hence our patient did not require de-sensitization prior to the transplant. The HLA region on chromosome 6 and its structure is depicted in the Figs. 3.1 and 3.2 [1, 2].

How Could We Have Prevented Our Patient from Becoming Sensitized?

Avoiding sensitization can be achieved by reducing blood transfusions through the use of erythropoiesis-stimulating agents, but when transfusion is absolutely necessary, the use of leukodepleted and irradiated blood products should be considered [3]. In high-income countries, where whole blood transfusions (unmodified blood from the donor), which are more likely to be immunogenic, are rarely used. In contrast, in upper-middle, lower-middle, and low-income countries, 0.2%, 16%, and 47% respectively of whole blood is transfused [4].

Table 3.3 Recipient and donor HLA class I and class II

Recipient:						
LOCUS	HLA-A*	HLA-B*	HLA-C*	HLA-DRB1*	HLA-DQB1 *	HLA-DPB1*
1	11:01:01	13:01:01	04:03:01	13:02:01	06:09:01	04:01:01
2	24:02:01	08:01:01	07:02:01	11:08:01	03:01: 01	04:01:01
Donor:						
LOCUS	HLA-A*	HLA-B*	HLA-C*	HLA-DRB1*	HLA-DQB1*	HLA-DPB1*
1	24:02:01	27:07:01	07:02:01	07:01:01	02:02:01	02:01:02
2	11:01:01	07:02:01	15:02:01	07:01:01	03:03:02	13:01:01

3/6 mismatch, 2/6 match

Table 3.4 Immune work up summary

Sensitization history	Multiple blood transfusion
CDC crossmatch	Negative
FLOW crossmatch	B cell positive
HLA typing donor	A*24:02:01, 11:01:01, B*27:07:01, 07:02:01, DRB1*07:01:01,07:01:01
HLA typing recipient	A*11:01:01,24:02:01, B*13:01:01,08:01:01, DRB1*13:02:01,11:08:01
Single antigen bead I	Low level anti-HLA with highest MFI 1766
Single antigen bead II	Multiple anti-HLA with highest MFI 3300

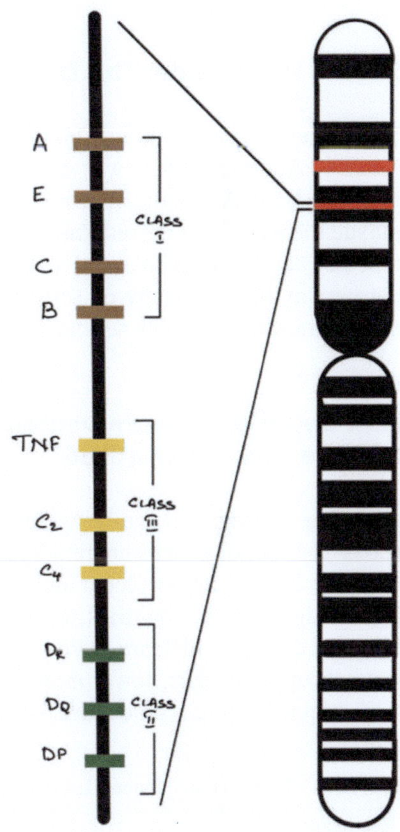

Fig. 3.1 Human Leukocyte Antigen region on chromosome 6

Technic of HLA Typing Using Modern Molecular Technics

The HLA system located on chromosome 6 contains >200 genes. The HLA system which is a polymorphic genetic system in humans comprises over 42,000 related alleles.

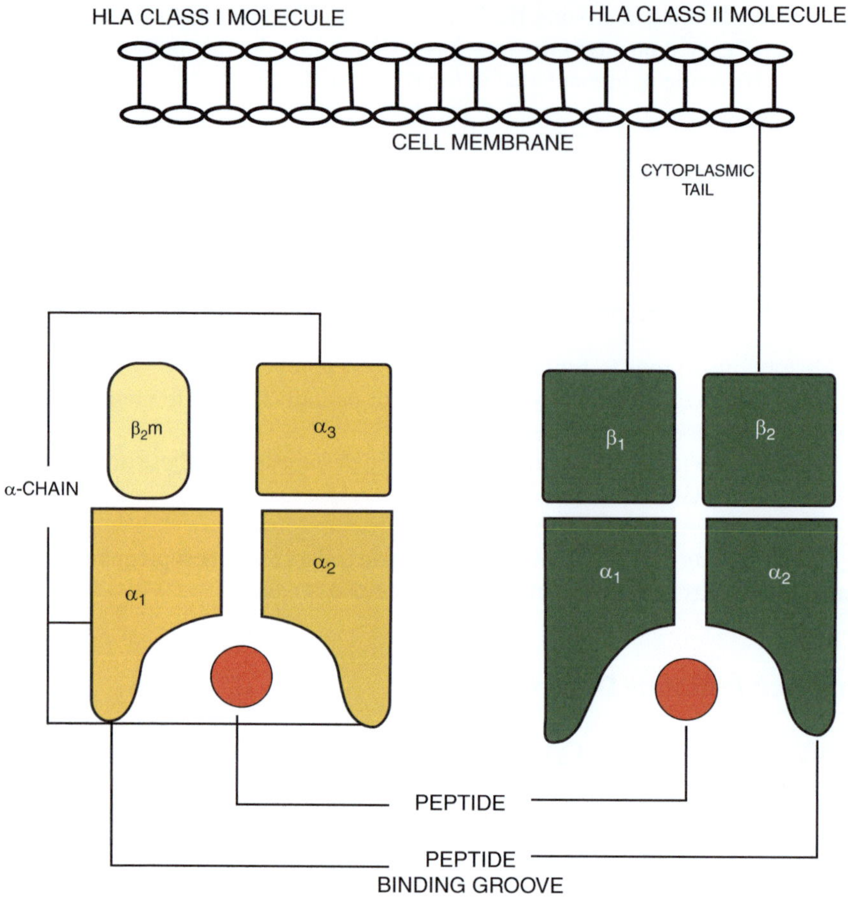

Fig. 3.2 Human leuckocyte antigen molecule structure

1. Sequence specific primer (SSP): This is a PCR based test which identifies specific DNA sequences using primers designed for known alleles, which is low to intermediate resolution.
2. Sequence specific oligonucleotide (SSO): This is hybridization based test, using specific probes for known alleles.
3. Next generation sequencing (NGS): This is based on High throughput DNA sequencing technology which reads entire HLA gene region. This has high accuracy because the entire gene region is sequenced.

High resolution uses primers specific for each subtype. Low resolution uses primers specific for one HLA type, but not specific for each subtype.

Intermittent resolution refers to a situation where HLA typing results are not consistently at the same level of detail across different HLA loci or samples leading to inconsistencies in matching and clinical interpretation.

While looking at missing HLA Alleles use of single antigen beads miss HLA alleles. Hence if positive crossmatch and HLA antibodies are present then ensure donor HLA alleles are included in the HLA panel.

Sensitization History

Sensitization means exposure to non-self HLA. It is a major barrier to transplantation by reducing the chances of finding an immunologically matched donor. A detailed history should help in identifying sensitization which include the following:

1. History of previous transplant
2. History of pregnancy: parity and number of partners leading to pregnancy
3. History of blood product transfusions
4. Relationship to the donor to elicit a history of exposure of these antigens, e.g., husband to wife, child to mother

All sensitizing events are not equal [5]. There is data to suggest pregnancy related sensitization may negatively impact transplant outcomes [6].

Anti HLA Antibody Detection

Complement Dependent Cytotoxicity cross match (CDC) detects the presence of complement binding antibody, but it could be IgG or IgM which can be identified by use of Dithiothreitol (DTT), DDT depletes IgM antibodies, improving the accuracy in cases where these antibodies are present. This method involves mixing donor cells with recipient serum, which contains potential HLA antibodies. If antibodies bind to the donor cells, the complement system is activated, leading to cell lysis, which is visualized using a fluorescent stain. When AHG (Antihuman Globulin) is added, it increases sensitivity for detecting low-affinity antibodies. The following table summarizes different crossmatch tests, their respective goals, and the clinical scenarios in which they are useful (Table 3.5).

The **flow crossmatches** are more sensitive than the CDC based assay for the detection of anti-donor antibodies and is capable of detecting both complement-binding and non-complement-binding antibodies; however it does not discriminate between cytotoxic and non-cytotoxic antibodies. In this test, donor T and B cells are mixed with recipient serum. If antibodies are present, they bind to the donor cells, and fluorescence is measured using flow cytometry. The amount of fluorescence correlates with the presence of antibodies and the strength of the crossmatch. Positive T-cell Flow crossmatch with negative CDC crossmatch is beneficial in predicting risk of rejection. Reasons for positive B cell only flow crossmatch could be donor specific antibody to Class II or low level to class I or false positive "sticky B cells."

Table 3.5 Following table summarizes different crossmatch tests, their respective goals, and the clinical scenarios in which they are useful

Method	Goal	Use
CDC crossmatch	Identifies cytotoxic antibodies	First-line test; detects complement-binding antibodies
AHG crossmatch	Enhances the sensitivity of CDC crossmatch	Improves detection of low-affinity antibodies
DTT crossmatch	Depletes IgM antibodies	Removes IgM antibodies in autoimmune or infection cases
Flow crossmatch	Sensitive detection of anti-donor antibodies	More sensitive; detects non-complement-binding antibodies
T-cell crossmatch	Identifies Class I donor-specific antibodies (DSA)	Detects Class I DSA
B-cell crossmatch	Identifies Class I and II donor-specific antibodies (DSA)	Detects Class I and II DSA
Pronase crossmatch	Removes Fc/background in Rituximab-treated patients	Refines crossmatch in Rituximab-treated patients
Endothelial cell crossmatch	Identifies non-HLA antibodies	Detects non-HLA antibodies
Autologous crossmatch	Detects autoantibodies	Identifies autoantibodies from the recipient
Virtual crossmatch	Most sensitive method for donor-recipient compatibility	Non-invasive, uses HLA typing and patient antibodies

Pronase Crossmatch: In patients treated with Rituximab (a B-cell depleting agent), the presence of Rituximab may interfere with crossmatch results by removing CD20 molecules on B cells. To address this, Pronase is used to remove CD20 from B cells, allowing for a more accurate crossmatch assessment.

Virtual crossmatch does not involve mixing patient serum with donor cells. Instead, it uses the donor's HLA typing data and analyzes the recipient's antibodies against HLA. This method is non-invasive and provides an efficient way to assess compatibility, particularly when physical crossmatching is not possible.

Panel Reactive Antibodies (PRA)

PRA testing is a measure of the level of sensitization of the recipient's immune system to donor HLA antigens. A high PRA indicates that the recipient has a considerable number of antibodies against potential donors, increasing the risk of transplant rejection. PRA is typically performed using complement-dependent cytotoxicity (CDC) or more advanced techniques like Luminex assays.

- Traditional PRA: Historically, PRA testing was performed using CDC, which detects antibodies by mixing patient serum with a panel of donor cells and

observing complement activation. The percentage of donor cells that are lysed indicates the level of sensitization (i.e., the PRA percentage).

- Modern PRA: Newer, more sensitive techniques, such as Luminex-based assays and flow cytometry, allow for the detection of a broader range of antibodies, including those that may not activate complement. These tests identify specific donor HLA antigens to which the recipient is sensitized, helping predict the likelihood of graft rejection more accurately. High PRA levels, particularly above 50%, increase the risk of rejection and may make finding a compatible donor more challenging.

Luminex Assay

The Luminex assay is used to detect antibodies against HLA. Unlike traditional methods such as CDC, the Luminex assay uses bead-based flow cytometry to detect the presence of antibodies that target specific donor HLA antigens. In this assay, beads are coated with specific HLA antigens. When recipient serum is mixed with these beads, antibodies (if present) will bind to the corresponding HLA antigens. A secondary fluorescently-labeled antibody binds to any bound antibodies, and the presence of fluorescence is detected by a Luminex analyzer. This allows for the detection of donor-specific antibodies (DSA) and provides a quantitative assessment of the strength of the antibody response [7].

- Advantages of Luminex: Highly sensitive and specific compared to traditional PRA and CDC; can detect non-complement-fixing antibodies, which are missed by CDC-based methods; allows for detection of donor-specific antibodies (DSA), which are critical in predicting graft rejection; provides detailed information about specific HLA mismatches between the donor and recipient.

Conclusion

The immunological evaluation prior to transplant plays a significant role in predicting the long-term graft survival rates. The first step would be HLA matching. The HLA typing helps determine our induction and maintenance immunosuppression. Polymorphism in A, B, and DR is most relevant with regard to organ transplants. Crossmatch testing plays a vital role in determining donor-recipient compatibility in kidney transplantation. By detecting anti-donor antibodies, these tests help predict the risk of rejection and guide transplant decisions. While traditional methods such as CDC crossmatch remain valuable, newer techniques like flow cytometry, Luminex, and virtual crossmatch provide more sensitivity and precision. The choice of crossmatch method should be tailored to the individual patient's clinical needs, including prior sensitization, the use of Rituximab, and the presence of non-HLA antibodies.

Table 3.6 This table describes terminologies in immunological work-up

Immunogenicity	It is the ability of an antigen to mount an immune response. It could be humoral, or cell mediated
Antigenicity (immunological reactivity)	It is the ability of the antigen to react with the products formed as a part of the response
Haptens	They are molecules with antigenicity without immunogenicity
Complete antigen	Molecules with both immunogenic and antigenic properties
Epitope	The smallest discrete part of the antigen that reacts with the adaptive immune system. It is called the immunodominant part of the antigen. An antigen can have multiple epitopes

- The order of risk of rejection is as follows:
  Positive CDC cross match, positive flow cross match/positive DSA, high panel reactive antibodies, high HLA mismatch, previous transplant, African-American decent.

The different terminologies for immunological explanation are given in Table 3.6.

This patient underwent a successful kidney transplantation with the proposed donor. He recieved single dose thymoglobulin (1mg/kg body weight) as induction agent and is currently on triple immunosuppression. Combination of CDC, flow crossmatch, and Single Antigen Bead Assay currently gives the best and most comprehensive immunological risk assessment for a donor recipient pair.

References

1. Abbas K, Mubarak M. Expanding role of antibodies in kidney transplantation. World J Transplant. 2025;15(1):99220. https://doi.org/10.5500/wjt.v15.i1.99220.
2. Galliford J, Game DS. Modern renal transplantation: present challenges and future prospects. Postgrad Med J. 2009;85:91–101.
3. Jackman RP, Deng X, Bolgiano D, Utter GH, Schechterly C, Lebedeva M, Operskalski E, Luban NL, Alter H, Busch MP, Slichter SJ, Norris PJ. Leukoreduction, UV, and HLA antibody. Transfusion. 2014;54:672–80. https://doi.org/10.1111/trf.12317.
4. World Health Organization. Global status report on blood safety and availability. 2021. https://www.who.int/publications/i/item/9789240051683. Accessed 10 March 2025.
5. Anwar IJ, Jackson AM, Locke JE, Kwun J. Editorial: sensitization and desensitization in organ transplantation. Front Immunol. 2021;12:784472. https://doi.org/10.3389/fimmu.2021.784472.
6. Nellore A, Killian JT Jr, Porrett PM. Memory B cells in pregnancy sensitization. Front Immunol. 2021;12:688987. https://doi.org/10.3389/fimmu.2021.688987.
7. Solez K, et al. Luminex crossmatch and the prediction of kidney graft survival. Transplantation Proceedings. 2009;41(9):3950–52.

HLA Loci and Typing in Solid Organ Transplantation: A Case-Based Approach to HLA Typing and Applications in Clinical Practice

4

Shruti Tapiawala, Dolly Daniel, and Suchita Jogale

Introduction

The human leukocyte antigen (HLA) loci are essential for solid organ transplantation because of their ability to modulate immune responses and achieve donor-recipient matching. "Self" and "non-self" discrimination is achieved by highly polymorphic proteins called HLAs, which are encoded by genes located on chromosome 6. The ability of HLA molecules to present peptides to T cells dictates immune responses, making them pivotal in transplant compatibility and immunological tolerance and hence their crucial roles in antigen presentation and immune response regulation [1, 2].

Evolution of the techniques to identify the HLA molecules and delineate the non self has helped in field of transplantation—both solid and hematopoietic. The detailed knowledge helps in choosing well matched donors and immune stratify the recipient based on the differences. This has resulted in better outcomes.

This review article provides a comprehensive overview of the evolution of human leukocyte antigen (HLA) typing methodologies, tracing the progression from early serological techniques to high-resolution molecular and next-generation sequencing (NGS)-based platforms. It highlights how advances in HLA genotyping have significantly enhanced our understanding of histocompatibility and immune recognition. The article also explores the critical role of accurate HLA typing in clinical

S. Tapiawala
Institute of Renal Sciences, Gleneagles Hospitals, Mumbai, Maharashtra, India

D. Daniel (✉)
Transfusion Medicine and Immunohaematology, Christian Medical College, Vellore, Tamil Nadu, India
e-mail: dollyd@cmcvellore.ac.in

S. Jogale
ATDI Lab Private Limited, Mumbai, Maharashtra, India

G. Abraham et al. (eds.), *Immunological Challenges in Organ Transplantation*, https://doi.org/10.1007/978-981-95-1533-2_4

transplantation, emphasizing its impact on donor-recipient matching, graft survival, and the prevention of immune-mediated rejection.

History

The Major Histocompatibility Complex (MHC) and its human counterpart, the Human Leukocyte Antigen (HLA) system, have been central to immunology, transplantation biology, and disease susceptibility research. These highly polymorphic gene clusters play crucial roles in antigen presentation and immune response regulation. The discovery and subsequent characterization of the MHC and HLA have significantly advanced our understanding of immunogenetics and human disease [1].

Early Discoveries and Initial Concepts

The foundations of MHC research were laid in the early twentieth century when it was observed that tissue grafts were rejected between genetically non-identical individuals. The first major breakthrough occurred in the 1930s when Peter Gorer identified a group of genes in mice, later termed the Histocompatibility (H) antigens, which were responsible for transplant rejection [3]. Subsequent studies by George Snell further elucidated these genes, leading to the identification of the murine MHC, termed H-2 system [4].

Discovery of the Human MHC: HLA System

The discovery of the human MHC began in the 1950s when Jean Dausset, working with transfusion patients, identified leukocyte-specific antibodies. These antibodies led to the recognition of the first human histocompatibility antigens, forming the basis of the HLA system. In parallel, Rose Payne and colleagues confirmed that leukocyte antigens were inherited, paving the way for HLA typing and its applications in transplantation [5].

By the 1960s and 1970s, serological and genetic studies refined the HLA classification into major loci—HLA-A, HLA-B, and HLA-C (Class I) and HLA-DR, HLA-DQ, and HLA-DP (Class II). The emergence of molecular biology in the 1980s allowed researchers to sequence HLA genes, revealing their extreme polymorphism and structural complexity [5, 6].

The HLA system is a critical component of the immune system that facilitates self-recognition and immune surveillance. Discovered in the mid-twentieth century, HLA molecules are central to adaptive immunity and are particularly relevant in organ and tissue transplantation. The ability of HLA molecules to present peptides to T cells dictates immune responses, making them pivotal in transplant compatibility and immunological tolerance [7].

Structure of the HLA System

HLA genes are located on the short arm of chromosome 6 and are categorized into three classes (see Fig. 4.1 for Genetic location of HLA system):

Class I HLA (HLA-A, HLA-B, HLA-C): Found on almost all nucleated cells, these molecules present endogenous peptides to CD8+ cytotoxic T cells. They consist of a heavy chain associated with beta-2 microglobulin [2, 7]. See Fig. 4.2 for structure of HLA Class I molecule.

Class II HLA (HLA-DR, HLA-DQ, HLA-DP): Expressed mainly on antigen-presenting cells (APCs) such as dendritic cells, macrophages, and B cells, these molecules present exogenous peptides to CD4+ helper T cells. They consist of an alpha and a beta chain [2, 7]. See Fig. 4.2 for structure of HLA Class II molecule.

Class III HLA: Encodes components of the complement system and other immune-related proteins but is not directly involved in antigen presentation. HLA genes are highly polymorphic, contributing to individual variability in immune responses and transplant outcomes [2].

Function of the HLA System

HLA molecules play a fundamental role in antigen presentation and immune activation: Self versus Non-Self Recognition: HLA molecules present peptides to T cells, enabling the immune system to differentiate between self and foreign antigens. Adaptive Immunity: Through peptide presentation, HLA molecules initiate and regulate immune responses, influencing susceptibility to infections, autoimmune diseases, and transplant rejection. Immune Tolerance: Regulatory mechanisms, such as central and peripheral tolerance, prevent autoimmunity by ensuring that self-antigens do not elicit immune responses [8, 9].

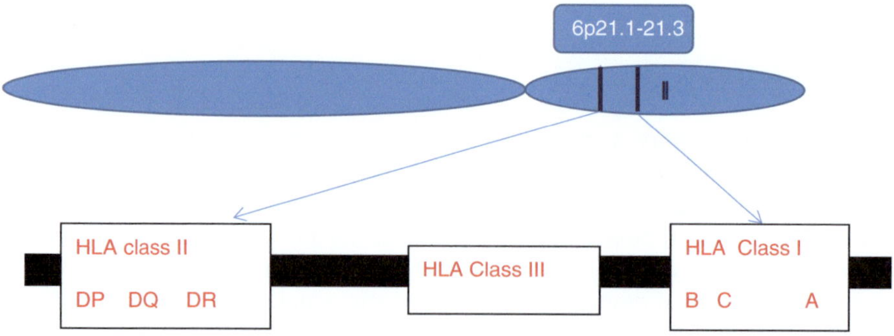

Fig. 4.1 Genetic location of HLA system

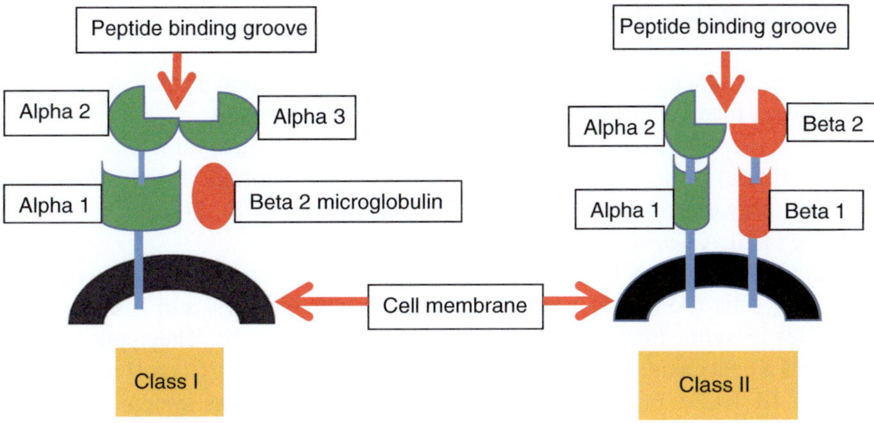

Fig. 4.2 Structure of HLA Class I and Class II molecule

HLA Inheritance

Is Mendelian inheritance and co-dominant in expression, meaning that both maternal and paternal alleles are expressed simultaneously. Unlike dominant or recessive inheritance patterns, each individual inherits two complete sets of HLA genes, one from each parent. Each set is called a haplotype. HLA genes are inherited together as part of a haplotype, which is a set of linked genes on a single chromosome. Each offspring receives one haplotype from each parent, leading to extensive genetic diversity. Figure 4.3 shows the possible combinations/matches that can occur [2].

So every child will be a haplomatch with each of their parents. There is a 25% chance that two siblings will be HLA identical, a 25% chance that they will be completely mismatched and a 50% chance that they will be a haplomatch.

Nomenclature

The HLA nomenclature is an intricate system that has evolved over time to accommodate the increasing number of discovered alleles and to standardize the classification of HLA genes. The system is maintained and updated by the World Health Organization (WHO) Nomenclature Committee for Factors of the HLA System, ensuring consistency in the identification and classification of HLA variants [10]. Molecular methods of typing are used now which include the PCR based SSP and SSOP techniques, and sequence based typing including the NGS.

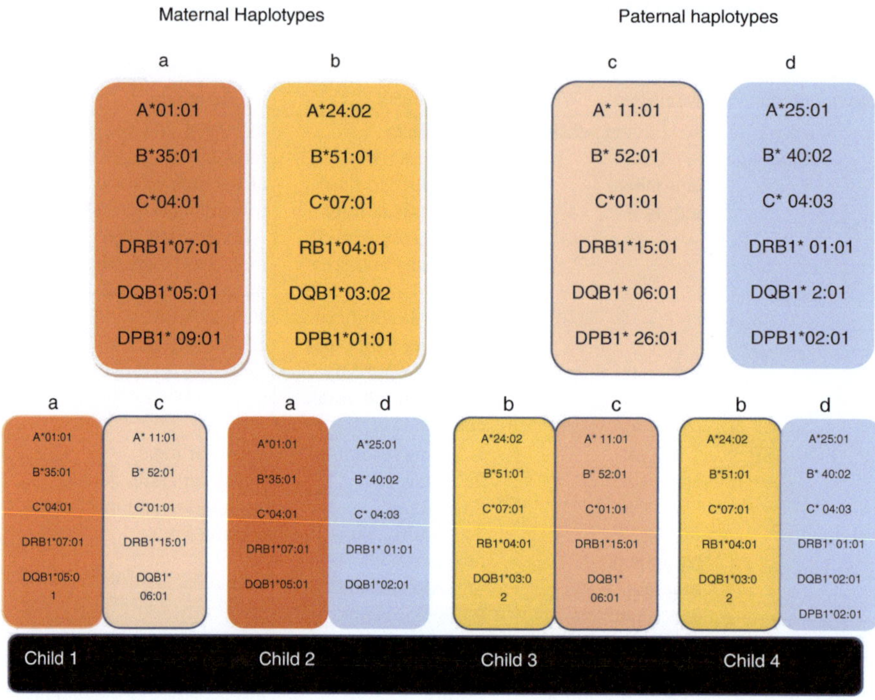

Fig. 4.3 Inheritance of HLA genes to offspring from parents

HLA Typing Methods

Evolution of HLA Typing Technologies

Initially, HLA typing relied on two primary techniques: Serological typing—Based on antigen-antibody reactions using antisera and Mixed lymphocyte culture (MLC)—Used to determine compatibility between donor and recipient lymphocytes. With technological advancements, more precise DNA-based molecular methods have been developed, including Sequence-Specific Oligonucleotide Probe Hybridization (SSOP), Sequence-Specific Primer Amplification (SSP), Sequencing-Based Typing (SBT), and Next Generation Sequencing (NGS).

These methods have significantly improved HLA matching and transplantation outcomes [11]. Table 4.1 demonstrates the comparison of all HLA typing methods.

The HLA system can be tested to give differing levels of resolution.

- The low resolution typing identifies only the allele family.
  Identifies HLA alleles at a broad level, grouping them into families based on shared serological reactivity. For example, HLA-A*02:01 and HLA-A*02:05 would both be reported as HLA-A*02.

Table 4.1 Comparison of different HLA typing methods

No	HLA typing methods	Advantages	Disadvantages	Applications
1	Serology method	– Less expensive – Less equipment required	1. Due to cross-reactivity possibility of false positive reactions 2. Less sensitive 3. Many antigens are unidentified due to unavailability of detecting antibodies and low reactivity 4. Required viable cells	Till 2004, it was used for renal transplantation as well as for bone marrow transplantation (BMT). After introduction of molecular methods, this method is not used as a standard method to identify HLA antigens
2	PCR-SSP	1. Less expensive 2. Less expensive equipment required 3. Rapid test	1. Low resolution 2. Lower throughput 3. Ambiguous results 4. Alpha chain typing kit is not available.	Used for renal transplantation, disease association and in transfusion
3	PCR-SSOP	1. Intermediate to high resolution typing 2. High throughput 3. Robust 4. Ideal for virtual crossmatch	1. Ambiguous results	Used for transplantation, disease association and in transfusion
4	SBT	1. Highest resolution 2. Full allele typing 3. Direct identification new allele	1. Lowest throughput 2. More technically demanding 3. Expensive	Used for organ transplantation, BMT or hematopoietic stem cell transplantation and for resolution of ambiguous results from other methods
5	NGS	1. Highest resolution 2. Full allele typing 3. Direct identification new allele	1. More technically demanding 2. Expensive for small volume	Used for organ transplantation, BMT or hematopoietic stem cell transplantation and for resolution of ambiguous results from other methods

- The intermediate resolution gives a list of the most probable alleles.
  Provides more detailed allele information than low-resolution typing but does not reach the full specificity typing. It typically distinguishes groups of alleles that differ at the protein level but may not identify each specific allele variant. For example, it reports HLA-A*02:01 or 02:02 specifying which exact allele is present.
- The high resolution typing defines the specific allele identified. The latter is most commonly done currently on the NGS platform.
  Identifies HLA alleles at the most specific level by determining the exact nucleotide sequence of the gene's antigen recognition site. For example, it can reports up to six or eight digits HLA-A**02:01:01:01 or HLA-A**02:01:01.

Serological Methods—One of the earliest and most foundational phenotyping techniques in the field of histocompatibility testing was the complement-mediated microlymphocytotoxicity assay, pioneered by Paul Terasaki and colleagues in the 1960s. This method utilized live lymphocytes—T cells for HLA class I antigen detection and B cells for class II—isolated from peripheral blood [12]. Sera containing anti-HLA antibodies, often derived from sensitized individuals such as multiparous women or multi-transfused patients, were used as reagents due to their broad HLA alloantibody repertoire [12]. The assay was typically performed using Terasaki trays, which are 60- or 72-well plates designed to facilitate testing of multiple antigen-serum combinations in small volumes.

The test relies on the principle of complement-dependent cytotoxicity (CDC), wherein the binding of specific antibodies to surface HLA molecules on lymphocytes triggers the classical complement cascade, resulting in cell lysis. The presence of cell death, indicated by uptake of vital dyes such as trypan blue or eosin, signified a positive reaction. This technique became the gold standard for HLA typing for several decades and was instrumental in the early development of transplantation immunology. Importantly, the same CDC principle was later adapted for cross-matching, allowing for the detection of preformed donor-specific antibodies (DSAs) in recipient serum, thereby providing critical risk assessment prior to transplantation [13, 14]. See Fig. 4.4 for illustration of serology typing method.

While the microlymphocytotoxicity assay has largely been supplanted by more sensitive and specific DNA-based and solid-phase immunoassays in modern practice, it laid the groundwork for understanding HLA immunogenicity and compatibility assessment. It remains a valuable historical and practical reference point, especially in centers where resource limitations preclude the use of advanced platforms.

Molecular HLA Typing Methods

The advent of DNA-based methods has revolutionized the field of HLA typing, marking a significant departure from traditional serological approaches. This transformation was largely enabled by the development of the polymerase chain reaction

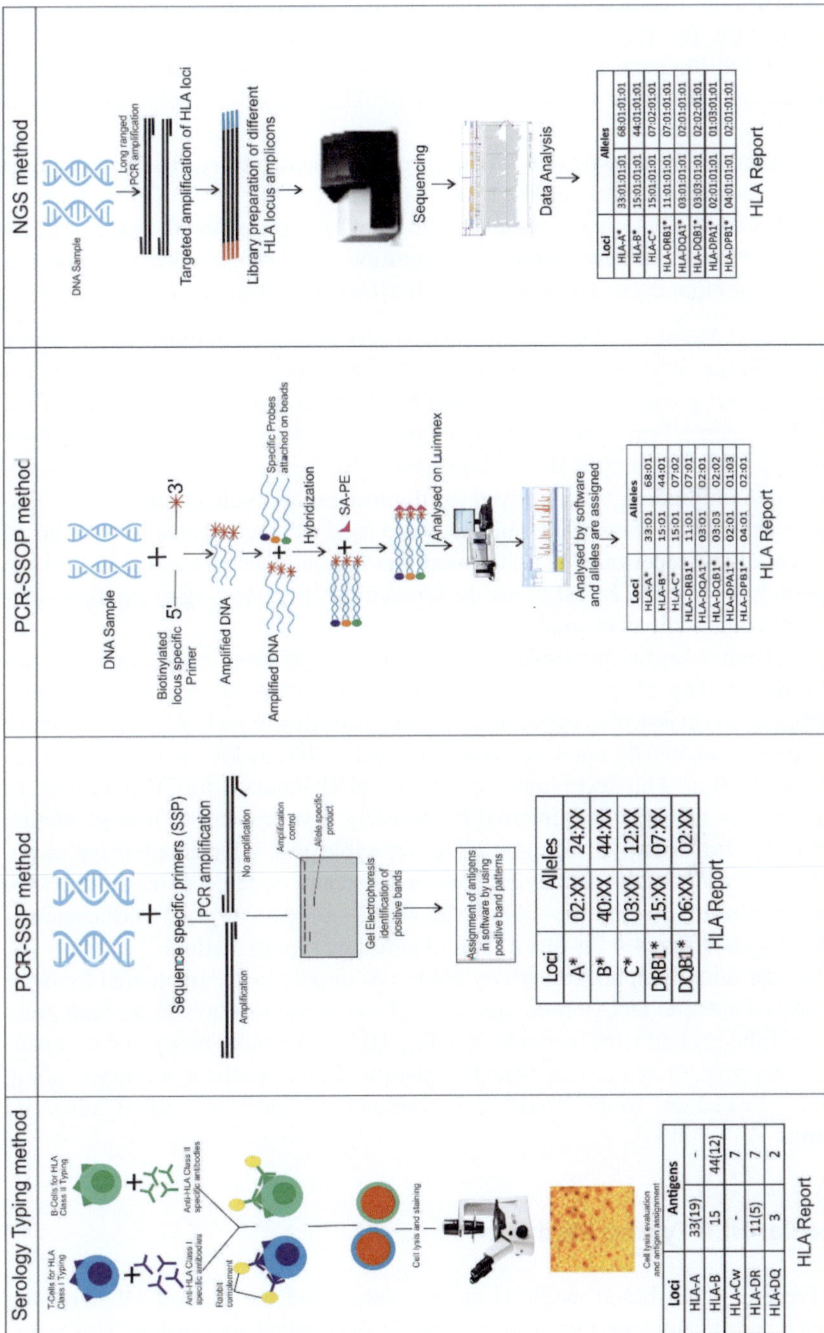

Fig. 4.4 Illustration of Different HLA typing methods [15]

(PCR), which allows for the exponential amplification of specific DNA sequences, thereby enabling the analysis of minute quantities of genomic material with high accuracy [16]. PCR-based techniques provide enhanced precision, reproducibility, and sensitivity, and have the added advantage of detecting both expressed and non-expressed alleles, which serological methods often miss. These methods offer greater precision, reproducibility, sensitivity, and resolution compared to traditional serological techniques. Some of the key molecular HLA typing approaches include PCR-sequence-specific primers (SSP), PCR-sequence-specific oligonucleotide probes (SSO), real-time PCR, sequencing-based typing (SBT), and next-generation sequencing (NGS). See Fig. 4.4 for illustration of molecular typing methods.

PCR—Sequence-Specific Primers (PCR-SSP)

In 1989, Newton et al. introduced the use of sequence-specific primers for DNA amplification using the Amplification Refractory Mutation System (ARMS), a method that significantly advanced the molecular analysis of polymorphic genes such as those of the HLA complex [17]. The ARMS technique exploits the high specificity of PCR by designing primers that are complementary only to specific alleles at their 3′ termini. A mismatch at this critical position prevents the primer from annealing effectively, thereby blocking the initiation of amplification. This allows for the discrimination of single nucleotide polymorphisms (SNPs) with high fidelity, enabling precise allele identification.

In the context of HLA typing, ARMS-based PCR (commonly referred to as PCR-SSP) became a powerful tool for genotyping due to its capacity to detect minute differences between highly homologous HLA alleles. The process involves setting up multiple parallel PCR reactions, each with primers designed to amplify only if a specific allele is present. Following amplification, the PCR products are separated using agarose gel electrophoresis, and the presence or absence of specific DNA bands is interpreted according to a predefined allele-calling pattern.

This method offered several advantages over conventional serological typing. Unlike serological methods, which rely on the availability and specificity of antisera and are often limited by cross-reactivity and subjectivity, PCR-SSP provides an objective, reproducible, and highly sensitive molecular approach. It also enables detection of null or rare alleles that might be missed in serological assays [18]. Moreover, ARMS-PCR can be completed in a relatively short time frame (often within a few hours), making it suitable for clinical settings, including deceased donor transplant matching, where rapid and accurate typing is essential.

PCR—SSOP Method

The sequence-specific oligonucleotide (SSO) hybridization technique is a widely used molecular method for HLA typing that enables intermediate to high-resolution allele identification. The process begins with the amplification of genomic DNA

using locus-specific primers designed to target specific HLA genes, such as HLA-A, HLA-B, or HLA-DRB1. These primers amplify conserved regions within the highly polymorphic HLA loci, ensuring robust and consistent PCR products suitable for downstream analysis [19].

Following amplification, the PCR products are hybridized with a panel of oligonucleotide probes that are complementary to allele-specific sequences within the HLA genes. These probes are immobilized on fluorescently labeled microspheres—each with a unique spectral signature—facilitating the simultaneous interrogation of multiple alleles in a single reaction. A reporter molecule, typically a streptavidin-conjugated phycoerythrin (PE), binds to biotinylated PCR products, and fluorescence signals are then detected and analyzed using the Luminex xMAP technology [20] (see Fig. 4.4).

The Luminex platform enables high-throughput, multiplexed testing by analyzing up to 100 distinct bead sets in a single reaction well. This system significantly reduces turnaround time and allows for batch testing of multiple samples, making it particularly suitable for donor registries, large-scale HLA screening, and clinical laboratories. Moreover, the method provides intermediate (two-digit) to high-resolution (four-digit) allele typing depending on the probe design and panel coverage. At the four-digit resolution level, alleles are distinguished at the protein sequence level, which is especially valuable in clinical settings such as virtual crossmatching (VXM), where precise donor-recipient allele matching is required to predict compatibility and avoid antibody-mediated rejection [21].

SSO typing, particularly when integrated with the Luminex system, offers a balance of speed, accuracy, and resolution. Although it may lack the ultra-high resolution of sequencing-based methods, its scalability, cost-effectiveness, and ability to detect most clinically relevant polymorphisms make it an indispensable tool in histocompatibility laboratories worldwide.

Sequence-Based Typing (SBT)

Among the various PCR-based HLA typing methods, sequence-based typing (SBT) is widely regarded as one of the most accurate and reliable techniques for high-resolution allele identification. Unlike hybridization-based methods such as sequence-specific oligonucleotide (SSO) or sequence-specific primer (SSP) assays, which rely on the presence or absence of probe or primer binding to infer allele identity, SBT provides the actual nucleotide sequence of the targeted HLA gene regions. This direct sequencing approach allows for precise and unambiguous assignment of HLA alleles, often at the four-digit or higher resolution level, making it particularly valuable in both clinical and research settings [22, 23].

SBT typically involves the amplification of specific exons within HLA loci—most commonly exons 2 and 3 for class I genes and exon 2 for class II genes—followed by bidirectional Sanger sequencing [24]. The resulting chromatograms are analyzed and compared against reference databases, such as the IMGT/HLA database, to determine the most likely allele match. Because many HLA alleles differ by

only one or a few nucleotides, the ability to read exact base sequences is crucial for resolving closely related alleles that cannot be distinguished by SSP or SSO methods [2].

One of the major strengths of SBT is its capacity to resolve ambiguities that often arise from probe-based typing. In many cases, multiple alleles may share the same hybridization pattern, leading to ambiguous or multiple possible allele calls. SBT overcomes this limitation by revealing the specific nucleotide variations responsible for allelic diversity, allowing for definitive typing results. Moreover, as the method does not rely on predefined panels of probes or primers, it is particularly well-suited for the identification of novel or rare alleles, thereby contributing to the continuous expansion of the known HLA allele repertoire [24].

Despite its high resolution and accuracy, SBT does have limitations, including relatively high costs, longer turnaround times, and the potential for phasing ambiguities—especially in heterozygous individuals—due to the inability to determine which nucleotide sequence belongs to which allele when multiple polymorphisms are present in a single sequencing read. Nonetheless, advancements such as allele-specific sequencing strategies and the incorporation of complementary methods (e.g., cloning or next-generation sequencing) have been employed to address these challenges [25].

Next-Generation Sequencing (NGS)

The advent of next-generation sequencing (NGS) has revolutionized HLA genotyping by providing unparalleled depth, resolution, and throughput compared to traditional methods such as Sanger sequencing. NGS-based HLA typing combines locus-specific PCR amplification with massively parallel sequencing, enabling comprehensive analysis of both class I (HLA-A, -B, -C) and class II (HLA-DR, -DQ, -DP) genes with high precision. This technology allows for allele-level resolution up to six digits, distinguishing not only coding variations but also intronic and untranslated region polymorphisms that may influence gene expression and immune responses [26, 27].

One of the most significant advantages of NGS over earlier genotyping methods is its ability to accurately phase sequencing reads. In heterozygous individuals, where two different alleles are present at the same locus, NGS can assign nucleotide variants to their respective alleles by sequencing long contiguous stretches of DNA. This phasing capability is critical for resolving ambiguities that are commonly encountered in Sanger-based sequence-based typing (SBT), especially in regions with high polymorphism [2, 28].

NGS is also inherently scalable, capable of processing dozens to hundreds of samples simultaneously, making it ideally suited for large-scale donor registries, population genetics studies, and high-throughput clinical laboratories. The platform flexibility—ranging from benchtop sequencers for smaller volumes to high-capacity instruments for bulk sequencing—further supports its adoption across diverse laboratory settings [29].

In addition to its high resolution and scalability, NGS enables the detection of novel alleles and rare variants that may be missed by probe- or primer-based approaches. This is particularly important for patients from underrepresented populations, where reference allele databases may be incomplete, and for personalized transplantation strategies, where even subtle HLA mismatches can impact graft survival and immune compatibility [30].

NGS has now emerged as the gold standard in HLA typing and is increasingly being integrated into both clinical and research workflows. Its robustness, reproducibility, and comprehensive data output make it an essential tool for advancing precision medicine in transplantation, immunogenetics, and disease association studies. An example of a high resolution typed allele by NGS is shown in Fig. 4.5.

Updates and Revisions

- The nomenclature is periodically revised to integrate new allele discoveries, refine existing allele classifications, and incorporate genomic sequencing data.
- These updates enhance accuracy in HLA matching, particularly in transplantation medicine.
- The extremely high polymorphism at the HLA locus is reflected in the number of alleles described at each locus as shown in the Table 4.2 below.

The clinical impact of HLA typing on renal transplant has been evident in many studies. The better the match, the better the outcome. With the high degree of polymorphism in HLA, it can be rather challenging to get a perfect match.

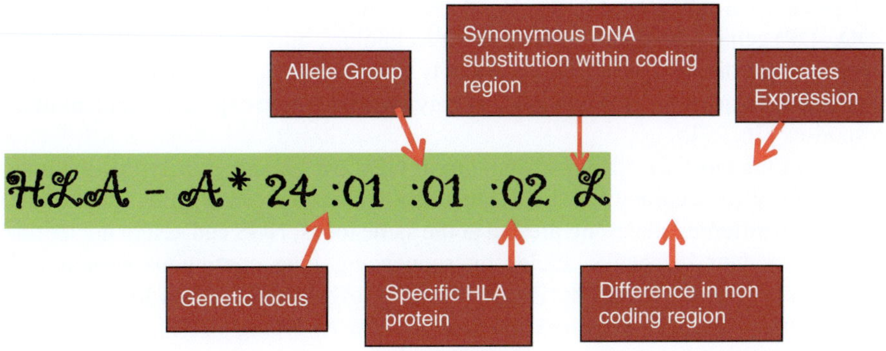

Fig. 4.5 Nomenclature of HLA allele

Table 4.2 Total number of HLA alleles identified

Gene	A*	B*	C*	DRB1*	DQA1*	DQB1*	DPA1*	DPB1*
No of alleles	8012	9573	7995	4530	688	2491	619	2393

Clinical Application in Organ Transplantation

HLA typing plays a pivotal role in the field of organ transplantation by guiding donor-recipient matching, reducing the risk of rejection, and improving long-term graft survival. MHC is the most polymorphic region in the human genome. It impacts the graft outcome by mediating the discrimination between self and non-self. The non self represents the mismatched antigens/alleles from the donor (immunizer). HLA typing plays an important role in immune stratification of a prospective recipient. The HLA typing of a recipient and the prospective donor helps identify the mismatches with the donor (immunizer) and hence identify the foreign antigens the recipient's immune system will be exposed to. Further the mismatched antigens and alleles help pre transplant identification of donor specific antibodies and post-transplant monitoring of DSAs and identification of De Novo DSAs [31].

The clinical application of HLA typing in organ transplantation has evolved from low-resolution, serological methods to increasingly sophisticated high-resolution molecular techniques, enabling more precise immunogenetic profiling. Low-resolution typing, often referred to as antigen-level or two-digit typing, identifies broad HLA antigen groups (e.g., HLA-A*02), which are typically sufficient for solid organ transplantation in urgent clinical scenarios, particularly kidney and liver transplants. In such settings, the emphasis is often placed on avoiding major incompatibilities at HLA-A, -B, and -DR loci, which are most strongly associated with acute rejection and poor graft outcomes [31].

However, with advancements in molecular technologies, high-resolution typing (four- to six-digit level) has become increasingly important in refining donor-recipient compatibility. High-resolution typing identifies specific allelic variants (e.g., HLA-A02:01 vs. HLA-A02:07) and can reveal subtle mismatches that may not be detected with low-resolution methods. This level of detail is particularly critical for highly sensitized patients—those with preformed donor-specific antibodies (DSAs)—where allele-level mismatches can trigger antibody-mediated rejection. In such cases, virtual crossmatching using high-resolution HLA data helps predict compatibility and avoid transplant failure [32].

Role of HLA Typing in Transplant Compatibility

HLA typing is used to assess compatibility between organ donors and recipients, helping to minimize the risk of immune rejection and identify donor-specific antibodies (DSAs).

Following the first successful kidney transplant between identical twins in 1954, research demonstrated that HLA-identical kidney transplants had significantly better graft survival rates compared to HLA-mismatched transplants [33]. Historically, the primary focus has been on matching HLA-A, HLA-B, and HLA-DR due to their high levels of polymorphism and expression.

A zero antigen mismatch was initially defined as the absence of any HLA-A, HLA-B, or HLA-DR antigen in the donor's HLA profile that differed from the recipient's. However, early limitations in typing technology and reagent availability restricted the assessment of other loci, such as HLA-C, HLA-DP, and HLA-DQ.

With advancements in polymerase chain reaction (PCR)-based technologies, particularly reverse sequence-specific oligonucleotide (R-SSO) and sequence-specific primer (SSP) methods, HLA typing has evolved from serological approaches to molecular-based techniques, improving accuracy in allele identification and mismatch detection.

Allelic Variants and Antibody Formation

Serological assays, which rely on whole-antigen antibody reactions, lack the ability to distinguish between allelic variants with high specificity.

Previously, a zero antigen mismatch was defined based on the absence of HLA-A, -B, or -DR antigens in the donor that were different from those in the recipient. However, modern molecular typing methods have increased the number of possible mismatches significantly, allowing for up to two allele or antigen mismatches at each of the following loci: HLA-A, HLA-B, HLA-C*, HLA-DRB1*, HLA-DRB3/4/5*, HLA-DQA1*, HLA-DQB1*, HLA-DPA1*, and HLA-DPB1*.

HLA molecules such as HLA-Cw, HLA-DP, and HLA-DQ are now recognized as important antigenic targets for immune responses. Studies indicate that pre-transplant or de novo antibodies against HLA-C, DP, and DQ are associated with poor graft outcomes, increased rates of acute rejection, and immunologically driven graft loss. Notably, HLA-DQ antibodies have emerged as the most commonly detected class II donor-specific antibodies (DSAs). As a result, HLA typing for these loci has been integrated into organ allocation algorithms in the United States and Europe.

A complete HLA typing of all clinically relevant alleles allows for assessing mismatches in a more objective way. Apart from defining allelic mismatches, epitope/eplet mismatches, PIRCHE scores, electromagnetic mismatch scores, can be calculated, which have been shown to impact on graft outcome, de novo antibody formation and thereby impacting graft longevity.

Impact of HLA Matching and Graft Outcomes

Numerous evidence-based single- and multicenter studies have consistently demonstrated that reducing the number of HLA mismatches between donor and recipient correlates with improved long-term graft survival in renal transplantation. These findings underscore the immunological importance of HLA compatibility and have informed both clinical practice and organ allocation policies worldwide.

A pivotal study by Opelz et al., derived from the Collaborative Transplant Study (CTS) database, provided robust evidence of the impact of HLA mismatching on long-term graft outcomes. This study revealed that patients receiving kidney transplants with a full HLA mismatch at the HLA-A, -B, and -DR loci (defined as a 2:2:2 mismatch) experienced a 17% lower graft survival rate at 10 years compared to those with a complete HLA match (0:0:0). This substantial difference highlights the importance of minimizing HLA mismatches when feasible, particularly in younger patients who may require repeat transplants over their lifetime [34].

Among the HLA loci, mismatching at HLA-DR has been shown to have the most pronounced negative effect on graft survival, followed by HLA-B, and finally HLA-A. The hierarchy of immunogenicity aligns with the central role of HLA-DR in initiating alloimmune responses, given its high expression on antigen-presenting cells and strong involvement in CD4+ T-cell activation. Consequently, many organ allocation algorithms prioritize matching at the HLA-DR locus to enhance long-term outcomes [35].

Interestingly, while the impact of each locus appears to be distinct in the early post-transplant period, long-term studies have shown that the influence of HLA-A, -B, and -DR mismatches on graft survival tends to converge beyond the 10-year mark. At this stage, their effects appear additive, suggesting that cumulative alloimmune exposure over time—regardless of the specific locus—plays a significant role in chronic allograft injury and eventual graft loss [36].

This evolving understanding supports a more nuanced approach to donor selection, in which HLA matching is considered alongside other critical factors such as recipient sensitization status, donor quality, and recipient comorbidities. Furthermore, with advances in immunosuppressive therapies, the relative impact of HLA mismatching may be mitigated in some contexts; however, the fundamental association between HLA compatibility and improved graft longevity remains well established.

Data from both retrospective registries and prospective studies strongly advocate for the minimization of HLA mismatches to optimize transplant outcomes. The enduring influence of HLA disparities, even in the context of modern immunosuppression, affirms the value of immunogenetic matching as a cornerstone of transplant success [37].

While high-resolution typing is the standard of care in hematopoietic stem cell transplantation due to its stringent matching requirements, its role in solid organ transplantation is expanding. Advances in next-generation sequencing (NGS) now allow for rapid, high-throughput, and cost-effective generation of high-resolution HLA data, making it increasingly feasible for routine use in transplant centers [28].

As transplant immunology moves toward precision medicine, incorporating high-resolution HLA typing into clinical decision-making is expected to improve personalized risk stratification, optimize immunosuppression, and enhance graft outcomes across various organ types.

Here we can see the relevance of the intermediate resolution typing. From a zero mismatch he moved into the category of a haplomatch. Transplantation is not a contraindication, these differences can most certainly cause alloimmunisation and subsequent graft injury, and hence must be viewed with caution, particularly when there is intent to taper immunosuppression. Further if patient has DSAs against the mismatched HLA antigens and alleles, they can make the patient high risk for rejection. This can only be identified when we have a molecular intermediate level typing.

The availability of high-resolution data has enhanced the use of eplet-based matching algorithms, such as HLAMatchmaker, which assess compatibility at the structural epitope level rather than the traditional antigen level. This approach enables a more nuanced assessment of immunologic risk and facilitates better donor selection, especially in re-transplant or pediatric patients where long-term outcomes are a priority [38].

In this cohort study done by Thiago Abramo Alves et al., of 159 renal transplant recipients, 17.6% experienced allograft loss during the follow-up period, while rejection episodes were observed in 37.1% of patients. Among these, antibody-mediated rejection (ABMR) accounted for 13.6% of all cases. The HLA-DR eMM was most strongly associated with a worse immunologic outcome than eMM mismatches for HLA-AB [39].

Aleksandar Senev et al. evaluated 926 kidney recipients for the effect of number of eplet mismatches (mismatch load) on de novo formation of donor-specific HLA antibodies (DSAs) and transplant outcomes. De novo donor-specific antibodies (DSAs) developed in 43 patients (4.6%). Multivariable analysis revealed that the eplet mismatch load, when verified through antibody reactivity, was independently and significantly associated with both the development of de novo DSAs and subsequent graft failure. This association was predominantly driven by mismatches at the HLA-DQ locus, highlighting the immunologic relevance of DQ-specific antibody-verified eplet mismatches [40].

Principles of HLA-Based Relationship Testing and DNA Profiling

Each individual inherits one set of HLA alleles from each parent, forming a unique haplotype. By comparing HLA haplotypes among family members, genetic relatedness can be established.

Haplotype Sharing: In parent-child testing, the child must inherit one haplotype from each biological parent, making HLA typing a reliable tool for confirming parentage.

Case Study 1

A 49-year-old male patient with a diagnosis of chronic kidney disease (CKD) was evaluated for a pre-emptive living donor renal transplantation. The potential donor was his spouse. The patient had no prior history of blood transfusions, organ transplantation, or sensitizing events. In order to establish and confirm the biological relationship between the donor and recipient, HLA typing of their child was performed. A family haplotype analysis was subsequently constructed to assess the segregation and inheritance of HLA alleles, supporting the parentage and genetic relationship between the donor (wife) and recipient (husband). See Fig. 4.6 for haplotype segregation and inheritance.

HLA typing was performed on their daughter. The analysis demonstrated that the daughter inherited haplotype ***'b'*** from her father and haplotype ***'d'*** from her mother. A pedigree-based haplotype analysis was subsequently constructed, confirming the segregation of parental HLA alleles and thereby substantiating the familial relationship and haplotype inheritance pattern.

Case Study 2

A 31-year-old male patient with end-stage renal disease on maintenance hemodialysis (MHD) was evaluated for a living donor renal transplantation. His sister was identified as the prospective donor. Initial HLA typing of the patient and donor revealed that they were HLA-unidentical, with complete mismatching at all loci. To

Recipient (Husband)

Haplotype –'a'	Haplotype 'b'
HLA-A*02:01	HLA-A*11:01
HLA-B*40:06	HLA-B*07:05
HLA-DRB1*15:01	HLA-DRB1*15:01

Donor (Wife)

Haplotype –'c'	Haplotype –'d'
HLA-A*01:01	HLA-A*01:01
HLA-B*55:01	HLA-B*57:01
HLA-DRB1*11:03	HLA-DRB1*04:04

Daughter

Haplotype –'b'	Haplotype –'d'
HLA-A*11:01	HLA-A*01:01
HLA-B*07:05	HLA-B*57:01
HLA-DRB1*15:01	HLA-DRB1*04:04

Fig. 4.6 Showing the haplotype inheritance from parents to child

verify the biological relationship and further evaluate the unexpected HLA disparity, HLA typing of both parents was performed. The results demonstrated that the recipient had inherited haplotype **‘a’** from the mother and haplotype **‘c’** from the father, while the donor had inherited haplotype **‘b’** from the mother and haplotype **‘d’** from the father (see Fig. 4.7 for inheritance of haplotypes). This pattern confirmed that the siblings shared no common haplotypes, consistent with Mendelian inheritance.

Such findings are in line with classical genetic expectations, where there is a 25% probability of siblings being HLA-unidentical (i.e., sharing no HLA haplotypes), despite being full biological siblings.

Case Study 3

A 35-year-old patient Mr. DSD with ESRD comes with his sibling ACD as a prospective donor for a pre transplant immunological work-up. He has a HLA report that shows him to be HLA identical to his brother. This is his second transplant. The first transplant was 5 years ago which he lost due to a suspected acute antibody mediated rejection. His mother was his previous donor.

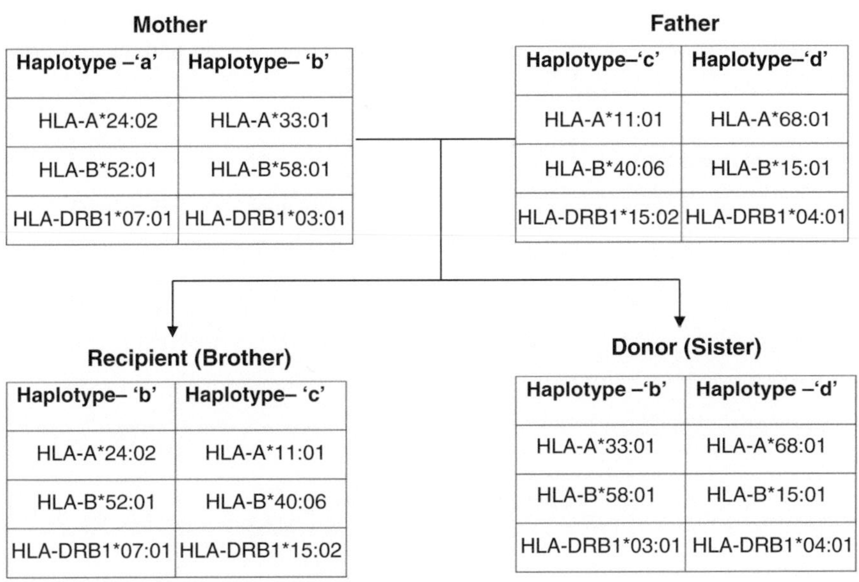

Fig. 4.7 Showing the haplotype inheritance from parents

Locus	A*	B*	C*	DRB1*	DQB1*	Inference
Mr DSD	02	52	04	09	03	10/10 match-HLA identical
	24	35	07	04	05	
Sibling ACD	02	52	04	09	03	
	24	35	07	04	05	

He was evaluated with a single antigen bead (SAB) assay. Virtual crossmatch (VXM) could not be performed to identify DSAs as HLA typing available with the patient was a two digit low resolution typing. An intermediate resolution HLA typing was recommended.

His high-resolution HLA typing results were as below:

Locus	A*	B*	C*	DRB1*	DQB1*	Inference
Mr.	02:01	52:03	04:01	09:01	03:02	Haplomatch (5/10)
	24:02	35:01	07:01	04:01	05:01	
	02:02	52:03	04:01	09:01	03:02	
	24:02	35:03	07:02	04:03	05:03	

DNA Profiling for Relationship

DNA profiling helps in providing a robust method for assessing genetic relatedness between individuals and proving relationship. Since HLA genes are highly polymorphic and inherited in a Mendelian codominant pattern, analyzing HLA markers can confirm biological relationships with high accuracy.

Case Study 4

A 35-year-old male patient was evaluated for a second living donor renal transplantation. His younger brother was identified as a willing and medically eligible donor. However, HLA typing did not demonstrate any shared haplotypes between the donor and recipient, raising concerns regarding their biological relationship.

Given the complete HLA mismatch and in accordance with legal and ethical requirements for living organ donation, DNA profiling (short tandem repeat [STR] analysis) was performed to establish the familial relationship. The DNA profiling confirmed a full sibling relationship between the donor and recipient, thereby satisfying legal and regulatory criteria for proceeding with the transplant. (Example of DNA profiling report is shown in Fig. 4.8.)

This case underscores the utility of DNA-based identity testing in scenarios where HLA typing alone may not be sufficient to confirm biological relationships, particularly in the context of living-related organ donation. Extended Kinship Analysis: HLA profiling is also used in cases requiring verification of more distant

DNA PROFILING REPORT

Test Type: Sibling Relationship Analysis (STR Profiling)
Method: Capillary Electrophoresis-based Short Tandem Repeat (STR) Genotyping
Markers Tested: 21 autosomal STR loci + Amelogenin (sex determination)

Sample Details:

Individual	Sample ID	Relationship Claimed
Individual A (Recipient)	R-001	Brother
Individual B (Donor)	D-001	Brother

Locus	Individual A (R-001)	Individual B (D-001)
D3S1358	15, 16	15, 17
vWA	17, 18	18, 19
FGA	22, 25	22, 24
D8S1179	12, 13	12, 14
D21S11	29, 30	28, 30
D18S51	14, 17	14, 16
D5S818	11, 12	11, 13
D13S317	10, 11	10, 11
D7S820	8, 10	9, 10
CSF1PO	11, 12	11, 12
TH01	6, 9	6, 9
TPOX	8, 8	8, 8
D16S539	11, 12	11, 12
D2S1338	19, 20	19, 21
D19S433	13, 14	13, 15
D12S391	18, 21	18, 22
D1S1656	12, 14	12, 13
D10S1248	13, 15	13, 16
D22S1045	14, 15	14, 15
D2S441	10, 12	11, 12
SE33	20, 24	22, 24
Amelogenin	X, Y	X, Y

Interpretation:

Based on the analysis of 21 autosomal STR loci, there is a high likelihood of a full sibling relationship between Individual A and Individual B. The statistical probability (Sibling Index) exceeds the threshold for full-sibling relatedness as per accepted guidelines (SI > 100). The profies show a high degree of allele sharing consistent with Mendelian inheritance patterns expected in full siblings.

Fig. 4.8 Example of DNA profiling report

familial connections, such as grandparent-grandchild or avuncular (aunt/uncle-niece/nephew) relationships.

Clinical and Forensic Applications

Parentage Testing: HLA typing is utilized alongside other genetic markers to confirm biological parentage, especially in cases where traditional short tandem repeat (STR) analysis may be inconclusive.

Forensic Identification: In cases involving missing persons, disaster victim identification, or disputed inheritance claims, HLA profiling can help establish genetic links between individuals.

Legal and Immigration Cases: HLA typing is sometimes employed in legal disputes, including immigration cases requiring proof of family relationships.

HLA-based DNA profiling, combined with modern molecular techniques such as next-generation sequencing (NGS) and sequence-based typing (SBT), enhances the precision and reliability of relationship verification in clinical and forensic settings.

HLA Bioinformatics and KIR Genotyping

The advent of next-generation sequencing (NGS) has significantly advanced the field of histocompatibility and immunogenetics by enabling high-resolution HLA genotyping. For clinicians involved in transplantation, autoimmunity, and pharmacogenetics, understanding the bioinformatics framework that underpins NGS-based HLA typing is essential for interpreting results and applying them in clinical practice.

Bioinformatics Strategies

NGS-based HLA genotyping relies on complex bioinformatics pipelines that transform raw sequencing reads into clinically meaningful allele assignments. Key components include:

Reference-based Alignment: Sequencing reads are aligned to curated HLA databases such as the IPD-IMGT/HLA database. These references contain thousands of known HLA alleles, and accurate alignment is critical to distinguish between closely related variants.

Phasing Algorithms: Since HLA genes are highly polymorphic and often co-amplified, phasing (i.e., determining which variants reside on the same chromosome) is essential for accurate allele pairing. Algorithms use read overlap or statistical modeling to phase alleles.

Variant Calling and Genotyping Software: Tools such as Omixon HLA Twin, TypeStream, ATHLATES, and NGSengine are commonly used to analyze sequence

data and assign allele calls. These programs account for mismatches, insertions/ deletions, and sequencing errors [41].

Killer-Cell Immunoglobulin-Like Receptors (KIRs) Typing

Killer-cell Immunoglobulin-like Receptors (KIRs) are a family of receptors expressed primarily on natural killer (NK) cells and some T-cell subsets. KIRs play a critical role in the regulation of innate immune responses by interacting with HLA class I molecules, influencing NK cell education, activation, and inhibition. KIR genes are highly polymorphic and variably present across individuals [42].

The presence or absence of specific KIR genes, along with their corresponding HLA ligands, has important implications in:

Hematopoietic stem cell transplantation (HSCT): Certain KIR-HLA mismatches (especially in haploidentical or unrelated donor settings) can enhance graft-versus-leukemia (GVL) effects without increasing graft-versus-host disease (GVHD). Reproductive immunology: KIR-HLA interactions influence implantation and pregnancy success; specific maternal KIR genotypes have been linked to preeclampsia and recurrent miscarriage. Infectious and autoimmune diseases: KIR gene content and interactions with HLA types are implicated in diseases like HIV, hepatitis, rheumatoid arthritis, and ankylosing spondylitis.

KIR gene typing is emerging as an important tool in immunogenetics, especially in transplantation and reproductive immunology. While current clinical applications are primarily limited to gene-content analysis using kit-based PCR methods, advances in NGS and bioinformatics are poised to unlock deeper insights and enhance clinical utility [42].

Recommendations for HLA Typing in Organ Transplantation

Accurate HLA typing remains essential for immunological risk stratification and successful donor-recipient matching in kidney transplantation. International transplant societies have developed detailed recommendations to standardize HLA typing practices, facilitate equitable organ allocation, and improve long-term graft outcomes. Key organizations contributing to these recommendations include the European Federation for Immunogenetics (EFI), the American Society for Histocompatibility and Immunogenetics (ASHI), the British Society for Histocompatibility and Immunogenetics (BSHI), and global policy-setting bodies such as The Transplantation Society (TTS) and Kidney Disease: Improving Global Outcomes (KDIGO) [43–45].

Consensus guidelines universally recommend molecular-based HLA typing at a minimum of intermediate resolution for the HLA class I (HLA-A, -B, -C) and class II (HLA-DRB1, -DQB1) loci in all transplant candidates. High-resolution (allele-level) typing is increasingly advocated, particularly in sensitized patients and for living donor-recipient pairs. Additionally, typing for HLA-DQA1 and HLA-DPB1

loci is strongly recommended in contexts where donor-specific antibody (DSA) testing includes these specificities or in regions where DP mismatches are routinely considered in allocation algorithms [46].

EFI and ASHI guidelines emphasize that high-resolution DNA-based methods such as next-generation sequencing (NGS) or sequence-based typing (SBT) should be the standard for pre-transplant recipient typing [EFI Standards Version 8.0; ASHI Accreditation Manual, 2022].

Conclusion

HLA typing is an important immune stratification tool in the armamentarium of evaluation of a recipient–donor pair. An intermediate- or high-resolution HLA typing at A, B, C, DRB, DP, and DQ typing helps in determining the compatibility of the donor and the recipient. The advances in HLA typing technologies can help in increasing the accuracy of HLA typing and avoid ambiguities, further adding to precision in immunestartifying a recipient and hence helping in tailoring the immunosuppression. DNA profiling of individuals can be performed to established for legal purposes. HLA informatics help in developing algorithms for constructing databases for HLA sequences, accurate HLA allo identification and creating software for matching donors and recipients.

References

1. Klein J, Sato A. The HLA system. First of two parts. N Engl J Med. 2000;343:702–9.
2. Erlich HA. HLA DNA typing: past, present, and future. Tissue Antigens. 2012;80(1):1–11.
3. Gorer PA. The genetic and antigenic basis of tumour transplantation. J Pathol Bacteriol. 1937;44:691–7.
4. Thorsby E. A short history of HLA. Tissue Antigens. 2009;74(2):101–16.
5. Bodmer W. A historical perspective on HLA. Immunother Adv. 2023;3:1–3.
6. van Rood JJ, van Leeuwen A. Leukocyte grouping: a method and its application. J Clin Invest. 1963;42:1382–90.
7. Sheldon S, Poulton K. HLA typing and its influence on organ transplantation. In: Methods in molecular biology, Transplantation immunology: methods and protocols, vol. 333. Totowa, NJ: Humana Press; 2006. p. 157–74.
8. Parham P. Function and polymorphism of human leukocyte antigen-A,B,C molecules. Am J Med. 1988;85(6A):2–5.
9. Kostyu DD, Hannick LI, Traweek JL, Ghanayem M, Heilpern D, Dawson DV. HLA class I polymorphism: structure and function and still questions. Hum Immunol. 1997;57(1):1–18.
10. Marsh SG, Albert ED, Bodmer WF, Bontrop RE, Dupont B, Erlich HA, et al. Nomenclature for factors of the HLA system, 2010. Tissue Antigens. 2010;75:291–455.
11. Jung HL. Shedding a new light on the HLA matching. Korean J Hematol. 2011;46(1):1–2.
12. Terasaki PI, McClelland JD. Microdroplet assay of human serum cytotoxins. Nature. 1964;204(4962):998–1000.
13. Patel R, Terasaki PI. Significance of the positive crossmatch test in kidney transplantation. N Engl J Med. 1969;280(14):735–9.
14. Mittal RD, et al. Significance of complement-dependent cytotoxicity crossmatch in renal transplantation: past and present perspectives. Indian J Nephrol. 2008;18(1):2–6.

15. Tapiawala S, Jogale S. Human leukocyte antigen typing in organ transplantation: methods, clinical relevance, and practical applications. Indian J Kidney Dis. 2024;3:4–11.
16. Saiki RK, et al. Enzymatic amplification of beta-globin genomic sequences and restriction site analysis for diagnosis of sickle cell anemia. Science. 1985;230(4732):1350–4.
17. Newton CR, Graham A, Heptinstall LE, Powell SJ, Summers C, Kalsheker N, Smith JC, Markham AF. Analysis of any point mutation in DNA. The amplification refractory mutation system (ARMS). Nucleic Acids Res. 1989;17(7):2503–16.
18. Olerup O, Zetterquist H. HLA-DR typing by PCR amplification with sequence-specific primers (PCR-SSP) in 2 hours: an alternative to serological DR typing in clinical practice. Tissue Antigens. 1992;39(5):225–35.
19. Hollenbach JA, et al. Molecular-based HLA typing using sequence-specific oligonucleotide probe hybridization. Curr Protoc Immunol. 2001;9(5):1–9.
20. Fleischhauer K, Maiers M. Principles of molecular methods for HLA typing. In: Histocompatibility testing. Springer; 2005. p. 63–84.
21. Guidicelli G, et al. Virtual crossmatch in kidney transplantation: where do we stand? Hum Immunol. 2007;68(1):1–9.
22. Cereb N, Kim C, Yang SY. Rapid and high-resolution DNA typing of HLA-A and -B loci by oligonucleotide hybridization using immobilized sequence-specific probes. Tissue Antigens. 1995;45(1):1–11.
23. Bottger EC, et al. High-resolution typing of HLA class I genes by PCR amplification with sequence-specific primers followed by sequencing. Tissue Antigens. 1999;53(4):295–302.
24. Cereb N, Maye P, Lee S, Kong Y, Yang SY. Locus-specific amplification of HLA class I genes from genomic DNA: sequencing-based typing of HLA-A, -B, and -C alleles. Tissue Antigens. 2001;57(4):273–83.
25. Milius RP, et al. Genotype list strings: a grammar for describing HLA and KIR genotyping results in a text string. Tissue Antigens. 2015;85(5):361–72.
26. Hosomichi K, Shiina T, Tajima A, Inoue I. The impact of next-generation sequencing technologies on HLA research. J Hum Genet. 2015;60(11):665–73.
27. Lind C, Ferriola D, Mackiewicz K, et al. Next-generation sequencing: the solution for high-resolution, unambiguous human leukocyte antigen typing. Hum Immunol. 2010;71(10):1033–42.
28. Bentley G, Higuchi R, Hoglund B, et al. High-resolution, high-throughput HLA genotyping by next-generation sequencing. Tissue Antigens. 2013;81(3):141–50.
29. Knecht H, Albrecht V, Henke M, et al. Implementation of next-generation sequencing in clinical HLA typing: challenges and solutions. Int J Mol Sci. 2020;21(18):6467.
30. Petersdorf EW, et al. HLA matching in unrelated hematopoietic cell transplantation. Curr Opin Hematol. 2014;21(6):451–7.
31. Opelz G, Döhler B. HLA matching in kidney transplantation: a retrospective analysis of 135,985 recipients. N Engl J Med. 2007;357(6):605–12.
32. Tambur AR, Campbell P, Claas FHJ, et al. Sensitization in transplantation: assessment of risk (STAR) 2017 working group meeting report. Am J Transplant. 2018;18(7):1604–14.
33. Dash SC. Ranjith Nair, Vineet Behera, kidney transplantation: the journey across a century. Med J Armed Forces India. 2023;79(6):631–7.
34. Opelz G, Wujciak T, Dohler B, Scherer S, Mytilineos J. HLA compatibility and organ transplant survival. Rev Immunogenet. 1999;1:334–42.
35. Dyer PA, Johnson RW, Martin S, et al. Evidence that matching for HLA antigens significantly increases transplant survival in 1001 renal transplants performed in the northwest region of England. Transplantation. 1989;48(1):131–5.
36. Festenstein H, Doyle P, Holmes J. Long-term follow-up in London transplant group recipients of cadaver renal allografts. The influence of HLA matching on transplant outcome. N Engl J Med. 1986;314(1):7–14.
37. Ayoub G, Terasaki P. HLA-DR matching in multicenter, single-typing laboratory data. Transplantation. 1982;33(5):515–7.
38. Duquesnoy RJ. A structurally based approach to determine HLA compatibility at the epitope level. Hum Immunol. 2007;68(6):563–76.

39. Alves TA, Nascimento E, de Castro LB, Fabreti-Oliveira RA. Impact of HLA eplet mismatch load in immunological outcomes after living donor kidney transplantation. Transpl Immunol. 2023;80:101908.
40. Senev A, Coemans M, Lerut E, Van Sandt V, Kerkhofs J, Daniëls L, Driessche MV, Compernolle V, Sprangers B, Van Loon E, Callemeyn J, Claas F, Tambur AR, Verbeke G, Kuypers D, Emonds MP, Naesens M. Eplet mismatch load and *De novo* occurrence of donor-specific anti-HLA antibodies, rejection, and graft failure after kidney transplantation: an observational cohort study. J Am Soc Nephrol. 2020;31(9):2193–204.
41. Klasberg S, Surendranath V, Lange V, Schöfl G. Bioinformatics strategies, challenges, and opportunities for next generation sequencing-based HLA genotyping. Transfus Med Hemother. 2019;46:312–32.
42. Hou L, Chen M, Steiner N, Kariyawasam K, Ng J, Hurley CK. Killer cell immunoglobulin-like receptors (KIR) typing by DNA sequencing. Methods Mol Biol. 2012;882:431–68.
43. KDIGO Transplant Work Group. KDIGO clinical practice guideline on the evaluation and Care of Living Kidney Donors. Transplantation. 2020;104(4S1):S7–S13.
44. European Federation for Immunogenetics. Standards for histocompatibility and immunogenetics laboratories, version 8.0. EFI; 2021.
45. American Society for Histocompatibility and Immunogenetics. Accreditation manual, 2022 edition. ASHI; 2022.
46. Claas FHJ, Witvliet MD, Duquesnoy RJ, Persijn GG, Doxiadis IIN. The acceptable mismatch program as a fast tool for highly sensitized patients awaiting a cadaveric kidney transplant. Am J Transplant. 2004;4(5):826–32.

Eplets, Epitopes, and Antibody Verified Eplets in Clinical Transplantation

5

Feroz Aziz, Divyesh Engineer, Subho Banerjee,
Shruti Tapiawala, Meenakshi Singh, and Suchita Jogale

Eplets and Epitopes in Transplantation

Feroz Aziz, Divyesh Engineer, and Subho Banerjee

Introduction

Molecular typing methods have led to incessant identification of newer HLA (Human Leukocyte Antigen) alleles with latest estimates around forty thousand and still counting [1]. HLA being a highly polymorphic molecule, except for having an HLA-identical sibling, it is very difficult to transplant without crossing the immunological barrier of HLA. Despite the polymorphism there is a lot of homologies between the HLA molecules leaving few islands of polymorphic amino acid residues called epitopes through the course of HLA molecules that confers the differential immunogenicity. The immune cells like T lymphocytes, B lymphocytes, or NK cells identify and interact with these epitopes to mount an immune response. The

F. Aziz (✉)
Iqraa International Hospital and Research Centre, Aster MIMS Hospital,
Calicut, Kerala, India

D. Engineer · S. Banerjee
Department of Nephrology, IKDRC-ITS, Ahmedabad, Gujarat, India

S. Tapiawala (✉)
Institute of Renal Sciences, Gleneagles Hospital, Mumbai, Maharashtra, India

M. Singh
Transplant Immunology and Immunogenetics Laboratory, ACTREC, Tata Memorial Center,
Mumbai, Maharashtra, India

S. Jogale
ATDI Lab Pvt Ltd, Mumbai, Maharashtra, India

G. Abraham et al. (eds.), *Immunological Challenges in Organ Transplantation*,
https://doi.org/10.1007/978-981-95-1533-2_5

epitope has a geographical layout identified as structural epitope in which the hot spots of polymorphisms are inscribed and identified as functional epitopes or eplets [2]. The principal responsibility of the HLA is to present the foreign antigens to the host immune system and the polymorphisms around the peptide binding groove is of utmost importance. The ever increasing HLA alleles can be broken down into varying combinations of around 500 identified epitopes [3]. Analogously epitopes are shared across different HLA antigens and can account for cross-reactivity between HLA molecules [4]. Epitopes can conceptually challenge the current matching algorithms which are dependent on matching at antigen level [5, 6]. In this article we try to present an overview of epitopes with respect to its clinical utility.

Clinical Case Scenario

Mr. R had undergone living donor kidney transplantation with mother-in-law as donor 8 years ago after a negative immunological work-up. His native kidney disease was unclear. After transplantation he developed antibody medicated rejection which despite treatment resulted in suboptimal graft function. During the ensuing years, he developed chronic graft dysfunction and had to be started on hemodialysis. His immunological work-up showed the presence of Class II antibodies (Fig. 5.1) causing crossmatch positivity with multiple donors. He has negative class I anti-HLA antibodies. HLA typing of recipient, prior donor, and multiple new donors are shown in Table 5.1. The question is how to identify a safe donor with epitope matching? The analysis of the clinical case is done during the course of this chapter.

In the science of histocompatibility, epitope as a concept is nothing new. It had a relevant past, meaningful present, and probably hopeful future. The transition of histocompatibility laboratories from the serological era to the era of next-generation sequencing has resulted in a substantial increase in data generation, fostering renewed optimism for developing solutions to the complex challenges encountered in contemporary clinical practice.

Defining Epitope and Eplet

The immune system recognizes and forms antibodies against epitopes on foreign HLA antigens. Eplet or functional epitope is single or small numbers of polymorphic amino acids within 3 Å (Angstrom) radius on or near surface of HLA molecule. Eplet is functional part of epitope and is responsible for antibody specificity by binding with complementary-determining region 3 (CDR 3) on antibody paratope (Fig. 5.2) [7]. Structural epitope (also called just epitope) is larger part of HLA antigen comprising of around 25 amino acids with eplet or functional epitopes at its center and serve as both recognition and binding site for antibody (Fig. 5.3). Both polymorphic as well as non-polymorphic amino acids in a radius of 15 Å (angstrom) from eplets serve as anchor point for antibody binding [8]. There are B cell

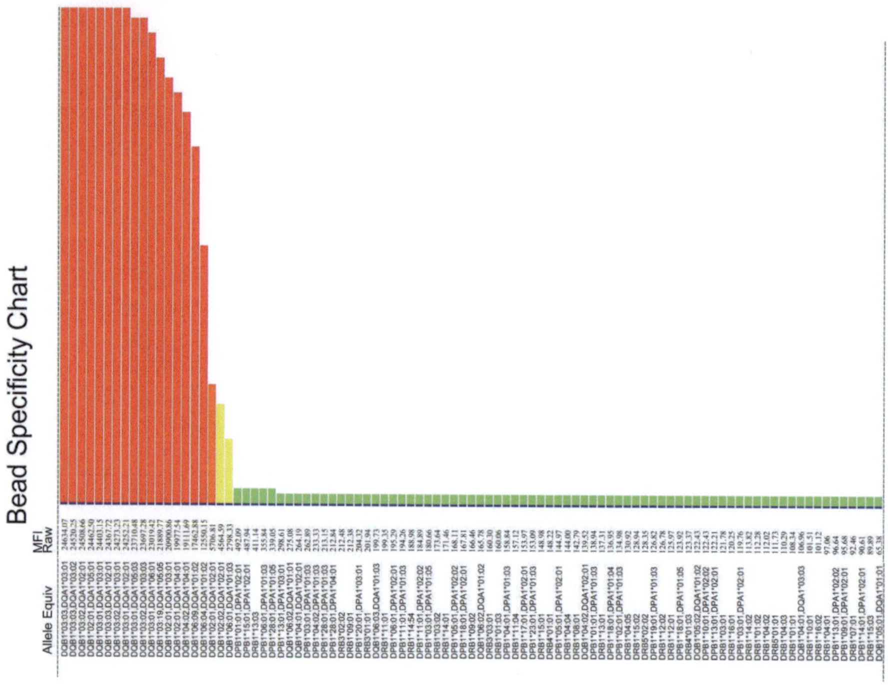

Fig. 5.1 Class II SAB analysis of recipient

and T cell epitopes. B cell epitopes are recognized by specific antibodies and serve as binding site on HLA molecule. B cell receptors will recognize foreign B cell epitopes on HLA molecule and will bind with it. Foreign HLA molecules coupled with B cell receptor are then internalized and processed. This processed donor HLA molecule is then presented to T cell with recipient class II MHC molecule. T cell epitopes are specific areas on donor HLA molecule that are presented by recipient antigen presenting cell (APC) as processed peptide in the binding groove of recipient MHC class II antigen to CD4 T cell. T cell receptor (TCR) recognizes T cell epitope and gets activated. TCR on CD4 T cells also recognizes T cell epitopes on foreign HLA antigen presented by recipient B cell with MHC class II molecules and send activating signals to B cells for antibody production. So both T and B cells epitopes are crucial for cellular and humoral alloimmune response [9, 10].

Epitopes: Relevant Past

During the serological era of 1960–1970s, both Van Rood and Paul Terasaki had reported serological patterns that reacted with a particular group of HLA antigens which were subsequently called Cross-Reactive Antigens (CREGS) [11, 12]. More patterns of serological reactivity emerged in ensuing years which along with advancements in molecular HLA typing added new dimension to histocompatibility.

Table 5.1 High-resolution HLA typing of recipient, prior donor, and three potential donors for retransplantation

	Relation	HLA-A*	HLA-A*	HLA-B*	HLA-B*	HLA-C*	HLA-C*	HLA-DRB*	HLA-DRB*	HLA-DQA*	HLA-DQA*	HLA-DQB*	HLA-DQB*
Recipient		01:01	03:01	07:06	15:17	07:01	07:02	01:01	04:10	01:01	03:03	04:02	05:01
Prior Donor	Mother in law	03:01	11:01	35:03	40:06	04:01	04:01	11:01	13:01	01:03	05:09	03:01	06:01
New donor 1	Wife	03:01	23:01	40:06	41:01	04:01	17:01	07:01	11:01	02:01	05:09	02:01	03:01
New donor 2	Sister	01:01	02:11	15:17	57:01	07:01	06:02	01:01	07:01	01:01	02:01	03:03	05:01
New donor 3	PKE donor	01:01	02:11	15:01	40:06	03:03	15:02	14:04	15:01	01:03	01:04	05:03	06:01

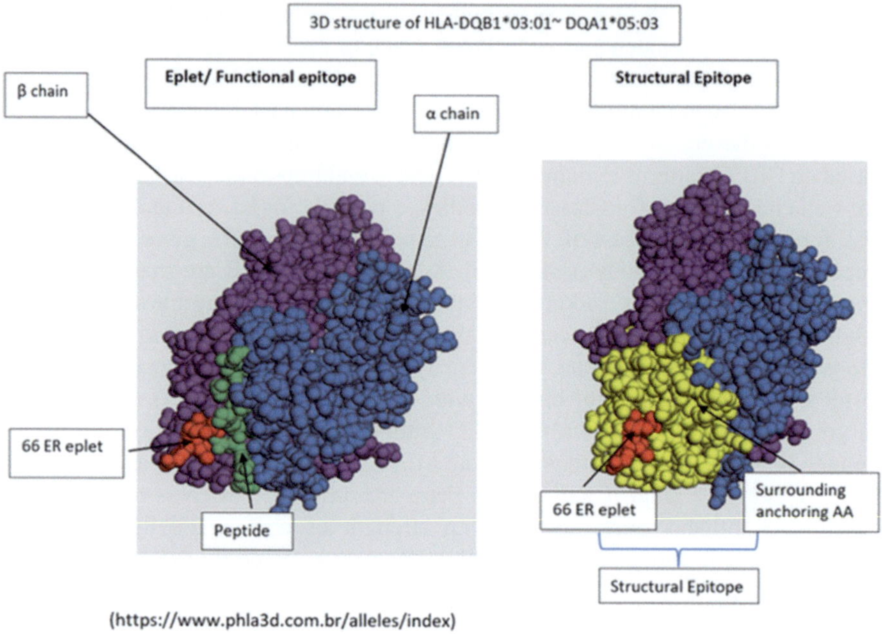

Fig. 5.2 Structural and functional epitope (Reproduced with permission from pHLA3D (https://www.phla3d.com.br); The pHLA3D database provides predicted 3D structures of HLA molecules and is a useful resource in immunogenetics (Rigo et al. [7])

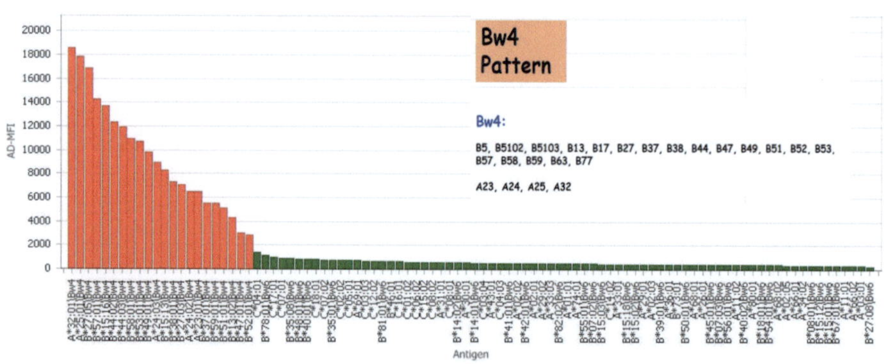

Fig. 5.3 Single antigen bead profile of Bw4 epitope profile. The constituent antibodies's MFI range from 18,000 to 2500 but all carry the same value from an alloreactive standpoint, meaning the pattern is more important than MFI value

Studies on alloreactive sera from sensitized patients have shown that majority of sera showed reactivity patterns befitting CREGS sharing public epitopes rather than narrow reactivity patterns against private epitopes [13]. Retrospective analysis of registries in the USA and Europe was done to analyze whether CREG matching had any improvement in allograft survival. The studies were not able to tease out

the advantage of CREG matching amidst already existing matching algorithms like "0"antigen mismatch and "0" DR mismatch organ allocation policies [14, 15]. Table 5.2 shows the usually identified CREG patterns and Table 5.3 shows few epitopes that has been identified as contributing to certain CREG patterns.

Even in the era of Single Antigen Bead testing, analysis of antibody profiles based on CREG patterns remains highly relevant and important. The CREG patterns are well characterized for Class I antibodies, especially for HLA-A and HLA-B and less for HLA-C and Class II. Being a part of CREG pattern gives a functional importance to the antibody profile [16]. As the beads of SAB are sharing the epitopes, the complementary antibody to a particular epitope may have to share its binding capacity to a set of beads. As the antibody gets shared across the different beads having the same epitope, the intensity or quantity of binding may get weaker result in varying intensity of binding with varying MFI values [17]. Due to these varying nature of reactivity within CREG groups, we consider the pattern of reactivity may be more important than absolute MFI readings of the beads as depicted in Fig. 5.3. Within a major CREG group there are subgroups of cross-reactivity and constituent antigens may have varying strength of cross-reactivity (Fig. 5.4). Vendors of single antigen beads provide software that can easily discern the CREG pattern present if any.

Epitopes: Meaningful Present

In the current molecular era, the focus of transplant clinical practice has shifted from the management of acute rejection to the pursuit of improved long-term outcomes. Clinicians are increasingly aiming not only to prevent the development of de novo donor-specific antibodies (DSAs) but also to expand transplantation opportunities for sensitized patients. These evolving clinical goals have renewed interest in the concept of epitope matching, a field further propelled by advances in laboratory sciences, including the elucidation of the three-dimensional structure of HLA molecules and the advent of high-resolution HLA typing through next-generation sequencing (NGS).

HLA mismatch and Epitope mismatch: Large registry based retrospective data has shown that more HLA mismatches are associated with inferior graft survival [18, 19]. Nevertheless, the inferences from these macro-studies have been debated due to inherent limitation of being retrospective and method of outcome analysis. The difference in outcome associated with HLA antigen mismatches was not appreciable in small single center studies with limited number of recipients. From a qualitative perspective, it has been proven that certain mismatches compared to others are more likely to develop DSA and result in inferior outcome. High resolution typing and unravelling of 3D spatial arrangement of HLA molecules has facilitated to understand the immunogenicity of various mismatches. Experts are increasingly identifying immunogenic epitopes by studies that include human-monoclonal antibodies, absorption/elution studies and pattern analysis of serum antibody reactivity [20–23]. Each HLA allele can be considered as a combination of immunogenic

Table 5.2 Cross reactive groups (CREGS) associated with different HLA specificities

CREG groups							
A1$_{CREG}$	A1, A3, A11, A23, A24, A29, A30, A31, A36, A80						
A2$_{CREG}$	A2, A23, A24, A68, A69, B57, B58						
A10$_{CREG}$	A11, A25, A26, A32, A33, A34, A43, A66, A68, A69, A74						
Bw4$_{CREG}$	A23, A24, A25, A32, B13, B27, B37, B44, B47, B38, B49, B51, N52, B53, B57, B58, B59, B63, B77						
Bw6$_{CREG}$	B7, B8, B18, B27:08, B35, B39, B40, B4005, B41, B42, B45, B48, B50, B54, B55, B56, B60, B61, B62, B64, B65, B67, B70, B71, B72, B75, B76, B78, B81, B82						
B5$_{CREG}$	B18, B35, B46, B49, B50, B51, B52, B53, B57, B58, B62, B63, B71, B72, B73, B75, B76, B77, B78, B7						
B7$_{CREG}$	B7, B8, B13, B27, B41, B42, B47, B48, B54, B55, B56, B59, B60, B61, B67, B81, B82						
B8$_{CREG}$	B8, B18, B38, B39, B59, B64, B65, B67						
B12$_{CREG}$	B13, B37, B41, B44, B45, B47, B49, B50, B60, B61						
C1$_{CREG}$	Cw1, Cw7, Cw8, Cw9, Cw10, Cw12, Cw14, Cw16, Bw46, B73						
C2$_{CREG}$	Cw2, Cw4, Cw5, Cw6, Cw15, Cw17, Cw18						
DR1$_{CREG}$	DR1, DR10, DR103	DR51$_{CREG}$	DR51,DR15,DR16	DR52$_{CREG}$	DR52,DR11,DR12, DR13, DR14, DR17,DR18	DR53$_{CREG}$	DR53,DR4,DR7, DR9
DQ1$_{CREG}$	DQ5, DQ6	DQ2$_{CREG}$	DQ2	DQ3$_{CREG}$	DQ7,DQ8,DQ9	DQ4$_{CREG}$	DQ4
DP1c$_{CREG}$	DP2, DP3, DP4, DP6, DP9, DP10, DP11, DP14, DP17, DP18, DP20, DP28			DP2c$_{CREG}$	DP1, DP5, DP13, DP15,DP19,DP23		

Table 5.3 Certain epitopes identified with CREG groups

Epitope	Antigens sharing the epitope
17 RS	A30
127K	A2,A23,A24,A68,A69
76 ESN	B7, B8, B14,B18,B*27:08,B35,B39,B41,B42,B45,B48, B50,B54,B55,B56,B60,B61, B62,B64, B65, B67,B70, B75, B76, B78,B81,B82
80K	Cw2, Cw4,Cw5,Cw6,Cw15,Cw17,Cw18
80 I	A23,A24,A25,A32,B38,B49,B51,B52,B53,B57,B58,B59,B63,B77
163 EW	A*66:02,B7,B13, B27, B47, B48,B60, B61,B73, B81,Cw2,Cw17

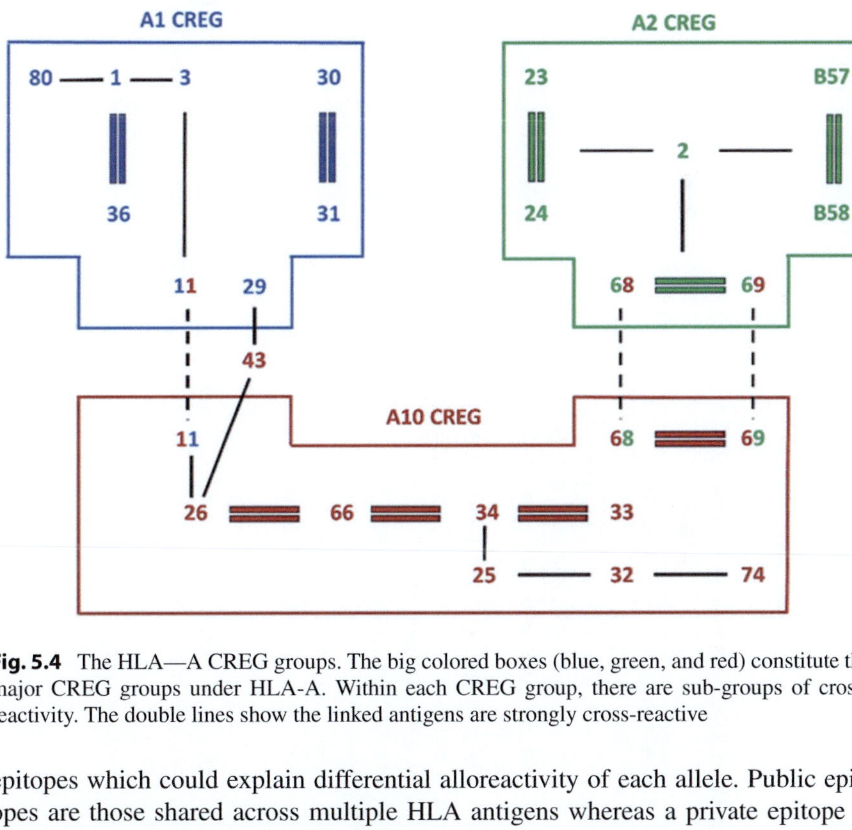

Fig. 5.4 The HLA—A CREG groups. The big colored boxes (blue, green, and red) constitute the major CREG groups under HLA-A. Within each CREG group, there are sub-groups of cross-reactivity. The double lines show the linked antigens are strongly cross-reactive

epitopes which could explain differential alloreactivity of each allele. Public epitopes are those shared across multiple HLA antigens whereas a private epitope is seen uniquely to a particular HLA allele.

Rene Duquesnoy was pioneer in pushing forth the concept of epitope matching and came up with a computational software-based platform called "HLA matchmaker" which could calculate the epitope mismatches between recipient and donor on populating two field high resolution HLA typing data into the software tool.

Many centers have retrospectively analyzed the eplet mismatches in their cohort and have tried correlate the transplant outcome with respect to eplet mismatch load. Significant proportion of studies have seen an association with increased eplet

Table 5.4 The most used software tools for epitope matching in available literature

Software	Basis	Concept	
HLA Matchmaker	B cell epitope	Eplet mismatch between HLA alleles	R. Duquesnoty University of Pittsburg, USA
Electrostatic mismatch score	B cell epitope	Electrostatic charge of polymorphic amino acids in mismatched HLA alleles	V. Kosmoliaptsis University of Cambridge, UK
PIRCHE	T cell epitope	Predict T cell rejection by identifying indirectly recognizable HLA epitopes	Eric Spierings & Team, Private Company
HLA EMMA	B cell epitope	Mismatch of polymorphic amino-acid residues at solvent-accessible positions	S.Heidt and C. Kramer University of Leiden, NLD
HLA epitope registry	B cell epitope	Eplet mismatches between HLA alleles	Federal University of Piauí

mismatch especially class II with development of de novo DSA and inferior graft outcome [19, 24, 25]. Recipients with increased eplet mismatch have also been associated with increased biopsy proven cellular rejections implying that increased eplet mismatch is an indicator of increased alloreactivity in general [26]. A problem with eplet mismatch is that it considers all eplets same and no importance is given to qualitative aspect of eplet mismatch like physiochemical properties of amino acid substitution and presence or absence of required T cell epitope. Some eplet mismatches are more immunogenic than others. So though higher number of eplet mismatches increase risk of de novo DSA but even single eplet mismatch may be sufficient to generate de novo DSA [10, 27].

To address this issue, different centers have come up with software tools similar to HLA Matchmaker as shown in Table 5.4, based on their definition of important epitopes and their consideration of which interface of allorecognition pathway is important from an immunogenicity standpoint [28–31]. Fundamentally all these tools use the same principle but with respective unique approaches. The mismatched epitopes at T cell epitope and B Cell epitope interface are not similar and analogously information readout is also uniquely different across tools. Studies have been done correlating and integrating these different algorithms [32–34]. All these tools are based on the genetic variations of ever-expanding domain of HLA set with the proviso that only a limited number of epitopes have undergone intensive validation process to confirm their potential to mount an antibody response. The gaps and unknown in the science of epitopes may be more than we presume.

To consider impact of physiochemical property of amino acid mismatch (surface charge, polarity, hydrophobic nature), three dimensional Electro Static mismatch score (3D-EMS) was developed by Vasilis Kosmoliaptsis. 3D-EMS can be calculated on the basis of two field HLA typing of donor and recipient with use of software. (https://mohorianulab.org/shiny/kosmoliaptsis/HLA_Algorithms/). This scoring system permits quantitative assessment of surface electrostatic potential difference between recipient and donor HLA molecule at three-dimensional level structure. Interlocus comparison is done with each mismatched donor HLA class I

and II molecule compared electrostatically to each class I and class II molecule respectively to derive electrostatic similarity distance(ESD) and the minimum ESD value is taken to represent EMS-3D score. In study of 191 healthy females immunized with lymphocyte from male partner, there was strong association between 3D EMS of donor with development of denovo DSA. Similarly in clinical transplant cohort, 3D EMS of HLA-DR and HLA-DQ of donor independently correlated with denovo DSA formation [35].

B cells require activation from CD4 T cells for antibody generation. CD4 T cell allo-recognition and activation of the cellular response can be predicted using the Predicted Indirectly Re-Cognizable HLA Epitopes (PIRCHE-II) algorithm developed by Geneugelijk and Spiering. PIRCHE-II algorithm can estimate the number of mismatched HLA peptides that can activate CD4 T cells through the indirect pathway. PIRCHE-II scores have been associated with de novo DSA and graft failure [36].

HLA epitope mismatch algorithm (HLA-EMMA) is a software program developed by Leiden University group, Netherlands. It compares simultaneously the HLA class I and class II amino acid sequences of the donor with the HLA amino acid sequences of the recipient and determines the polymorphic solvent accessible amino acid mismatches that are readily accessible to B cell receptors. In study of 191 healthy females immunized with lymphocyte from male partner, incidence of denovo DSA was >80% in group with highest number of solvent accessible amino acid mismatches at HLA-A, HLA-B, and HLA-DQ mismatches [31].

Utility of Epitope Tool

In the ever-existing situation of organ shortage, any algorithm should be backed with scientific evidence and beneficial to majority of patients. HLA typing and anti-HLA antibody assessment are the two time-tested histocompatibility tests practiced across the globe. Whether epitope matching can be used instead of current system based on HLA typing is the fundamental question. With all the limitations we have understood about epitopes, there could be three clinical scenarios where epitope matching can be very useful. The scenarios are as follows

- *Calculate and compare the epitope mismatches with the donor or multiple* donors.
- *Analyze the reactivity pattern of an antibody profile of a SAB readout.*
- *Assess the eplet mismatch from prior immunizers to see how much memory has to be crossed.*

Aforementioned clinical scenarios may provide a template on which the epitope data can be superimposed for clinical utility. The jury is still out whether the available epitope matching tools can be utilized in prospective transplants.

Calculate and compare the epitope mismatches with the donor or multiple donors: Many retrospective studies have shown that more HLA eplet mismatches, especially in class II has been associated with inferior outcome [26, 37, 38]. The

registry-based studies have modelled out a particular number of mismatches being associated with inferior outcome in their population but that cannot be generalized to be used in prospective algorithms. In a study by C. Wiebe et al, HLA-DR and HLA-DQ epitope matching performs better than traditional low resolution HLA antigen matching in minimizing risk of de novo donor specific antibody (DSA) formation in kidney transplantation. Threshold eplet mismatch sum (sum of mismatched donor eplets for a given locus) of 10 for HLA-DR and 17 for HLA-DQ was associated with minimal development of de novo DSA [6]. Single molecule eplet mismatch (number of eplet mismatch for each specific molecule in locus) appears to be more precise than eplet mismatch sum especially when patients with only one out of two molecules mismatched. In another study by Wiebe C et al, single molecule eplet mismatch allowed patients to be divided into low (HLA-DR <7 and HLA-DQ < 9), intermediate (HLA-DR ≥ 7 and HLA-DQ ≤14 or HLA-DR 0–6 and HLA-DQ 9–14), and high-risk categories (HLA-DR 0–22 and HLA-DQ ≥ 15). These risk categories were significantly correlated with Banff ≥1A T cell mediated rejection, HLA-DR/DQ de novo DSA formation, antibody mediated rejection and all cause graft loss [26]. But all eplet are not same, some may be more immunogenic than others. Some are verified using monoclonal antibodies, while others remain theoretical. A large registry based study involving around 118,313 patients looked into the immunogenicity of mismatched eplets and identified that only fewer subsets of mismatches were associated with inferior allograft outcome, raising the concern of giving equal importance for all eplet mismatches [39]. Though antibody verified eplets have been shown to be more detrimental with inferior graft outcome, there is a lag in identifying Class II antibody verified epitopes [40]. It has also been reported that certain combination of epitopes may be more immunogenic than when present in isolation [41]. We need more data on quality of epitope mismatch from different populations before we can attribute any importance to quantitative epitope mismatch for prospective decision making in clinical solid organ transplantation. In our understanding, utmost we may initiate a discussion with prospective transplant recipients about the risk inferred from available literature and counsel on importance of adherence to immunosuppression.

Eplet match and tacrolimus level: Though high eplet mismatch increases risk of de novo DSA formation, tacrolimus trough level is important determinant of whether recipient with high eplet mismatch develops de novo DSA. In study by Chris Wiebe et al with 596 renal transplant recipients, both eplet mismatch and trough tacrolimus level were independent predictor of de novo DSA formation. In recipients at high alloimmune risk (>11 HLA-DR or DQ eplet mismatch) who developed de novo DSA, higher proportion had tacrolimus trough level < 5 ng/mL compared to high-risk patient who do not develop de novo DSA. Recipients with low alloimmune risk (≤11 HLA-DR or DQ eplet mismatch) tolerated trough tacrolimus level < 5 ng/mL without development of de novo DSA. So eplet mismatch can help to tailor maintenance immunosuppression. In situations requiring reduction of maintenance immunosuppression like BKV nephritis, those with high eplet mismatch are the high risk subset for development of de novo DSA so they should undergo monitoring for de novo DSA with single antigen bead test(SAB) [42].

Table 5.5 Eplet analysis of different A*02 alleles

A*02:01	62GK	62GE	79GT	107W	127K	142TKH	144K	144KHA	149AH	150AAH	193AV	253Q
A*02:02	62GK	62GE	79GT	107W	127K	142TKH	144K	144KHA	149AH	150AAH	193AV	253Q
A*02:05	62GK	62GE	79GT	107W	127K	142TKH	144K	144KHA	149AH	150AAH	193AV	253Q
A*02:11	62GK	62GE	79GT	107W	127K	142TKH	144K	144KHA	149AH	150AAH	193AV	253Q

Analyze the reactivity pattern of an antibody profile of a SAB Readout: Single antigen bead analysis is a very sensitive tool that can make clinician overloaded with myriads of data making clinical decision making complicated. Data from good quality laboratories have reported 25% to 60% prevalence of anti-HLA antibodies in apparently non-sensitized individuals [43, 44]. A clinically relevant alloimmune response is typically directed against immunogenic epitopes. While not mandatory, the actual alloreactive single antigen bead (SAB) profile can often be explained by mapping it to a corresponding panel of immunogenic epitopes. Epitope analysis of the reactive pattern is a simple tool to test the reactivity of some questionable beads. Epitope analysis also helps when a particular mismatched allele of interest is absent on single antigen bead panel. In the present case, donor 2 and 3 have A*02:11 as mismatched allele. A*02:11 allele bead is absent in current SAB panel from both vendors. In this scenario, eplet analysis can be used to decide which bead to be considered for deciding donor specific antibody assignment. Antibody verified eplets of A*02:11 were identical to that of A*02:01, A*02:02 and A*02:05 which were present in SAB panel of both vendors (Table 5.5).

Assess the eplet mismatch from prior immunizers to see how much memory has to be crossed: Assessing the memory response would be where epitope analysis would be most useful. None of the current day histocompatibility tests like cell-based crossmatches or bead-based analysis can efficiently identify the alloimmune memory response. The eplet mismatches with the prior known immunizer (like prior donor) against which the recipient has reacted to is something that has to be avoided in next transplantation. Many a times the DSA profile may not be a complete mirror of the potential memory response.

Epitope Matching Software: How to Use It in Clinical Practice

There are many tools than can be used to calculate epitope matching. Although different calculators use repository data bases that may be different, the approach and technique is similar. Studies done using multiple calculators in the same population and have given congruent results.

Prerequisites for Epitope Analysis

- Identify the software calculator—there are fee based or license based or free to use software tools. In this review we would be using HLA eplet registry –a free to use, B cell epitope-based program *HLA Eplet Registry* (epregistry.com.br).
- Need two field high resolution typing. Imputation from one field based typing can have inherent deficiencies.
- SAB profile of sensitized patients.

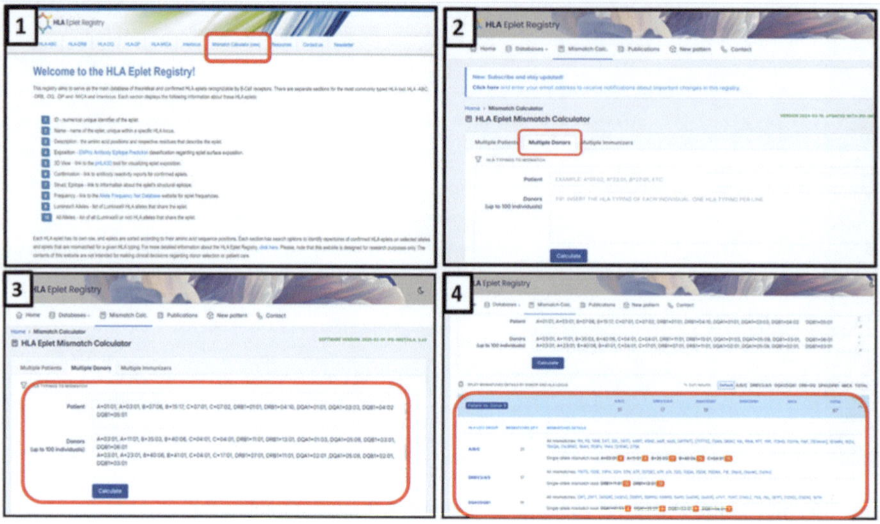

Fig. 5.5 Steps 1 to 4 describes how to use the HLA Eplet registry software tool. The details are briefed in the text (Reproduced with permission from the HLA Eplet Registry, LIB, UFPI (www.epregistry.com.br) (Adapted from Duquesnoy et al. [3])

Using the software in analyzing the eplet mismatch between recipient and donor: Steps of using the HLA Eplet registry software for eplet mismatch is being briefed in Fig. 5.5.

- Step 1: Open the website *HLAEpletRegistry* (epregistry.com.br). Choose the option mismatch calculator as shown in red box.
- Step 2: Select the option of multiple donors.
- Step 3: Type in the recipient and donor HLA typing, two field high resolution as mentioned in the software with proper spacing. Even hundred different donors can be evaluated against a recipient.
- Step 4: On entering calculate—the software would populate all the number of mismatched eplets for both classes. Class I (ABC) DR, DQ and DP are calculated separately. The name of mismatched eplets would also be detailed and the eplets in parenthesis are the antibody verified ones.
- Step 5: Analyze SAB data to detect donor mismatched eplet-specific antibodies: This step involves mapping the HLA alleles corresponding to positive Single Antigen Bead (SAB) reactivities to their respective eplets (Fig. 5.6). A donor eplet-specific antibody is defined as one that targets a mismatched eplet uniquely present on SAB beads corresponding to the donor's mismatched class I (HLA-A, -B, -C) or class II (HLA-DR, -DQ, -DP) antigens. If the same mismatched eplet is also found on negative SAB beads, the antibody is not considered donor-specific [45]. Each eplet entry, when prompted, provides details on the associated Luminex alleles that express the epitope. This allows a bidirectional analysis: determining whether the antibody profile corresponds to an immunogenic epitope pattern, or vice versa. Figure 5.6 shows eplet analysis of SAB class II profile

Sero	Allele	DQB- EPLETS	DQA-EPLETS	MFI
DQ9	DQB1*03:03 DQA1*03:01	130R, 135D, 167R, 182N, 185I, 45G, 45GV, 46VY, 55PP, 66ER, 74EL, 77T, 84QL	129H, 40E, 40ERV, 55R, 61FT, 75I, 76V,2D	24634
DQ9	DQB1*03:03 DQA1*03:02	130R, 135D, 167R, 182N, 185I, 45G, 45GV, 46VY, 55PP, 66ER, 74EL, 77T, 84QL	129H, 40E, 40ERV, 55R, 61FT, 75I, 76V,2D	24520
DQ8	DQB1*03:02 DQA1*02:01	130R, 135D, 167R, 182N, 185I, 45G, 45GV, 46VY, 55PP, 66ER, 74EL, 77T, 84QL	129H, 40E, 40ERV, 47KHL, 55R, 61FT, 75I,2D	24509
DQ2	DQB1*02:01 DQA1*05:01	130R, 135D, 167R, 182S, 45G, 52LL, 66DR, 77R, 84QL	129H, 40GR, 55R, 61FT,2D	24463
DQ7	DQB1*03:01 DQA1*03:01	130R, 135D, 182N, 45EV, 46VY, 55PP, 66ER, 74EL, 77T, 84QL	129H, 40E, 40ERV, 55R, 61FT, 75I, 76V,2D	24403
DQ9	DQB1*03:03 DQA1*02:01	130R, 135D, 167R, 182N, 185I, 45G, 45GV, 46VY, 55PP, 66ER, 74EL, 77T, 84QL	129H, 40E, 40ERV, 47KHL, 55R, 61FT, 75I,2D	24368
DQ8	DQB1*03:02 DQA1*03:01	130R, 135D, 167R, 182N, 185I, 45G, 45GV, 46VY, 55PP, 66ER, 74EL, 77T, 84QL	129H, 40E, 40ERV, 55R, 61FT, 75I, 76V,2D	24273
DQ7	DQB1*03:01 DQA1*02:01	130R, 135D, 182N, 45EV, 46VY, 55PP, 66ER, 74EL, 77T, 84QL	129H, 40E, 40ERV, 47KHL, 55R, 61FT, 75I,2D	24252
DQ7	DQB1*03:01 DQA1*05:03	130R, 135D, 182N, 45EV, 46VY, 55PP, 66ER, 74EL, 77T, 84QL	129H, 40GR, 55R, 61FT,2D	23710
DQ8	DQB1*03:02 DQA1*03:03	130R, 135D, 167R, 182N, 185I, 45G, 45GV, 46VY, 55PP, 66ER, 74EL, 77T, 84QL	129H, 40E, 40ERV, 55R, 61FT, 75I, 76V,2D	23697
DQ7	DQB1*03:01 DQA1*06:01	130R, 135D, 182N, 45EV, 46VY, 55PP, 66ER, 74EL, 77T, 84QL	129H, 40GR, 55R, 61FT, 75I,2D	23019
DQ7	DQB1*03:19 DQA1*05:05	130R, 135D, 182N, 185I, 45EV, 46VY, 55PP, 66ER, 74EL, 77T, 84QL	129H, 40GR, 55R, 61FT,2D	21890
DQ2	DQB1*02:01 DQA1*03:01	130R, 135D, 167R, 182S, 45G, 52LL, 66DR, 77R, 84QL	129H, 40E, 40ERV, 55R, 61FT, 75I, 76V,2D	20901
DQ2	DQB1*02:01 DQA1*04:01	130R, 135D, 167R, 182S, 45G, 52LL, 66DR, 77R, 84QL	129H, 40GR, 55R, 61FT, 75I,2D	19978
DQ4	DQB1*04:02 DQA1*04:01	130R, 135D, 167R, 182N, 185I, 45G, 45GV, 46VY, 55R, 56L, 74S, 77T, 84QL	129H, 40GR, 55R, 61FT, 75I,2D	19112
DQ6	DQB1*06:09 DQA1*01:02	125G, 135D, 167R, 182S, 45G, 45GV, 46VY, 52PQ, 55R, 57V, 66ER, 74EL, 77T, 87Y	129QS, 40E, 52SK, 75I,2D	17463
DQ6	DQB1*06:04 DQA1*01:02	125G, 135D, 167R, 182S, 45G, 45GV, 46VY, 52PQ, 55R, 57V, 66ER, 74EL, 77T, 87Y	129QS, 40E, 52SK, 75I,2D	12550
DQ2	DQB1*02:01 DQA1*02:01	130R, 135D, 167R, 182S, 45G, 52LL, 66DR, 77R, 84QL	129H, 40E, 40ERV, 47KHL, 55R, 61FT, 75I,2D	5707
DQ2	DQB1*02:02 DQA1*02:01	130R, 167R, 182S, 45G, 52LL, 66DR, 77R, 84QL	129H, 40E, 40ERV, 47KHL, 55R, 61FT, 75I,2D	4565
DQ6	DQB1*06:01 DQA1*01:03	125G, 130R, 135D, 182S, 3P, 45G, 45GV, 46VY, 52PQ, 55R, 66DR, 74EL, 77T, 86A, 87F	129H, 130A, 40E, 52SK, 75I,2D	2798
DQ6	DQB1*06:02 DQA1*01:01	125G, 130R, 135D, 167R, 182S, 45G, 45GV, 46VY, 52PQ, 55R, 67VG, 70GT, 74EL, 77T, 86A, 87F	129QS, 40E, 52SK, 75I,2D	275
DQ4	DQB1*04:01 DQA1*02:01	130R, 135D, 167R, 182N, 185I, 23L, 45G, 45GV, 46VY, 55R, 56L, 74S, 77T, 84QL	129H, 40E, 40ERV, 47KHL, 55R, 61FT, 75I,2D	264
DQ6	DQB1*06:03 DQA1*01:03	125G, 130R, 135D, 167R, 182S, 45G, 45GV, 46VY, 52PQ, 55R, 67VG, 70GT, 74EL, 77T, 86A, 87F	129H, 130A, 40E, 52SK, 75I,2D	200
DQ6	DQB1*06:02 DQA1*01:02	125G, 130R, 135D, 167R, 182S, 45G, 45GV, 46VY, 52PQ, 55R, 67VG, 70GT, 74EL, 77T, 86A, 87F	129QS, 40E, 52SK, 75I,2D	166
DQ4	DQB1*04:02 DQA1*02:01	130R, 135D, 167R, 182N, 185I, 45G, 45GV, 46VY, 55R, 56L, 74S, 77T, 84QL	129H, 40E, 40ERV, 47KHL, 55R, 61FT, 75I,2D	140
DQ5	DQB1*05:02 DQA1*01:02	116I, 130R, 135D, 167R, 182S, 45G, 45GV, 46VY, 52PQ, 55R, 67VG, 74S, 77R, 86A, 87Y	129QS, 40E, 52SK, 75I,2D	122
DQ4	DQB1*04:01 DQA1*03:03	130R, 135D, 167R, 182N, 185I, 23L, 45G, 45GV, 46VY, 55R, 56L, 74S, 77T, 84QL	129H, 40E, 40ERV, 55R, 61FT, 75I, 76V,2D	107
DQ5	DQB1*05:01 DQA1*01:01	116I, 125SQ, 130R, 135D, 167R, 182S, 45G, 45GV, 46VY, 52PQ, 55R, 57V, 67VG, 74S, 77R, 86A, 87Y	129QS, 40E, 52SK, 75I,2D	65

Fig. 5.6 Step 5: Eplet analysis depicting potential eplets against immunizer

of current case. Mismatched eplet against prior donor like 45 EV, 55PP, 66DR, 66ER, and 40 GR are present only in positive SAB beads (MFI >1000) confirming that antibody profile corresponds to this mismatched eplets. So if these eplets are present in any other prospective donor, then their may be potential for interaction of anti-HLA antibodies with these mismatched eplets.

Discussion of the clinical case: Mr. R is being worked up for a second transplant with three potential donors (wife, sister, and PKE donor). In the hypothetical scenarios of Mr. R having negative anti-HLA antibody, different HLA molecular mismatch scores can be used to select a donor with least alloimmune risk. Eplet mismatch load, HLA-EMMA score, PIRCHE score, EMS score of the recipient with different donors are shown below (Tables 5.6, 5.7, 5.8, 5.9, 5.10, and 5.11). Recipient has least total eplet mismatch [46] and HLA-DR/DQ eplet mismatch [12] with donor 2 (sister) among available donors. So she may be best suitable donor from immune point of view. HLA-EMMA score conveys the same result with Mr. R has least total and solvent accessible AA mismatch with donor 2 [41, 46] but as per PIRCHE II analysis, donor 3 is ideal as Mr. R has least total PIRCHE II score (150.96). Result of 3D EMS score cannot be used in this case as EMS 3D has 99.94% HLA coverage and certain uncommon alleles in recipient and donors are not covered (Fig. 5.7).

But in current scenario Mr. R exhibited an alloimmune response to the initial donor (his mother-in-law), leading to the development of anti-HLA antibodies predominantly directed against epitopes on the DQ loci. This indicates that specific mismatched eplets within the DQ antigens were immunogenic and triggered the allo-response. The single antigen bead (SAB) assay revealed an antibody profile consistent with this sensitization. As shown in Table 5.6, multiple eplet mismatches were identified between Mr. R and the first donor, several of which are likely to be

Table 5.6 Eplet mismatch load with different donors

Recipient vs.	Relation	ABC	DRB1/345	DQA/ DQB	Total	DQ Eplet mismatches (EpM)	DQ EpM correlating with SAB	Crossmatch
Prior donor	MIL	31	17	19	67	3P,40GR, 45EV,55PP,66DR, 66ER, 74EL,87F,125G,130A	3P,40GR, 45EV, 55PP,66DR,66ER	Positive
Donor 1	Wife	35	23	19	77	40GR,45EV, 47KHL, 52LL, 55PP, 66DR, 66ER, 74EL	40GR,45EV, 52LL, 55PP, 66DR, 66ER	Positive
Donor 2	Sister	27	14	12	53	47KHL, 55PP, 66ER, 74EL	55PP, 66ER	Positive
Donor 3	PKE	40	17	12	69	3P, 66DR, 74EL, 87F, 125G,130A	3P, 66DR	Negative

Table 5.7 HLA–EMMA scores with different donors

	HLA Allele–Donor	Total AA mismatch	Solvent accessible AA mismatch	Allele mismatch	
Prior donor	HLA-A	2	0	1	Total AA mismatch = 69 Total Solvent accessible AA mismatch = 48
	HLA-B	20	11	2	
	HLA-DR	14	12	2	
	HLA-DQB1	15	11	2	
	HLA-DQA1	18	14	2	
Donor 1 (Wife)	HLA-A	20	15	1	Total AA mismatch = 97 Total Solvent accessible AA mismatch = 71
	HLA-B	14	8	2	
	HLA-DR	24	21	2	
	HLA-DQB1	20	12	2	
	HLA-DQA1	19	15	2	
Donor 2 (Sister)	HLA-A	18	16	1	Total AA mismatch = 53 Total Solvent accessible AA mismatch = 42
	HLA-B	7	3	1	
	HLA-DR	16	14	1	
	HLA-DQB1	6	4	1	
	HLA-DQA1	6	5	1	
Donor 3(PKE)	HLA-A	18	16	1	Total AA mismatch = 73 Total Solvent accessible AA mismatch = 56
	HLA-B	15	8	2	
	HLA-DR	21	19	2	
	HLA-DQB1	13	9	2	
	HLA-DQA1	6	4	2	

the immunological targets of the observed antibody response. Among these, 55PP, 66DR, and 66ER emerged as key eplet mismatches. Of these, 55PP and 66ER appear to be the most highly alloreactive, as both are shared by Donor-1 (wife) and Donor-2 (sister)—correlating with the positive crossmatch results seen with these donors. This suggests that these eplets played a central role in the sensitization process and should be avoided in future donor selections. Although 66DR was also a mismatched eplet, its contribution to the antibody response appears to be limited. This is supported by the negative crossmatch with Donor-3, despite the presence of 66DR. Therefore, antigens containing 66DR may be considered clinically acceptable in this context, as they did not elicit a strong alloantibody response. In contrast, Donor-3, identified through a paired kidney exchange (PKE) program, did not carry the highly alloreactive eplets 55PP or 66ER, and this absence aligns with the negative crossmatch result observed.

Table 5.8 3D EMS score with different donors

	HLA locus wise 3D EMS against different donors							
Index	A	A	B	B	DRB1	DRB1	DQA1_DQB1	DQA1_DQB1
Prior donor(Immunizer)	hla_match	0.141	0.237	0.276	0.272	0.249	0.214	0.429
Donor 1—Wife	hla_match	0.261	0.276	0.21	0.249	0.272	Unknown	Unknown
Donor 2—Sister	hla_match	Unknown	hla_match	0.195	hla_match	0.249	hla_match	0.361
Donor 3—PKE donor	hla_match	Unknown	0.214	0.276	0.249	0.21	0.1	Unknown

Note: 3D EMS score is missing of certain rare allele not covered in EMS3D as it has 99.94% HLA coverage. Missing HLA combinations for EMS3D in current case: A*02:11,A*03:01, DQA1*02:01_DQB1*03:01,DQA1*01:01_DQB1*05:01_DQA1*05:09_DQB1*02:01,DQA1*03:03_DQB1*04:02, DQA1*01:04_DQB1*06:01 DQA1*03:03_DQB1*04:02

Table 5.9 CDC scoring of dead cells

Score	% dead cells	Interpretation
1	0–10	Negative
2	11–20	Doubtful positive
4	21–50	Weakly positive
6	51–80	Positive
8	81–100	Strongly positive

Table 5.10 SAB class I results

Antigen specificity	Allele specificity	Normalized MFI values
Cw6	C*06:02	2405
Cw7	C*07:02	6458
Cw17	C*17:01	4206
Cw18	C*18:02	2765

Table 5.11 SAB class II results

Antigen specificity	Allele specificity	Normalized MFI values
DQ4	DQA1*02:01,-,DQB1*04:01	993
DQ4	DQA1*03:03,-,DQB1*04:01	1834
DQ4	DQA1*02:01,-,DQB1*04:02	1523
DP1	DPA1*02:01,-,DPB1*01:01	2112

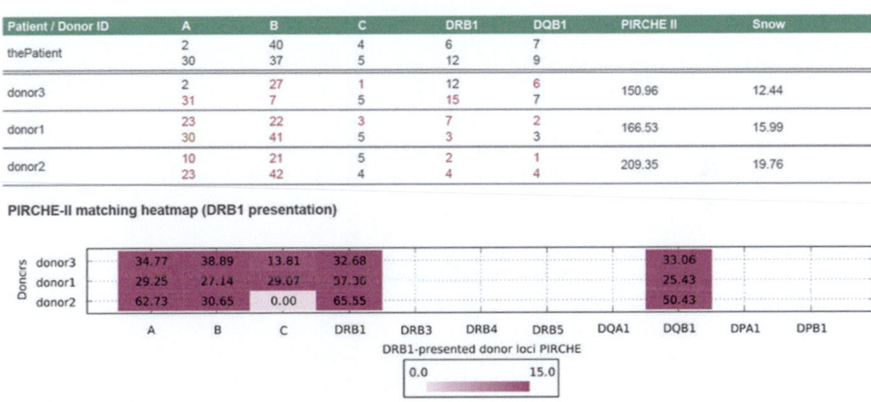

Fig. 5.7 PIRCHE-II scores with different donors

Furthermore, Donor-3 also lacked the majority of other immunogenic eplets shared with the prior immunizer, reinforcing the interpretation that Mr. Risun likely to mount a memory alloimmune response against this donor. In conclusion, among the mismatched eplets, 55PP and 66ER are identified as the primary alloreactive targets, and HLA antigens expressing these eplets should be avoided in future donor matching. Eplets such as 66DR, which did not trigger a clinically significant antibody response, may be considered permissible mismatches. These findings support

Donor-3 as an immunologically optimal match for Mr. R, minimizing the risk of antibody-mediated rejection and enhancing the prospects for successful transplantation.

Epitopes: Hopeful Future

There is a lot of hope that epitope matching could be a remedy for all deficiencies in histocompatibility testing strategy to avoid de novo DSAs and improve long-term allograft outcome. At this point Epitope matching is like xenotransplantation of histocompatibility with large knowledge gaps. Nevertheless, there is some encouragement for epitope enthusiasts due to some good work happening is this direction. Genome Canada consortium has been studying the scope of epitope matching in Canada.

High resolution HLA typing of nearly 1850 individual representative of multiethnic population across Canada was analyzed and was round that genotype of these 1850 individuals can be represented by 149 epitopes—59 class I and 90 class II epitopes [45, 47]. The practicality of analysis of the mismatches between a much lesser number of epitopes compared to ever expanding HLA alleles is promising. They also found that distribution of HLA alleles in the society had been skewed with some alleles very common and some very rare making it difficult find matched counterparts for wider population. The distribution of epitopes in the population was more even making it principally easy to find a matched donor. The also did simulation exercise of their national organ allocation data using epitope mismatches. With simulation exercise of 2000 patients who had undergone transplantation they showed that the average eplet MM load of 30 with existing allocation policy could be brought down to 10 if pre transplant epitope matching is instituted. Computational stimulation by inculcating T-cell epitope matching into Eurotransplant waitlist showed that theoretically better epitope matching can be done without causing any disadvantage to already waitlisted patients by denials of offer [38].

In immunological high-risk patients with elevated PRAs, adding epitope analysis would provide more specificity in donor selection. As mentioned previously epitope analysis can add meaning to the highly sensitive and occasionally non-specific single antigen bead profile in giving clarity on the immune reactivity profile which can ultimately facilitate donor selection [48, 49].

Pediatric transplant recipients in whom we desire the maximum longevity of graft life is a subset like sensitized patients where epitope matching may have a role. Kausman et al. reported a series of consecutive pediatric renal transplantation where they found 8 or 19 kids transplanted with eplet exclusion has less number of Class II eplet mismatches and less number of de novo DSAs [50].

Epitope matching is definitely a concept in the right direction but the practicalities have to evolve with more data. Prolongation of already extended waiting periods to get an organ offer is of primary concern [51]. Once are as on able algorithm has been identified, sensitized, and pediatric patients may be the two subgroups in which we may try the efficacy of the process.

Antibody Verified Eplets and Their Implications in Transplantation

Shruti Tapiawala, Meenakshi Singh, and Suchita Jogale

Antibody Verification of Eplets

As mentioned earlier in the chapter, HLA eplets resemble the functional epitope, which generally determines the specificity of the antibody through interaction with the complementarity determining region 3 (CDR3) of the heavy chain of the antibody [5–7].

Eplet mismatch analysis in retrospective studies seems to explain, better than antigen-mismatch analysis, antibody reactivity patterns observed in patients who are sensitized. Because not all eplet mismatches were shown to lead to antibody formation, the HLA Epitope Registry proposes some ("antibody-verified") eplets to be of higher immunogenicity. Hence eplets have to be validated as Antibody-verified eplets for their clinical implication. They are verified by analyzing interaction with polyclonal sera from multi- or uni-parous women, murine and human monoclonal antibodies (mAb) and with antibodies purified by adsorption and elution from sera of HLA sensitized individuals.

The experimental procedures carried out for antibody verification of HLA eplets are discussed below.

HLA-Specific Human Monoclonal Antibodies

Human mAbs which were previously produced by cloned B cell hetero-hybridomas derived from pregnancy immunized individuals were tested in luminex SAB assay and subsequently analyzed for their HLA specificity. IgG human mAbs were tested in the Lifecodes HLA class I or HLA class II SAB assay (Immucor, Stamford, CT, USA) according to the manufacturer's instructions.

Eplets with one or more references of A1 or A2 level were considered truly antibody-verified. Eplets with level B, C, or D were classified as provisionally antibody-verified, Human mAb data presented in the database that had not been published in a peer-reviewed journal were considered as not sufficient for antibody verification.

Lymphocytotoxicity

Lymphocytotoxicity data for mAbs is evaluated. In this test, a panel of more than 800 second-field HLA-typed cells are tested in complement dependent cytotoxicity (CDC) assays. Only cells with a single SAB reactive allele are included for analysis.

The average CDC score for each allele is calculated from the previously determined reactivity grades as shown in Tables 5.9 [46].

Reactivity Analysis of Human mAbs HLA Epitope Mismatch Algorithm (HLA-EMMA) version 1.05 [33] was used to determine the solvent accessible amino acid mismatches between the HLA of the antibody producer and the mismatched HLA allele of the immunizer.

Role of International Databases and Registries (e.g., EpRegistry)
The HLA Epitope Registry is an online database founded under auspices of the 16th International HLA and Immunogenetics Workshop in 2012, which has documented all theoretically defined eplets, as well as their antibody verification status with the aim to reflect the eplet repertoire incorporated in the widely used HLA Matchmaker software. The Registry has formed a pivotal source of information on eplets that have been defined on HLA as well as on MICA (Human Major Histocompatibility Complex Class I Chain-Related gene A) and has been of great benefit to the field of histocompatibility. Antibody-verified eplets in the Registry have been verified by analyzing reactivity patterns with polyclonal sera as well as monoclonal serum from sensitized individuals as well as recombinant antibodies.

Limitations and Ongoing Challenges in Verification

Although antibody-verified eplet mismatches have been demonstrated to correlate with DSA formation and graft survival [13, 14], recent reports have indicated that there are still clinically relevant eplets which have not been antibody verified yet.

Eplet verification using different methods described above are not uniform in the information they depict. In general, the reactivity pattern of a eplet using polyclonal sera in a Luminex based single antigen bead assay (SAB) cannot be compared with the response attributed using monoclonal sera. The antibodies purified using either adsorption or elution from a patient's serum cannot be correlated with SAB reactivity is caused by antibody reactivity against a single eplet. During the analysis of the antibody reactivity with a eplet, it must be always kept in cognizance that laboratory procedures of elution and purification may reveal non-pathogenic /natural anti-HLA antibodies. In addition, the usage of murine mAbs to extrapolate the immunogenicity in human settings is not completely accurate since the immunogenicity of mismatched HLA antigen is determined by recipients HLA type. The specificity of the murine ab may be directed against a different HLA epitope than human antibody. Consequently, only human mAbs accurately depict eplet antibody verification.

Another limitation is that the mAbs used for verification of eplets is generally directed against the epitope on common HLA alleles, whereas epitopes on rare HLA alleles remain unverified. Therefore, the ethnic minorities with rare alleles will find it difficult to be classified for antibody verified eplets posing a risk for the population.

Clinical Significance of Antibody-Verified Eplets

Correlation with Donor-Specific Antibody (DSA) Formation

Donor-specific antibodies (DSA) are formed in response to polymorphic amino acid mismatches on donor human leukocyte antigens (HLA), specifically targeting antigenic determinants known as eplets—clusters of amino acids on the surface of HLA molecules that are recognized by B-cell receptors and antibodies. These eplets represent the minimal structural elements required for antibody binding and are critical to the development of humoral alloimmune responses. The generation of DSA is a major immunological barrier in renal transplantation, as these antibodies are associated with antibody-mediated rejection (ABMR), chronic allograft injury, and eventual graft loss [52–54].

Importantly, not all eplets are equally immunogenic. Each eplet possesses unique physicochemical properties, spatial configurations, and degrees of structural disparity relative to the recipient's HLA type, all of which influence its potential to elicit an immune response. This concept—termed the nonself–self paradigm—proposes that the immunogenicity of a given eplet depends not only on its foreignness but also on its similarity to self-eplets. Eplets that differ subtly from self may be more readily recognized by the immune system as foreign and can effectively trigger DSA formation [28, 55, 56].

The variation in immunogenicity among eplets has significant clinical implications. Quantification of eplet mismatch load, using tools like HLAMatchmaker, has emerged as a powerful approach to assess alloimmune risk. Studies have demonstrated that a higher eplet mismatch burden correlates with increased likelihood of de novo DSA development and poorer graft outcomes [6, 57]. Furthermore, certain individual eplets, such as 62GE on HLA-B or 55PP on HLA-DR, have been identified as particularly immunogenic, underscoring the importance of qualitative, not just quantitative, evaluation of mismatches [58].

By integrating eplet-level analysis into donor-recipient compatibility assessments, transplant clinicians can move beyond traditional antigen-level matching to a more precise understanding of humoral sensitization risk, ultimately improving long-term graft survival.

Eplet Load and Immunogenicity

Various population studies have shown quantitative correlation between the number of epitope mismatches between donor and recipient, and DSA development. This reflects the effect of number of mismatches on the immunogenicity of an HLA molecule. Albeit even 1 or a few epitope mismatches can induce an antibody.

The immunogenicity of epitopes in clinical renal transplantation has been extensively studied. Across various studies, the HLA Matchmaker computer algorithm (Duquesnoy RJ et al) was used to identify the epitope load of specific HLA mismatches and its effect on the development of de novo DSA [58].

Traditional HLA antigen-level matching does not sufficiently predict the risk of donor-specific antibody (DSA) formation or antibody-mediated rejection (ABMR).

Instead, immunologic analysis at the epitope level offers a more nuanced approach to understanding and predicting immune sensitization. While total eplet count offers an important first measure, emerging data indicate that not all eplets are equally immunogenic. Some eplets consistently induce stronger alloantibody responses than others, likely due to factors such as:

- Surface accessibility and conformational stability of the eplet on the HLA molecule.
- Physicochemical disparity from the recipient's self HLA repertoire.
- The presence of T-cell help, often via indirect allorecognition, which enhances B-cell activation and class-switching.

This means that both quantitative (number of mismatched eplets) and qualitative (structural and immunologic properties) aspects must be considered in assessing immunogenicity [55].

It was observed that there is a significant correlation between the epitope mismatches between the recipient and the donor and development of de novo DSA (Marlies et al). Using HLA Matchmaker Wiebe et al. reported in a population study that recipients with 10 or less HLA-DR eplet mismatches and 17 or less HLA-DQ eplet mismatches did not produce donor specific antibodies. In addition, they observed a synergistic effect of non-adherence to immunosuppressive medication and HLA class II eplet load above these thresholds in recipients who underwent a transplant were more likely to develop DSA, acute rejection and graft failure compared with adherent recipients with an epitope load beneath the thresholds [59]. It is important to note that the HLA class II eplet threshold were not antibody-verified eplets and only potential eplets. The work published above was corroborated in pediatric patients based on the same HLA DQ eplet load and reported that recipients of a graft were at a low risk to develop both DR and DQ specific de novo DSA. This has led to possible hypothesis of the association between the number of HLA classes I and II epitope mismatches and development of de novo DSA. Therefore, the number of epitope mismatches may be a parameter for predicting the outcome of graft survival and long-term outcome of renal transplant recipients. Sapir-Pichhadze et al. even described an association between the transplant glomerulopathy and the number of HLA-DR and -DQ eplet mismatches they observed the additive effect of both HLA-DR and -DQ eplet mismatches on antibody mediated rejection and transplant glomerulopathy [60].

Besides renal transplants, epitope matching appears to be affect other types of solid organ transplants. It has been observed that the number of HLA eplet mismatches correlates with development of DSA and consequently the outcome of the transplantation in both lung transplant recipients and heart transplant recipients. It is important to comment that all the above studies indicate the impact of the combined effect of the number of HLA class I and II epitope mismatches on the risk of developing de novo DSA, without considering the immunogenicity of the individual epitopes.

Predictive Value in Primary and Repeat Transplantation

Recently, Kosmoliaptsis et al. compared the number of eplet mismatches along with intralocus as well as interlocus comparison of number of HLA class I and II amino acid mismatches between donor and recipient alleles as predictive value for De Novo DSA formation. There was no advantage of the eplet approach over the number of amino acid mismatches. Furthermore, the physiochemical properties of the amino acid substitution can predict immunogenicity. Each amino acid is unique based on the hydrophobicity and electrostatic charge of the side chains. Therefore in addition to the number and position of individual mismatched amino acids, the hydrophobicity and electrostatic charge determined for each mismatched amino acid resulted in a hydrophobicity mismatch score (HMS) and electrostatic mismatch score (EMS). The study showed that both HMS and EMS appeared to be strong and superior predictors of HLA classes I and II de novo DSA formation after kidney transplantation. Compared with eplet mismatches and amino acid mismatches [61].

Considering various studies reported, it may be concluded that HLA eplets instead of HLA antigens refines HLA matching strategies resulting in expanded donor pool for highly sensitized patients thus facilitating personalized immunosuppressive treatment based on immunological risk evaluation. This strategy results not only in decreased DSA formation but also less graft failure.

The following case study will demonstrate the importance of identifying eplet verified antibodies and treating the allograft in a timely manner.

Case Study

Case 1

This is a 28-year-old male who underwent cadaver donor renal transplant on 28th July 2013, then he underwent second living donor ABO incompatible renal transplant in May 2019, donor being his mother. He has history of blood transfusion in Sept 2018. The patient presented in October 2023 with rise in creatinine and proteinuria if 2 g/24 h. The renal allograft biopsy was suggestive of Chronic ABMR with C4D positive.

His post-transplant Single Antigen Bead Class I and Class II was performed to evaluate the cause of AMBR. SAB Class I and Class II results are shown in Tables 5.10 and 5.11 whereas recipient's and donor's HLA typing is shown in Table 5.12.

In the absence of detectable donor-specific antibodies (DSAs) post-transplant, an eplet mismatch analysis was undertaken to investigate potential alloimmune mechanisms contributing to allograft dysfunction. Fig. 5.8 shows the eplet analysis carried out using HLAMatchmaker.

Eplet-Level Analysis of Alloimmune Reactivity

To investigate potential alloimmune contributors to allograft dysfunction in the absence of detectable donor-specific antibodies (DSAs), a detailed eplet-level

Table 5.12 HLA typing by PCR SSOP method (intermediate resolution)

Locus	HLA A*	HLA B*	HLA C*	HLA DRB1*	HLA DQB1*	HLA DPB1*
Donor	11:01	35:03	14:02	12:01	03:01	02:01
	26:01	35:03	16:02	11:01	03:02	04:02
Recipient	11:01	35:03	14:02	12:01	03:01	02:01
	32:01	51:01	15:02	04:03	03:02	59:01
Mismatched Ags	26:01	–	16:02	11:01	–	04:02

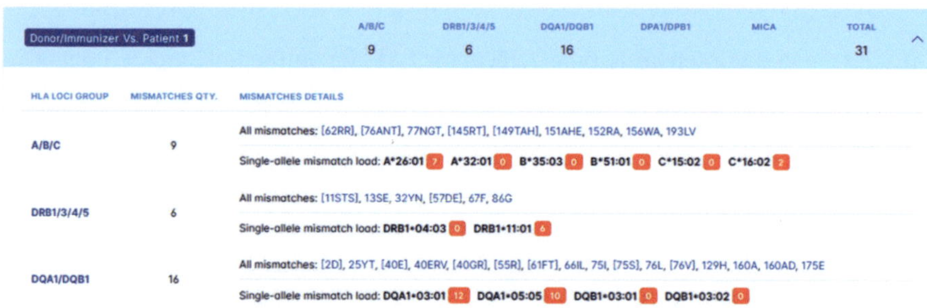

Fig. 5.8 Eplet Mismatching: (using HLA Eplet Registry: https://www.epregistry.com.br/)

comparison was performed. The mismatched eplets identified between the donor and recipient HLA alleles were systematically compared against the HLA antibody specificities detected by the single antigen bead (SAB) assay.

This comparison involved mapping the eplet compositions of each HLA antigen to which antibodies were present, and evaluating whether any of these eplets overlapped with those that were mismatched in the donor. Particular attention was paid to identifying shared eplets, i.e., those that were both:

- Present on the donor's mismatched HLA alleles, and
- Represented on one or more HLA antigens that elicited antibody binding in the SAB assay.

The presence of such shared eplets was interpreted as indirect evidence of potential cross-reactive alloimmune memory, suggesting that antibodies detected on the SAB panel may target epitopes structurally similar to donor eplets, even in the absence of donor-specific antibody as conventionally defined. This eplet-level correspondence helped refine the understanding of antibody specificity and supported the hypothesis of eplet-driven subclinical humoral alloimmunity as a possible contributor to allograft injury.

In this case, persistent antibody reactivity was observed on single antigen bead (SAB) assay; however, no donor-specific antibodies (DSAs) were detected at either the antigen or allele level against the donor HLA. A comprehensive non-HLA antibody screen, including antibodies against MICA, anti-endothelial cell antibodies

Table 5.13 Identification of antibody verified eplets

Antibody	Eplets	Exposition
C*06:02	62RR	High
	76ANT	High
	77NGT	High
	149TAH	High
	151AHE	High
	156WA	Intermediate
	193LV	High
C*07:02	152RA	High

(AECAs), and angiotensin II type 1 receptor (AT1R) antibodies, yielded negative results. Upon detailed eplet-level analysis, high and intermediate exposure antibody-verified eplets were identified that corresponded to mismatched donor eplets, as detailed in Table 5.13.

Despite regular immunological monitoring, the patient developed irreversible chronic antibody-mediated rejection (ABMR). The absence of DSAs at the antigen or allele level on SAB assay at the time of diagnosis suggested that the immunologic injury was likely mediated by de novo DSAs that emerged over time against the mismatched eplets. These eplet-level mismatches likely served as the immunogenic targets driving alloimmune injury in this case.

This case underscores the limitations of conventional HLA antibody detection methods that rely solely on antigen- or allele-level specificity. Although SAB assays did not reveal detectable donor-specific antibodies, the patient demonstrated persistent antibody reactivity and eventually developed chronic ABMR. The absence of non-HLA antibodies—including anti-MICA, AT1R, and AECAs—further directed attention to alternative mechanisms of alloimmune sensitization.

Eplet-level analysis using HLAMatchmaker revealed mismatches at verified immunogenic eplets, some of which were associated with high and intermediate antibody exposure risk. These findings support emerging evidence that certain eplet mismatches, particularly those with known immunogenic profiles, can induce de novo antibody formation post-transplant even when conventional SAB assays remain negative early in the course [6, 21].

The development of de novo DSAs targeting these mismatched eplets over time likely contributed to the chronic ABMR observed in this patient. This highlights the potential utility of integrating eplet mismatch analysis into pre-transplant risk stratification and post-transplant monitoring to identify subclinical alloimmune processes not captured by standard serological assays (Tambur et al., 2018). It also reinforces the notion that immunological risk may be underestimated when relying solely on traditional DSA assessments without considering epitope-level disparities [55].

Conclusion

Epitope matching has been an ever-evading challenge in histocompatibility. Advancements in high resolution typing and identification of 3D structure of HLA have facilitated a better understanding of importance of epitopes in immune recognition interface. There is an agreement among experts that we need more qualitative data to formulate a matching strategy based on epitopes for the future. Current literature which is dominated by association of incremental epitope mismatches and graft outcome is a crude estimate of real alloreactivity which in current form cannot be incorporated to any prospective matching algorithm. Although many advances have been made, only a limited number of epitopes have been validated to their immunogenic potential in large clinical cohorts. Advanced computational and machine-learning tools can be used to study the currently available epitopes and validate their immunogenicity. As the field advances, the integration of epitope data with clinical practice may offer more refined and personalized approaches to transplantation, ultimately benefiting patients.

Incorporating eplet-based analysis into pre- and post-transplant immunological evaluation may enhance our ability to predict and monitor alloimmune risk, especially in scenarios where conventional assays are insufficient. As precision in HLA immunogenetics continues to evolve, the integration of eplet immunogenicity profiles may inform donor selection, desensitization strategies, and personalized immunosuppressive regimens aimed at improving long-term graft outcomes. Post transplant monitoring for formation of DeNovo HLA antibodies and identifying if they are against the mismatched eplets may help prevent immune events by modification of immunosuppression and aid in diagnosis of cause of allograft injury.

The discovery of HLA specific human mAbs with new specificity will lead to advancements in the field of antibody verified eplets. In addition, the discovery of the newly verified eplets must be also regularly updated in the HLA Eplet registry to ensure revised antibody verification status of eplets. This will lead to up to date uniform database of antibody verified eplets aiding in the identification of immunogenic eplet mismatches or the level of eplet mismatch in prediction of transplant outcomes.

References

1. Barker DJ, Maccari G, Georgiou X, Cooper MA, Flicek P, Robinson J, et al. The IPD-IMGT/HLA database. Nucleic Acids Res. 2023;51(D1):D1053–60.
2. Cunningham BC, Wells JA. Comparison of a structural and a functional epitope. J Mol Biol. 1993;234(3):554–63.
3. Duquesnoy RJ, Marrari M, Sousa LCDDM, Barroso JRPDM, Aita KMDSU, Da Silva AS, et al. 16th IHIW: a website for antibody-defined HLA epitope registry. Int J Immunogenet. 2013;40(1):54–9.
4. Marrari M, Duquesnoy RJ. Why can sensitization by an HLA-DR2 mismatch lead to antibodies that react also with HLA-DR1? Hum Immunol. 2009;70(6):403–9.

5. Renaldo A, Roa-Bautista A, González-López E, López-Hoyos M, San SD. Epitope-level matching—a review of the novel concept of eplets in transplant histocompatibility. Transplantology. 2021;2(3):336–47.
6. Wiebe C, Pochinco D, Blydt-Hansen TD, Ho J, Birk PE, Karpinski M, et al. Class II HLA epitope matching—a strategy to minimize de Novo donor-specific antibody development and improve outcomes. Am J Transplant. 2013;13(12):3114–22.
7. Oliveira DMTE, Melo Santos De Serpa Brandão R, Claudio Demes Da Mata Sousa L, das Chagas Alves Lima F, Jamil Hadad Do Monte S, Sérgio Coelho Marroquim M, et al. pHLA3D: an online database of predicted three-dimensional structures of HLA molecules. Hum Immunol. 2019;80(10):834–41.
8. Mattoo A, Jaffe IS, Keating B, Montgomery RA, Mangiola M. Improving long-term kidney allograft survival by rethinking HLA compatibility: from molecular matching to non-HLA genes. Front Genet. 2024;15:1442018. https://doi.org/10.3389/fgene.2024.1442018/full.
9. Lanzavecchia A. Antigen-specific interaction between T and B cells. Nature. 1985;314(6011):537–9.
10. Kramer CSM, Roelen DL, Heidt S, Claas FHJ. Defining the immunogenicity and antigenicity of HLA epitopes is crucial for optimal epitope matching in clinical renal transplantation. HLA. 2017;90(1):5–16. https://doi.org/10.1111/tan.13038.
11. Mittal KK, Terasaki PI. Cross-reactivity in the HL-A system. Tissue Antigens. 1972;2(2):94–104.
12. Rood JJV, Leeuwen AV. Leukocyte grouping: a method and its application. J Clin Invest. 1963;42(9):1382–90.
13. Rodey GE, Neylan JF, Whelchel JD, Revels KW, Bray RA. Epitope specificity of HLA class I alloantibodies I. Frequency analysis of antibodies to private versus public specificities in potential transplant recipients. Hum Immunol. 1994;39(4):272–80.
14. Thompson J, Thacker L, Takemoto S. CREG matching for first kidney transplants performed by SEOPF centers between October 1987 and September 1995: an analysis of outcome and prospective benefit. Transplant Proc. 1997;29(1–2):1435–8.
15. Wujciak T, Opelz G. Evaluation of HLA matching for CREG antigens in Europe1. Transplantation. 1999;68(8):1097–9.
16. Rodey G. HLA beyond tears: introduction to human histocompatibility. 2nd ed. Durango, CO: De Novo: Distributed by Pel-Freez; 2000.
17. McCaughan J, Xu Q, Tinckam K. Detecting donor-specific antibodies: the importance of sorting the wheat from the chaff. HepatoBiliary Surg Nutr. 2019;8(1):37–52.
18. Opelz G, Wujciak T, Döhler B, Scherer S, Mytilineos J. HLA compatibility and organ transplant survival. Collaborative transplant study. Rev Immunogenet. 1999;1(3):334–42.
19. Williams RC, Opelz G, McGarvey CJ, Weil EJ, Chakkera HA. The risk of transplant failure with HLA mismatch in first adult kidney allografts from deceased donors. Transplantation. 2016;100(5):1094–102.
20. Bezstarosti S, Kramer CSM, Franke-van Dijk MEI, Vergunst M, Bakker KH, Uyar-Mercankaya M, et al. HLA-DQ-specific recombinant human monoclonal antibodies allow for in-depth analysis of HLA-DQ epitopes. Front Immunol. 2022;12:761893.
21. Duquesnoy RJ, Marrari M, Tambur AR, Mulder A, Da Mata Sousa LCD, Da Silva AS, et al. First report on the antibody verification of HLA-DR, HLA-DQ and HLA-DP epitopes recorded in the HLA epitope registry. Hum Immunol. 2014;75(11):1097–103.
22. El-Awar N, Jucaud V, Nguyen A. HLA epitopes: the targets of monoclonal and alloantibodies defined. J Immunol Res. 2017;2017:1–16.
23. Kramer C, Heidt S, Claas FHJ. Towards the identification of the relative immunogenicity of individual HLA antibody epitopes. Hum Immunol. 2019;80(4):218–20.
24. Charnaya O, Jones J, Philogene MC, Chiang PY, Segev DL, Massie AB, et al. Eplet mismatches associated with de novo donor-specific HLA antibody in pediatric kidney transplant recipients. Pediatr Nephrol Berl Ger. 2021;36(12):3971–9.

25. Kishikawa H, Kinoshita T, Hashimoto M, Fukae S, Taniguchi A, Yamanaka K, et al. Class II HLA eplet mismatch is a risk factor for De Novo donor-specific antibody development and antibody-mediated rejection in kidney transplantation recipients. Transplant Proc. 2018;50(8):2388–91.
26. Wiebe C, Kosmoliaptsis V, Pochinco D, Gibson IW, Ho J, Birk PE, et al. HLA-DR/DQ molecular mismatch: a prognostic biomarker for primary alloimmunity. Am J Transplant. 2019;19(6):1708–19.
27. Tambur AR. HLA-epitope matching or eplet risk stratification: the devil is in the details. Front Immunol. 2018;9:2010. https://doi.org/10.3389/fimmu.2018.02010/full.
28. Duquesnoy RJ. HLAMatchmaker: a molecularly based algorithm for histocompatibility determination. I. Description of the algorithm. Hum Immunol. 2002;63(5):339–52.
29. Geneugelijk K, Spierings E. PIRCHE-II: an algorithm to predict indirectly recognizable HLA epitopes in solid organ transplantation. Immunogenetics. 2020;72(1–2):119–29.
30. Kosmoliaptsis V, Mallon DH, Chen Y, Bolton EM, Bradley JA, Taylor CJ. Alloantibody responses after renal transplant failure can be better predicted by donor–recipient HLA amino acid sequence and physicochemical disparities than conventional HLA matching. Am J Transplant. 2016;16(7):2139–47.
31. Kramer CSM, Koster J, Haasnoot GW, Roelen DL, Claas FHJ, Heidt S. HLA-EMMA: a user-friendly tool to analyse HLA class I and class II compatibility on the amino acid level. HLA. 2020;96(1):43–51.
32. Aldea PL, Santionean MD, Elec A, Munteanu A, Antal O, Loga L, et al. An integrated approach using HLAMatchmaker and Pirche II for epitopic matching in pediatric kidney transplant—a Romanian single-center study. Children. 2023;10(11):1756.
33. Bedford A, Jervis S, Worthington J, Lowe M, Poulton K. Human leukocyte antigen epitope mismatch loads and the development of de novo donor-specific antibodies in cardiothoracic organ transplantation. Int J Immunogenet. 2022;49(1):30–8.
34. Crane C, Ingulli EG, Morris GP. Correlation of HLAMatchmaker and PIRCHE reveals T cell epitope mismatches high risk for donor specific antibodies (DSA) and antibody-mediated rejection (AMR): FR-OR61. J Am Soc Nephrol. 2023;34(11S):46.
35. Mallon DH, Kling C, Robb M, Ellinghaus E, Bradley JA, Taylor CJ, et al. Predicting humoral alloimmunity from differences in donor and recipient HLA surface electrostatic potential. J Immunol. 2018;201(12):3780–92.
36. Geneugelijk K, Niemann M, Drylewicz J, Van Zuilen AD, Joosten I, Allebes WA, et al. PIRCHE-II is related to graft failure after kidney transplantation. Front Immunol. 2018;9:321. https://doi.org/10.3389/fimmu.2018.00321/full.
37. Senev A, Coemans M, Lerut E, Van Sandt V, Kerkhofs J, Daniëls L, et al. Eplet mismatch load and De Novo occurrence of donor-specific anti-HLA antibodies, rejection, and graft failure after kidney transplantation: an observational cohort study. J Am Soc Nephrol. 2020;31(9):2193–204.
38. Walton DC, Cantwell L, Hiho S, Ta J, Wright S, Sullivan LC, et al. HLA class II eplet mismatch predicts De Novo DSA formation post lung transplant. Transpl Immunol. 2018;51:73–5.
39. Mohammadhassanzadeh H, Oualkacha K, Zhang W, Klement W, Bourdiec A, Lamsatfi J, et al. On path to informing hierarchy of eplet mismatches as determinants of kidney transplant loss. Kidney Int Rep. 2021;6(6):1567–79.
40. Sapir-Pichhadze R, Zhang X, Ferradji A, Madbouly A, Tinckam KJ, Gebel HM, et al. Epitopes as characterized by antibody-verified eplet mismatches determine risk of kidney transplant loss. Kidney Int. 2020;97(4):778–85.
41. McCaughan JA, Battle RK, Singh SKS, Tikkanen JM, Moayedi Y, Ross HJ, et al. Identification of risk epitope mismatches associated with de novo donor-specific HLA antibody development in cardiothoracic transplantation. Am J Transplant. 2018;18(12):2924–33.

42. Wiebe C, Rush DN, Nevins TE, Birk PE, Blydt-Hansen T, Gibson IW, et al. Class II eplet mismatch modulates tacrolimus trough levels required to prevent donor-specific antibody development. J Am Soc Nephrol. 2017;28(11):3353–62.
43. Gombos P, Opelz G, Scherer S, Morath C, Zeier M, Schemmer P, et al. Influence of test technique on sensitization status of patients on the kidney transplant waiting list. Am J Transplant. 2013;13(8):2075–82.
44. Morales-Buenrostro LE, Terasaki PI, Marino-Vázquez LA, Lee JH, El-Awar N, Alberú J. "Natural" human leukocyte antigen antibodies found in nonalloimmunized healthy males. Transplantation. 2008;86(8):1111–5.
45. Tran JN, Günther OP, Sherwood KR, Fenninger F, Allan LL, Lan J, et al. High-throughput sequencing defines donor and recipient HLA B-cell epitope frequencies for prospective matching in transplantation. Commun Biol. 2021;4(1):583.
46. Hamrick CW, Lebeck LK. Flow cytometric T and B cell crossmatching. In: Hahn AB, Land GA, Strothman RM, editors. ASHI laboratory manual. 4th ed. American Society for Histocompatibility and Immunogenetics; 2000. p. VI.B.4.1–5.
47. Sherwood KR, Tran J, Günther OP, Lan J, Aiyegbusi O, Liwski R, et al. Genome Canada precision medicine strategy for structured national implementation of epitope matching in renal transplantation. Hum Immunol. 2022;83(3):264–9.
48. Duquesnoy RJ, Howe J, Takemoto S. HLAMatchmaker: a molecularly based algorithm for histocompatibility determination. IV. An alternative strategy to increas the number of compatible donors for highly sensitized patients1. Transplantation. 2003;75(6):889–97.
49. Goodman RS, Taylor CJ, O'Rourke CM, Lynch A, Bradley JA, Key T. Utility of HLAMatchmaker and single-antigen HLA-antibody detection beads for identification of acceptable mismatches in highly sensitized patients awaiting kidney transplantation. Transplantation. 2006;81(9):1331–6.
50. Kausman JY, Walker AM, Cantwell LS, Quinlan C, Sypek MP, Ierino FL. Application of an epitope-based allocation system in pediatric kidney transplantation. Pediatr Transplant. 2016;20(7):931–8.
51. Affdal A, Ballesteros F, Malo MF, Sancho C, Cochran-Mavrikakis SL, Bryan S, et al. Canadian kidney transplant professionals' perspectives on precision medicine and molecular matching in kidney allocation. Transplant Direct. 2023;10(1):e1565.
52. Loupy A, Lefaucheur C. Antibody-mediated rejection of solid-organ allografts. N Engl J Med. 2018;379(12):1150–60.
53. Wiebe C, Gibson IW, Blydt-Hansen TD, et al. Evolution and clinical pathologic correlations of de novo donor-specific HLA antibody post kidney transplant. Am J Transplant. 2012;12(5):1157–67.
54. Everly MJ, Rebellato LM, Haisch CE, et al. Incidence and impact of de novo donor-specific alloantibody in primary renal allografts. Transplantation. 2013;95(3):410–7.
55. Tambur AR, Claas FHJ. HLA epitopes as viewed by antibodies: what is it all about? Am J Transplant. 2015;15(5):1148–54.
56. Kosmoliaptsis V, Bradley JA, Chaudhry AN, Taylor CJ. Predicting HLA class I alloreactivity: a comparison of methods for identifying permissible HLA mismatches in kidney transplantation. Am J Transplant. 2014;14(5):1275–82.
57. Sapir-Pichhadze R, Tinckam KJ, Quach K, et al. HLA-DR and HLA-DQ eplet mismatches and transplant glomerulopathy: a nested case–control study. Am J Transplant. 2015;15(1):137–48.
58. Duquesnoy RJ, Marrari M. HLAMatchmaker-based definition of structural epitopes and epitope-based matching for transplant compatibility. Curr Opin Organ Transplant. 2009;14(4):403–9.
59. Dankers MKA, Witvliet MD, Roelen DL, de Lange P, Korfage N, Persijn GG, Duquesnoy R, Doxiadis IIN, Claas FHJ. The number of amino acid triplet differences between patient and donor is predictive for the antibody reactivity against mismatched human leukocyte antigens. Transplantation. 2004;77(8):1236–9.

60. Sapir-Pichhadze R, Zhang X, Ferradji A, et al. Epitopes as characterized by antibody-verified eplet mismatches determine risk of kidney transplant loss. Kidney Int. 2020;97(4):778–85. https://doi.org/10.1016/j.kint.2019.10.028.
61. Kosmoliaptsis V, Bradley JA, Sharples LD, et al. Predicting the immunogenicity of human leukocyte antigen class I alloantigens using structural epitope analysis determined by HLAMatchmaker. Transplantation. 2008;85:1817–25.

Virtual Crossmatch in Kidney Transplantation

6

Sonia Mehrotra and Raj Kumar Sharma

Case 1 Shows how SAB analysis which estimated the existence of DSA against the donor by virtual crossmatch, before transplantation assessment, can help in the selection of a suitable donor in a sensitized recipient.

A 60-year-old male with chronic kidney disease on hemodialysis, a history of previous transfusion, no previous transplantation was evaluated. His blood group was B+. He underwent per-transplant evaluation for ABO-incompatible transplant with his wife as the donor, donor blood group was A+ (Table 6.1). CDC cross-match was negative with wife; Flow cross-match was negative with wife. Anti-HLA antibodies were identified against the listed HLA Class I antigens: (PRA 03%) MFI range for all Class I is 4661-2220 and anti-HLA antibodies were not identified against the listed HLA Class II antigens (PRA 0%) (Figs. 6.1 and 6.2). No donor-specific antibody (DSA) was identified against the wife as kidney donor but this was an ABO-incompatible transplantation (anti-A antibody 1:128).

Case 2 Shows that SAB analysis could be helpful to estimate the existence of DSA against the alleles present on donor from the SAB panel with MFI value evaluation. It is very useful if the SAB test is performed before transplantation. A 52-year-old female had chronic kidney disease on hemodialysis, with multiple pregnancies, history of transfusion but no history of previous transplantation. Her blood group was A+ (Table 6.1). She underwent per-transplant evaluation for ABOi transplant with her husband as donor. His blood group was B+ and baseline Anti-B titer was 1:128.

S. Mehrotra · R. K. Sharma (✉)
Department of Nephrology and Kidney Transplant Medicine, Medanta Kidney and Urology Institute, Lucknow, Uttar Pradesh, India
e-mail: Sonia.mehrotra@medanta.org; raj.sharma@medanta.org

G. Abraham et al. (eds.), *Immunological Challenges in Organ Transplantation*,
https://doi.org/10.1007/978-981-95-1533-2_6

Table 6.1 Case summary of crossmatch status

Cases	CDCXM	FCXM	SAB	Decision	Donor HLA antigens
Case-1	–	–	Abs in class I	XM presented both non-HLA antibodies and HLA-DSA, < sensitive	HLA typing—A*02, B*35, B*40, DRB1*15
			A*24:02, A*24:03, A*23:01	SAB DSA MFI ≥ 5000 projecting analyst of + XM & AR	
Case-1 with KDP donor	–	–	No DSA with above HLA antibodies		HLA typing—A*01 B*15, DRB1*13
Case-2 family donors	–	+	Multiple abs in class I & class II with DSA		HLA typing—A*01 B*15, DRB1*13
			B*15:16, B*58:01, B*57:01,B*27:05, B*27:03, A*11:02		
			DRB1*13:01,DRB1*13:05,DRB1*03:02, DRB1*03:01, DRB1*13:01 DPA1*02:01-DPB1*17:01, DRB1*03:03, DRB1*11:03, DRB1*03:01, DRB1*12:02, DRB1*12:01,DRB1*14:01,DRB1*11:01,DRB1*11:04		
One KDP donor	–	–	No DSA with above HLA antibodies		HLA typing—A*02, B*35, B*40, DRB1*15

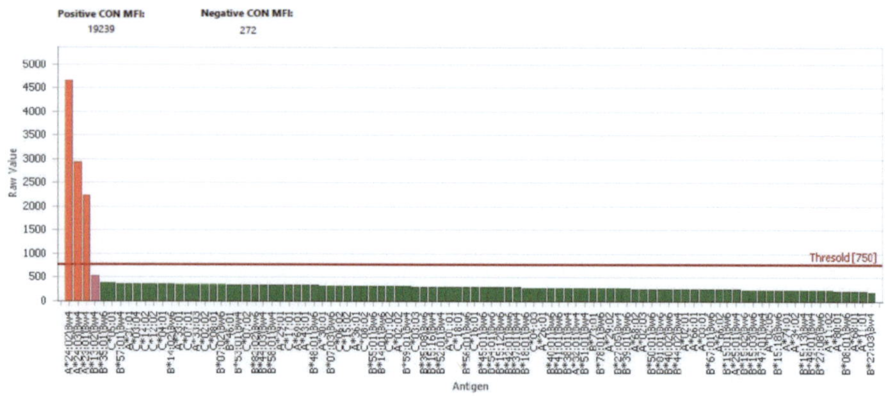

Fig. 6.1 HLA antibodies class I

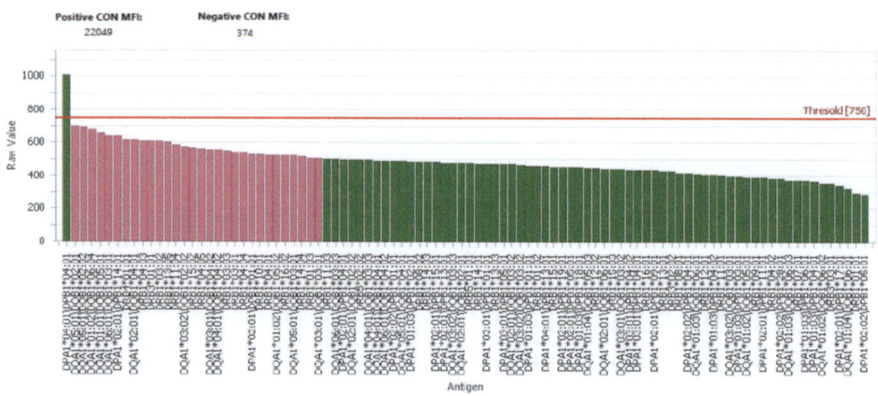

Fig. 6.2 HLA antibodies class II

Per-transplant evaluation of histocompatibility included CDC, FCXM between them. SAB assay of the recipient was performed. Their HLA typing of HLA-A, -B, -C, -DRB1, DQ loci were performed, CDCXM was negative with donor (husband), FCXM with the donor (husband) was positive for both T cell and B cell. The result of SAB is shown in Figs. 6.3 and 6.4 where there was reactivity against the donor's alleles. She possessed anti-HLA antibodies against HLA-B loci in class I and HLA-DR loci in class II. Anti-HLA antibodies were identified (PRA 6%) with MFI 7376 to 1606 and anti-HLA antibodies were identified against HLA Class II antigens (PRA 19%) with MFI 1061 to 759 (Figs. 6.3 and 6.4).

Both above pairs underwent evaluation of histocompatibility for Kidney Pair Donation with ABO compatible donor of case 1 by CDCXM and FCXM. This KPD donor was negative for both T cell and B cell physical crossmatch.

Case 1 also had negative flow Cross-match with ABO compatible KPD donor of case 2 (husband of case 2).

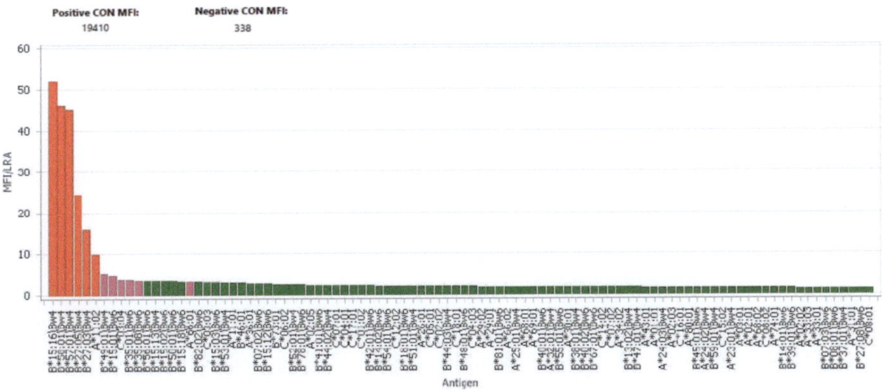

Fig. 6.3 HLA antibodies class I

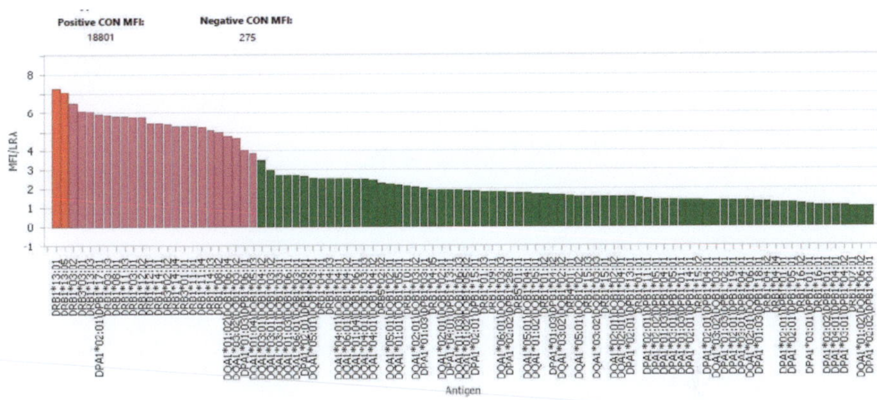

Fig. 6.4 HLA antibodies class II

Cell-based cross-match can detect both non-HLA antibodies and HLA-DSA. It is important to evaluate the positive cell-based crossmatch with SAB. In SAB, DSA MFI ≥5000 predicts positive cell-based Cross-match and post-transplant rejection.

These above two cases show the utility of SAB. The negative SAB assay or negative virtual crossmatch can help in the selection of a suitable donor.

If SAB assay detected anti-HLA antibodies but they are not donor-specific antibodies (DSA), then the patient can receive a kidney from ABO-compatible KPD donor for overcoming the ABO-incompatible of a family donor as is exemplified in these cases.

Definition

Virtual-crossmatch can predict a negative "physical" wet-crossmatch, a "compatible" crossmatch and "acceptable" outcomes. An assessment of immunologic compatibility based on the recipient's alloantibody profile compared to the donor's histocompatibility antigens is based on well-defined HLA antigens and antibodies.

Virtual-crossmatch is like a virtual reality (Web-based or paper-based crossmatch) which identifies the presence or absence of donor-specific HLA antibodies (DSA) in a prospective recipient going for transplant. Virtual crossmatching compares the patients' anti-HLA antibody specificity profile to the HLA type of the donors without actually performing a "wet" crossmatch (complement dependent cytotoxic (CDC) and flow cytometric crossmatch) [1].

History

The primary purpose of a pretransplant crossmatch is to predict the immunologic risk of rejection post-transplant. More than 50 years have passed since the first introduction of the complement-dependent cytotoxicity crossmatch by Terasaki and Patel called physical crossmatch (PXM), which remains one of the key final pieces of data used by clinicians to guide the compatibility acceptance of a donor organ. Physical crossmatch (PXM) has traditionally been used as the gold standard method because it reduces the risk of hyperacute and acute rejection in solid organ transplantation [2].

Physical crossmatches (PXM) show if a recipient is "pre-sensitized" to the prospective allograft donor with donor-specific antibodies (DSA). A positive physical crossmatch (CDC) is considered a contraindication for transplantation.

The parallel developments in antibody detection techniques on solid-phase assays and improved molecular HLA typing technologies have enabled the use of virtual crossmatch (VXM) as an accurate and specific test to identify DSA.

The virtual crossmatch (VXM) testing requires various prerequisites as listed:

1. **Mandatory directions**
2. **Technical directions**
3. **Clinical directions**

Mandatory Directions

The virtual crossmatch (VXM) pretransplant transplantation immunologic risk assessment is based on the patient's alloantibody profile compared to the donor's histocompatibility antigens. There is a requirement to perform the assessment of the full antibody profile of the patients to include all HLA loci—A, B, C, DR, DQA, DQB, DPA, and DPB with accurate complete HLA typing of the donor.

The outcome of the transplant procedure is based on a negative virtual crossmatch (VXM), which would predict the absent DSA, even if the donor's HLA is mismatched with the recipient or the recipient has a high calculated panel reactive antibody (cPRA) of 100%. The virtual crossmatch (VXM) can better define the understanding of the human major histocompatibility complex, predict outcomes in recipients with anti-HLA antibodies. Virtual crossmatch is prospectively utilized to select the most appropriate donor/recipient pair even among highly sensitized candidates [3–7].

In the USA, the new Kidney Allocation System (KAS) has also defined the geographical boundaries of kidney allocation. Laboratories are mandated to depend on virtual crossmatching. The advancements in VXM result in not transporting the blood samples to distant laboratories for PXM.

Technical Directions

VXM is an accurate technique. It has reduced the cold ischemia time by obviating the need to ship donor organs to long distances, but can have limitations to the assessment because live or deceased donor HLA typing is performed using SSP, SSO, or Real-Time PCR (RT-PCR), which provides an intermediate level of resolution. Patients can have allele-level antibodies which can be missed by the software (Table 6.2).

These ambiguities can hinder accurate VXM assessments. There is the possibility that the HLA typing of the donor is incorrect. UNOS has now mandated double entry of HLA typing. This has encouraged UNOS to plan authorizing HLA typing on two different samples by two methods to prevent dictation errors, like the current ABO blood group testing policy of regulatory bodies.

Clinical Directions

- Reduction in cold ischemic time (CIT).
- Avoidance of inappropriate organ refusals based on false positive PXM depends on the positive threshold established by each center. The prevalence of false positive PXM results can range from 5% to 20%.
- The accuracy of PXM interpretation can be further confounded by patient factors like the use of immunotherapies (autoimmune diseases or rituximab targeting anti-CD20).
- The protocol for the Indian setting is described in (Tables 6.3, 6.4, 6.5, 6.6, and 6.7).
- Transplantation across a positive PXM in the absence of donor-specific antibodies as confirmed by VXM does not appear to be associated with negative clinical consequences. Having a positive PXM in the setting of negative VXM can help in accepting the corresponding donor organ and avoid incorrect rejection of com-

Table 6.2 Summary of methods of imputation for low- to high-resolution HLA typing

	Single nucleotide polymorphism based imputation	Four-digit allele converter	Haplostats
Sponsor	Variable	HLA matchmaker	NMDP
Platform	Web-based	Excel spreadsheet	Web-based
Input	SNPs	Low-resolution HLA typing	Low-resolution HLA typing
Imputation of DQA1 and DPA1/DPB1 possible	Yes	No	No
URL	Variable SNP2HLA: http://software.broadinstitute.org/mpg/snp2hla	www.epitopes.net/downloads.html	www.haplostats.org
	Variable HIBAG: www.bioconductor.org/packages/release/bioc/html/HIBAG.html		
	Variable HLA*IMP:02: http://oxfordhla.well.ox.ac.uk		

Table 6.3 Comparing of various crossmatch risk assessment

CDC crossmatch	Flow crossmatch	Single-antigen bead (SAB)	Risk assessment
+	+	+	Hyperacute rejection due to DSA
–	+	+	Non-complement-fixing DSA
–	–	+	High background of beads, can be false positive or positive with low MFI
+	+	–	False-negative bead panel lacks representation of the donor antigen, prozone" effect, low-level anti-HLA antibody against a shared epitope, which is diluted with multiple beads
+	–	–	IgM antibody probably nonHLA Ab. can picked up by CDCXM
–	+	–	IgG antibody probably nonHLA Ab.

Table 6.4 Comparing of various crossmatch techniques

Crossmatch	CDCXM	FCXM	Luminex XM	VXM
Target and principle of test method	Recipient serum and donor's cells with complement target to anti-HLA antibodies	Recipient serum and donor's cells label with anti-human IgG antibody	Recipient serum and donor's cells Luminex beads coated with monoclonal IgG antibodies specific for HLA molecules	Interpretation of results from SAB (single antigen bead assay) and donor high resolution HLA molecular typing
Antigens	HLA (native) on cell membrane	HLA (native) on cell membrane	HLA (native) on cell membrane on Luminex beads	Recombinant HLA proteins SAB (single antigen bead assay)
Anti-HLA antibodies specific	**No**	**No**	**No**	**Yes**
Interpretation	Subjective: Operator-dependent technique	Objective: Based on the difference between test serum MFI and pool negative control serum	Objective: Based on the MFI of specific antigens	Objective: Based on the MFI of specific antigens
Sensitivity	**Low**	**Intermediate**	**Intermediate**	**High**
Predicting the clinical outcome of transplants	Hyperacute rejection	Acute rejection	Acute/chronic rejection	Acute/chronic rejection

CDCXM complement-dependent cytotoxicity, *HLA* human leukocyte antigen, *MFI* mean fluorescence intensity, *SAB* single antigen bead, *VXM* virtual crossmatch

Table 6.5 Proposed algorithm for pretransplant immunological risk assessment (for living-donor renal transplantation) in Indian setting

Pretransplant immunological risk assessment					
Stage	CDCXM	FCXM	SAB	HLA typing	
Stage I	CDCXM is negative go to stage II				
	CDCXM is borderline positive go to stage II for FCXM				
	CDCXM is positive high titers change the donor				
Stage II		FCXM negative	SAB negative	SSOP HLA typing	Go for transplant
		FCXM positive low titers	SAB positive low titers	High resolution HLA typing of donor	Go for transplant with desensitization even DSA
		FCXM positive high titers	SAB positive high titers	SSOP HLA typing if DSA change donor or KPD	If DSA change donor or KPD
Stage III	Final CDCXM is negative before a week of transplant				Go for transplant
	Final CDCXM is positive before a week of transplant	Go for FCXM negative before a week of transplant			Go for transplant
		Go for FCXM positive before a week of transplant			Wait for weeks
			SAB repeat negative before a week of transplant		Go for transplant
					Wait for weeks

CDC-XM complement-dependent cytotoxicity crossmatch, *FC-XM* flow cytometry crossmatch, *SAB* single antigen bead assay

patible donors or the unnecessary use of increased immunosuppression by rational identification of the rejection risk of transplant [8–10].

VXM can reduce the growing logistical difficulties and is associated with increased organ sharing as the UNOS has now moved to a new kidney allocation policy (KAS 250).

Table 6.6 Various molecular HLA typing methods

Molecular HLA typing	
Have largely replaced serological methods, use DNA analysis to identify HLA alleles, utilizing techniques like PCR-SSO, PCR-SSP, and sequencing-based typing (SBT)	
RT PCR HLA typing	HLA typing is a rapid and reliable method for identifying HLA alleles, crucial for organ and bone marrow transplantation, using sequence-specific primers and probes to detect and differentiate target alleles
SSP HLA typing	Amplifies HLA DNA regions using PCR and then hybridizes the amplified DNA with sequence-specific oligonucleotide probes to identify specific alleles
SSOP HLA typing	Uses PCR primers designed to amplify only specific HLA alleles, allowing for direct identification of the amplified product
NGS HLA typing	Provides a high-throughput approach to sequencing, enabling rapid and accurate HLA typing
Third-generation sequencing	Nanopore sequencing (ONT) have received the most attention for HLA typing applications, this platform does not require expensive equipment and can achieve allele-level typing within the time required for typing of deceased donors

Table 6.7 Cell and bead base antibodies identification crossmatch tests

	Medium	Assessment	Explanation
Cell base antibodies identification crossmatch			
CDCXM	By complement proteins to detect antibody binding and cell lysis	Limited sensitivity and specificity Report as scoring % of cells death	Both HLA/non-HLA antibodies. Detects and qualitative assessment
FCXM	Assess the binding of antibodies to donor cells	Sensitive in detecting low-level antibodies Report as ratio and median channel shift	Detects both HLA/ non-HLA antibodies and quantitative to semi-quantitative
Bead base antibodies identification crossmatch by Luminex			
Anti HLA antibody (screening)	Mixture of most common antigens	Positive or negative	Shows sensitization
Panel reactive antibodies (PRA)	Each bead is a phenotype of an individual	Positive (%) or negative	Sensitization in percentage
Single-antigen bead (SAB)	One HLA antigen on each bead	Positive or negative (semi-quantitative, using MFI)	If patient has any antibody, is this DSA?

The deceased donor kidney allocation system (KAS250) was implemented in March 2021 in the USA. It prioritizes recipients within a 250-nautical mile radius of the donor hospital. KAS250 was implemented to reduce geographic disparities in access to kidney transplantation [11, 12].

A maximum of 2 proximity points can be awarded to the recipient candidates inside the 250 NM and located closest to the donor hospital, and the points decrease

in a linear fashion out to the 250 NM circle, where a recipient at that distance receives no additional proximity points.

Distance-Based KAS revision essentially consists of three main components:

1. The implementation of a 250-nautical-mile (NM) fixed distance concentric circle (measured from the hospital where the deceased donor originated) as units of allocation replacing OPO's DSA and OPTN region (which previously functioned as units for "local" and "regional" allocation, respectively).
2. The addition of "proximity points," which are calculated based on the distance between the donor hospital and the transplant hospital where the recipient candidate is listed, to the candidate's total allocation score profile.
3. Extra ranking of pediatric and prior living donor candidates for certain types of donor kidneys. To define allocation priority and classification, each waitlisted recipient candidate is given a kidney allocation score on a given kidney match run formulated with the attributes as summarized in Table 6.8.

   The allocation points for the level of HLA sensitization are on a sliding scale divided into many CPRA categories ranging from 0 to 202.10. No point is awarded until the recipient CPRA reaches 20% (0.08 points). A full point is awarded only when the recipient CPRA reaches 70% (1.09 points). At 95%, 99%, and 100% CPRA, the recipient will be awarded 10.82, 50.09, and 202.10 points, correspondingly [13–15].

Virtual crossmatch and SAB evaluation has thus improved the deceased donor organ sharing with better outcomes and reduction in cold ischemia time (CIT).

Puttarajappa et al. reported [15] that the use of VXM increased 2.4% per year since implementation of the Kidney Allocation System in 2014. This approval is likely to grow at a faster pace under KAS 250 as centers look to include the use of VXM to minimize the trend of longer cold ischemic time (CIT) and organ discard.

U.S. registry analysis has shown that kidney recipients transplanted using VXM had a mean reduction of 1.2 and 2.5 h in CIT for local versus import kidneys respectively, as compared to recipients transplanted mainly by using PXM.

Table 6.8 On March 15, 2021, the deceased donor kidney allocation system (KAS) of the Organ Procurement and Transplantation Network (OPTN) had undergone geographical boundaries in ensuring an equitable allocation of deceased donor organs

Criteria	Allocation points
Waiting time (from dialysis start time or glomerular filtration rate of ≤20 mL/min	1/365 points for each day (equivalent to 1 point/year)
0-ABDR HLA mismatch (age 0–10)	4 points
0-ABDR HLA mismatch (age 11–17)	3 points
Age 0–10 (for donor with KDPI of <35%)	1 points
A prior living donor	4 points
One HLA-DR mismatch	1 points
0 HLA-DR mismatch	1 points
HLA sensitization by CPRA	0–202.10 points
Proximity	0–4 points

Scope of VXM

The single-antigen bead (SAB) test defines the level of HLA antibodies which are expressed semi-quantitatively as mean fluorescence intensity (MFI). MFI numbers for some samples can differ between laboratories due to differences in test procedures, testing kits, and protocols.

The introduction of the single-antigen bead-based test has significantly improved the test accuracy of virtual crossmatching. In this test, purified recombinant HLA antigens are conjugated to an array of polystyrene beads. Each bead carries only one HLA antigen, which makes it easy to interpret test results.

How the antigens are loaded on the beads in different lots is controlled by the manufacturer, and the amount of antibody binding to the beads is detected by fluorescence-labelled secondary anti-human immunoglobulin G (IgG) antibody. A limitation of this test could be due to the prozone effect.

The interfering factors—denatured antigens and complement—are some of the major interfering factors of the solid-phase single-antigen bead assay, resulting in falsely low MFI levels. Studies have shown that serum treatment with dithiothreitol (DTT) or ethylenediaminetetraacetic acid (EDTA) improves the creation of the C1 complex and removes the interference from the complement [8].

Denatured antigens on the single antigen beads expose cryptic epitopes, which can result in "false positives" (Table 6.3).

Other limitations of accurate virtual crossmatching include the presence of multiple low-level donor-specific antibodies and the varying HLA expression level of the donor. Highly sensitized patients, in particular, can have multiple low-level donor-specific antibodies against a given donor, making it difficult to predict compatibility, especially if the donor is homozygous for those HLA antigens.

HLA antibody assignment in these patients can be difficult, and many factors must be taken into consideration, such as sensitization route, history of sensitization, epitope analysis, and serological serotypes when assigning unacceptable antigens.

For the most part, VXM assessments can be safely used in unsensitized patients and patients who are moderately sensitized. For the very highly sensitized patients, VXM may need more in-depth evaluation. Complement-dependent cytotoxicity crossmatch and the flow crossmatch were implemented to evaluate HLA antibodies. In cell-based, physical XM, the number of antibodies bound on the cell surface is defined by two factors:

1. First is the level of antibodies in sera, and
2. Second, the antigen expression on the cell surface

It is difficult to identify the antibody specificity using these two crossmatches as is possible in SAB in terms of MFI.

Cell-based crossmatches are at times not very useful for highly sensitized patients who display multiple antibodies (Tables 6.4 and 6.5).

It has been shown in many studies that the presence of donor-specific antibodies to HLA is associated with lower graft survival and increased antibody-mediated rejection in solid organ transplantation.

The crossmatch result accuracy for HLA antibodies can predict the risk of antibody-mediated rejection post-transplant.

Higher MFI thresholds for reporting antibodies may miss important antibody patterns and should be avoided. There may be epitope reactivity with very low MFI that can be explained by a sensitizing event, and these antibody patterns should be considered, as they may be relevant in the setting of a memory response.

The presence of eplet/epitope analysis in vendor software programs facilitates this analysis. The HLA Eplet Registry is also useful [16]. The Eplet Registry is always evolving, and there may be some cross-reactive epitope groups (CREGs) that are not yet captured in some registries. Attention to CREGs will also assist in accurate antibody interpretation, as shown in Tables 6.2 and 6.9.

Flow cytometry crossmatch methods can be correlated to solid-phase methods.

Flow crossmatch can be used to investigate antibodies detected in the antibody assays, but physical crossmatch may be of limited use in the setting of very high cPRA sera since finding the right donor may be very challenging, as shown by our case 2.

On the other hand, low MFI reactivity may not be detected by the cell-based methods.

Table 6.9 Immunologic methods to determine structural human leukocyte antigen (HLA) compatibility, the number of mismatched donor epitopes or eplets by computer algorithm

Algorithm	Principles (computer algorithm)
HLA matchmaker	Donor-recipient HLA compatibility can be determined by examining and determining the number of mismatched donor epitopes or eplets
Physicochemical concept	Amino acid polymorphisms (amino acid mismatch score. Electrostatic mismatch score [EMS]) for each mismatched donor HLA
Antibody-verified epitopes	Physiological Terasaki epitopes (TerEps) experimentally defined using monoclonal antibodies (mAb) or alloantibodies that were absorbed and eluted from recombinant HLA cell lines and analyzed with single-antigen bead (SAB) assays
PIRCHE-II based on T-cell epitopes	HLA epitopes (PIRCHE) computational algorithm is designed to predict indirectly recognizable HLA-derived donor peptides that are likely to induce the production of donor-specific anti-HLA IgG antibodies

HLA Typing

Fifty years ago, microlymphocytotoxicity methods were used for HLA antigen identification.

HLA typing has undergone a change from serological HLA typing using reagents and methods that were not standardized to rapid sequencing of HLA genes with high accuracy.

Today, nucleotide sequences of HLA genes can be used to determine amino acid sequences of HLA proteins, and HLA mismatches can be described at the level of amino acid differences or epitopes [17].

Tools developed for molecular biology made it possible to take advantage of hybridization and selective gene amplification, as in SSP and SSO methods, to develop typing systems that have been used for decades.

During the last 10 years, NGS methods have become the gold standard, but this is likely to be short-lived as third generation sequencing continues to improve (Table 6.6).

Accurate and complete definition of the HLA type of the donor is very important for evaluating the virtual crossmatch.

Take Home

HLA Antibody Analysis Concerns

- Consider past sensitizing events.
- Perform epitope analysis.
- Use multiple antibody detection methods.
- Perform or predict high-resolution typing.
- Consider CREG reactivity.
- Look for antibodies to the alpha chain also.
- Set MFI threshold with your lab environment.

Role of HLA Epitope Matching in Kidney Transplantation

- HLA epitope matching is an algorithm, a modern technology of virtual crossmatch for the prediction of alloreactivity between donor and recipient HLA alleles.
- Can predict the differences in amino acid sequences, structural and/or physicochemical properties.
- Pre-transplant role in organ allocation.
- Avoid sensitization therapy.
- Prevent post-transplant de novo DSA development, acute rejection, and graft failure.

Basic Guidelines for Anti-HLA Antibody Detection Analysis

- Know about CREGs, DQBI, DPB1 regions.
- High negative/low positive control can shield antibody specificity.
- Be informed about technical limitations of the SAB assay.
- For result estimation, one would need recipient typing, donor typing, previous transplant, sensitization history (pregnancies, transfusions, medications, vaccines), antibody history, previous XM results, and results from other antibody assays.

Accurate in-depth characterization of HLA antibody in SAB testing (virtual crossmatch) and complete HLA typing of proposed donors is crucial for donor-recipient evaluation.

Our clinical cases show that such an approach resulted in finding an ABO-compatible, DSA-negative KPD donor for a sensitized recipient and an ABO-compatible KPD donor in families which had only an option of ABO-incompatible transplantation or a sensitized recipient.

References

1. Rocha Y, Jaramillo A, Neumann J, Hacke K, Palou E, Torres J. Crossmatch assays in transplantation: physical or virtual? A review. Medicine (Baltimore). 2023;102(50):e36527. https://doi.org/10.1097/MD.0000000000036527.
2. Patel R, Terasaki PI. Significance of the positive crossmatch test in kidney transplantation. N Engl J Med. 1969;280:735–9.
3. Sullivan HC, Gebel HM, Bray RA. Understanding solid-phase HLA antibody assays and the value of MFI. Hum Immunol. 2017;78(7–8):471–80.
4. Tait BD, Süsal C, Gebel HM, et al. Consensus guidelines on the testing and clinical management issues associated with HLA and non-HLA antibodies in transplantation. Transplantation. 2013;95(1):19–47.
5. Baxter-Lowe LA. The changing landscape of HLA typing: understanding how and when HLA typing data can be used with confidence from bench to bedside. Hum Immunol. 2021;82(7):466–77.
6. Coates PT, Wong G. Current controversies in nephrology-how to crossmatch for transplantation? Kidney Int. 2020;97:662–3.
7. Clifford Sullivan H, Krummey SM, Gebel HM, Bray RA, Jackson AM. Physical crossmatching vs virtual crossmatching: the end of an era? Or why give up a good thing? Hum Immunol. 2020;81:401–6.
8. Wehmeier C, Hönger G, Schaub S. Caveats of HLA antibody detection by solid-phase assays. Transpl Int. 2020;33:18–29.
9. Bhaskaran MC, Heidt S, Muthukumar T. Principles of virtual crossmatch testing for kidney transplantation. Kidney Int Rep. 2022;7:1179–88.
10. Visentin J, Guidicelli G, Bachelet T, et al. Denatured class I human leukocyte antigen antibodies in sensitized kidney recipients: prevalence, relevance, and impact on organ allocation. Transplantation. 2014;98:738–44.
11. Friedewald JJ, Turgeon N. Early experience with the new kidney allocation system. Clin J Am Soc Nephrol. 2017;12(12):2060–2. https://doi.org/10.2215/CJN.07520717.

12. Bray RA, Brannon P, Breitenbach C, Bryan C, Chen D-F, Lai J, McRacken T, Kirk A, Kaplan B, Pearson T, Gebel HM. The new OPTN kidney allocation policy: potential for inequitable access among highly sensitized patients. Am J Transplant. 2015;15:284–5.
13. Friedewald JJ, Samana CJ, Kasiske BL, et al. The kidney allocation system. Surg Clin North Am. 2013;93:1395–406.
14. Adler JT, Husain SA, King KL, Mohan S. Greater complexity and monitoring of the new kidney allocation system: implications and unintended consequences of concentric circle kidney allocation on network complexity. Am J Transplant. 2021;21(6):2007–13. https://doi.org/10.1111/ajt.16441.
15. Puttarajappa CM, Hariharan S, Zhang X, et al. Early effect of the circular model of kidney allocation in the United States. J Am Soc Nephrol. 2023;34:26–39.
16. Tafulo S, Malheiro J, Santos S, Dias L, Almeida M, Martins LS, Pedroso S, Mendes C, Lobato L, Castro-Henriques A. HLA class II eplet mismatch load improves prediction of dn DSA development after living donor kidney transplantation. Int J Immunogenet. 2021;48:1–7. https://doi.org/10.1111/iji.12519.
17. Tafulo S, Malheiro J, Santos S, Dias L, Almeida M, Martins LS, Pedroso S, Mendes C, Lobato L, Castro-Henriques A. Degree of HLA class II eplet mismatch load improves prediction of antibody-mediated rejection in living donor kidney transplantation. Hum Immunol. 2019;80:966–75. https://doi.org/10.1016/j.humimm.2019.09.010.

Molecular Biomarkers to Assess Kidney Allograft Status

7

Dinesh Kannabhiran, Gayatri Veeramani Jayaraman, and Thangamani Muthukumar

Introduction

Individuals with end-stage kidney disease live longer when they undergo kidney transplantation than when they remain on dialysis [1]. The success of human kidney transplantation, which began in Boston on December 23, 1954, has continued to improve. In the United States, the 5-year graft survival rate of kidney transplant recipients increased by 12% points from 66% in 1996–99 to 78% in 2012–2015 for deceased donor kidneys and by 8% points from 80% to 88% for living donor kidneys [2]. Interventions at multiple levels, including those in potential donors and transplant candidates, socioeconomic interventions, pre-transplant immunological interventions, and post-transplant interventions, have likely contributed to this improvement [2]. Nevertheless, alloimmunity catches up, and many recipients eventually experience graft failure. Patients who return to dialysis after a failed transplant represent a unique subpopulation of dialysis patients with myriad management challenges [3]. Thus, assessing allograft status to determine prognosis and kidney failure trajectory is a key goal of the transplant community, and biomarkers are indispensable for achieving this goal [4].

D. Kannabhiran
Division of Kidney Disease and Hypertension, Donald and Barbara Zucker School of Medicine at Hofstra/Northwell, Manhasset, NY, USA
e-mail: dkannabhiran@northwell.edu

G. V. Jayaraman
Internal Medicine/Nephrology, Mass General Brigham Hospitals, Boston, MA, USA
e-mail: drgayatri@fourrts.com, gayatri.jayaraman@mgh.harvard.edu

T. Muthukumar (✉)
Nephrology, Hypertension, and Transplantation Medicine, Weill Cornell Medicine, Cornell University, New York, NY, USA
e-mail: mut9002@med.cornell.edu

G. Abraham et al. (eds.), *Immunological Challenges in Organ Transplantation*,
https://doi.org/10.1007/978-981-95-1533-2_7

What Is a Biomarker?

The United States Food and Drug Administration defines a biomarker as a characteristic that is measured as an indicator of normal biological processes, pathogenic processes, or responses to an exposure or intervention, including therapeutic interventions. Thus, the serum creatinine level, estimated glomerular filtration rate, and albuminuria or proteinuria may be considered biomarkers. Nonetheless, these are neither specific nor sensitive enough to assess kidney allograft status [5]. While allograft histology is the current reference standard and the Banff classification of transplant pathology has helped standardize its reporting, invasive biopsies are fraught with complications, logistical challenges, and variability in interpretation [6].

Why Are Molecular Biomarkers Required?

The current clinical practice typically involves assessing kidney allograft function using serum creatinine or proteinuria, and a biopsy is done when such laboratory tests are abnormal. This traditional, decade-old approach is more reactive in terms of diagnosis and treatment. Instead, personalized transplantation medicine integrates molecular and clinical data to achieve the best outcomes for individual transplant recipients and aims to be proactive—to identify and prevent [5]. Molecular biomarkers are the cornerstones of personalized or precision transplantation medicine. Allograft rejection or dysfunction can be visualized using a timeline model [7]. In such a formulation, molecular perturbations in the allograft precede histological changes, which in turn precede biochemical derangement. The hypothesis that early intervention is highly efficacious is important for developing molecular surveillance strategies [7]. Molecular biomarkers that are noninvasive have additional advantages, including the avoidance of invasive biopsies and the convenience of serial measurements to provide a sequential assessment of the allograft and recipient's immune status.

This reasoning has resulted in the development and clinical application of several molecular biomarkers over the last decade. In this chapter, we focus on the underlying principles and clinical utility of the common molecular biomarkers in kidney transplantation that are currently available (Table 7.1).

Case Vignette

A 45-year-old woman with chronic kidney disease due to immunoglobulin A nephropathy was being evaluated for kidney transplantation. The patient's husband was a potential donor for the transplantation. No other potential living donors were available and the patient did not wish to undergo dialysis. She had a human leukocyte antigen (HLA) mismatch at the A and B loci and two mismatches at the DR locus with her husband. The flow cytometry crossmatch (FCXM) test was negative, with a median channel shift (MCS) of six for T cells and eight for B cells. The Luminex single antigen bead (SAB) assay showed the presence of donor-specific

Table 7.1 Molecular tests for assessing kidney allografts

Specimen	Test	Output	Product name (company)	Methodology
Serum	Anti-HLA IgG antibodies	Mean fluorescence intensity	Luminex platform	
			(a) Luminex xMAP instruments (Luminex Corporation/Thermo Fisher Scientific)	Quantitative fluorescent flow cytometry
			(b) Multiplex solid phase assays	Color-coded microspheres
			– LABScreen Single Antigen beads (One Lambda/Thermo Fisher Scientific)	
			– LIFECODES Single Antigen assays (Immucor/Werfen)	
Plasma	dd-cfDNA	Fraction (%)	AlloSure® Kidney and AlloSeq cfDNA (CareDx)	NGS (266 SNPs)
		Fraction (%) and/or absolute (cp/mL)	Prospera™ (Natera)	mmPCR-NGS (13,926 SNPs)
		Fraction (%)	TRAC™ kidney (Eurofins/ transplant genomics)	WGS (>100,000 SNPs)
		Fraction (%) and/or absolute (cp/mL)	VitaGraft™ and GraftAssure™	ddPCR (48 SNPs)
Whole blood	Gene expression profile	TX or not-TX[a]	TruGraf® kidney (Eurofins/ transplant genomics)	Microfluidic PCR
Whole blood	ATP in CD4+ T cells	ng/mL	ImmuKnow (Eurofins-Viracor)	Bioluminescence assay

From CMS.gov, MolDX: Molecular Testing for Solid Organ Allograft Rejection, revised version April 1, 2025 (https://www.cms.gov/medicare-coverage-database/view/article.aspx?articleid=58019)

AlloSeq cfDNA is a decentralized NGS-based kit for analyzing dd-cfDNA in individual laboratories. It is available only outside the USA

GraftAssure is a decentralized ddPCR-based kit for analyzing dd-cfDNA in individual laboratories

NGS next generation sequencing, *PCR* polymerase chain reaction, *mmPCR* multiplex microsphere PCR, *WGS* whole genome sequencing, *ddPCR* droplet digital PCR, *RT-PCR* reverse transcription PCR

[a] The test output is binary: Transplant eXcellence (TX), negative for subclinical rejection, or Not-TX, positive for subclinical rejection

antibodies (DSA) against HLA-A2, with a mean fluorescence intensity (MFI) of 4400. Should we proceed with transplantation?

SAB assays detect the presence of antibodies but do not provide any information on the ability of these antibodies to bind to antigens or injure target cells. The FCXM test, on the other hand, identifies antibodies that bind to cell surface antigens. In this test, the recipient serum is mixed with donor lymphocytes. T and B cells are distinguished by the addition of cell lineage-specific antibodies. As implemented in most laboratories, this test detects only immunoglobulin G (IgG) alloantibodies. IgG alloantibodies, if present in the serum, bind to donor B and T cells. Fluorescein-labeled anti-human IgG antibody is then added to detect the bound alloantibody. FCXM result is reported as the MCS of the fluorescence intensity. A predetermined cutoff value for the MCS is considered a positive crossmatch.

Luminex Single Antigen Bead Testing for Detecting Anti-HLA Antibodies

In addition to blood group antibodies, antibodies directed against donor HLA are important for kidney transplantation. Humans do not have natural antibodies against HLA but may develop it after sensitizing events, such as pregnancy, blood transfusion, or organ transplantation. An assay in which at least one molecule under analysis (an antigen or an antibody) is bound to a solid surface while the other reactants are in solution is called a solid-phase assay. The Luminex platform combines the Luminex xMAP instruments and multiplex solid-phase assays designed for the detection of proteins.

Single antigen bead (SAB) assays tested on a Luminex platform are the current standard for detecting circulating anti-HLA antibodies. In this assay, a bead set is used for each serum sample. Each bead set contains multiple polystyrene beads, each coated with a single purified HLA protein. Bead sets cover the most prevalent HLA phenotype in the general population. For example, the LABScreen Single Antigen assay covers 97 HLA Class I and 95 HLA Class II antigens. The SAB output is reported as an MFI value and detects the presence of antibodies directed against one or more HLA. Current assays are designed to detect IgG antibodies. If an antibody is directed against a donor, it is called a donor-specific antibody (DSA) [8].

When interpreting the results of SAB assays, several caveats should be considered. The MFI value is a semiquantitative assessment that does not accurately reflect the concentration or titer of the antibody but gives the bead's relative fluorescence without reference to a standard [9]. Cutoff values have not been standardized across laboratories. When to call an antibody present/positive is based on a statistical approach in most laboratories—fluorescence level above a threshold, rather than practical (associated with physical crossmatch result) or clinical (associated with rejection or graft failure) [10]. A Clinical Trials in Organ Transplantation (CTOT) study concluded that an MFI value of 1000 to 1500 yielded a high level of agreement among HLA laboratories to determine the presence or absence of an antibody

[11]. When DSA is reported to be negative, it could be a false negative result. Some of the causes of false negative results include (i) the bead set may not contain beads for a particular HLA, (ii) the SAB manufacturing process could result in lot-to-lot variability [12]; (iii) HLA structure present in the SAB assays is different between the two currently available vendors (One Lambda/Thermo-Fisher and Immucor/Lifecodes); (iv) shared epitopes could result in dilution of the antibodies that bind to each antigen [10]; and (v) inhibitory factors in the serum could interfere with antibody detection (prozone effect) and require dilution or pretreatment of the serum with heat, dithiothreitol, or ethylenediamine tetra-acetic acid. Thus, failure to detect an antibody in the SAB assay does not imply the absence of the antibody in circulation [8]. In addition, the density of antigens on the beads, the fluorochrome detection antibody used, and the setup of the instrument may affect the MFI value [8].

Similarly, when DSA is reported to be positive, it could be a false positive result. Accepting any antibody reactivity in the SAB assay as positive and clinically relevant carries the risk of false assignment of unacceptable antigens and diminishes a patient's chance of receiving an organ. Another cause of false-positive results is the presence of denatured antigens. Initially reported in sera from healthy unsensitized males, these anti-HLA antibodies targeted cryptic epitopes on HLA molecules typically exposed to denatured HLA on the beads, but not in the native configuration on cell surfaces. Such natural antibodies, likely produced due to cross-reactive epitopes found in microorganisms and not directed against an intact HLA, do not have any clinical impact on the allograft.

Another caveat when interpreting SAB assay results is donor HLA typing. HLAs are cell surface glycoproteins and are the primary determinants of alloimmunity. HLA class I is expressed on all nucleated cells and HLA class II is expressed on antigen-presenting cells (B cells, dendritic cells, and macrophages) and endothelial cells. There are three HLA class I genes, HLA-A, HLA-B, and HLA-C. There are three HLA class II genes: HLA-DP (α-chain encoded by HLA-DPA1 locus and β-chain encoded by HLA-DPB1), HLA-DQ (α-chain encoded by HLA-DQA1 and β-chain encoded by HLA-DQB1), and HLA-DR (α-chain encoded by HLA-DRA and 4 β-chains encoded by HLA-DRB1, DRB3, DRB4, and DRB5 loci). HLA are highly polymorphic antigens; >28,000 class I alleles and >12,000 class II alleles have been identified so far [13]. An allele is a variant of a gene that represents different versions of the same genetic sequence at a specific location (locus) on a chromosome. Despite the large number of alleles, only one-third have been commonly reported in the general population [14].

HLA nomenclature uses a specific format to identify HLA alleles: there are four sets of digits (called fields) separated by colons. For example, when an HLA is typed as HLA-A*02:101:01:02 N, the first number (02) is field 1, and the second number (101) is field 2. Field 1 corresponds to the serologically defined HLA protein; HLA-A2 by serology corresponds to HLA-A*02 by DNA typing. Field 2 is the HLA allele that encodes the same protein sequence within the antigen-binding site; HLA-A*02:101 and HLA-A*02:102 are different proteins but cannot be differentiated by serologic typing) [15]. HLA typing is currently done by DNA-based

methods. Low-resolution typing is at the level of field 1 resolution. High-resolution typing is at the level of field 2 resolution. Allelic resolution is up to the level of field 4 resolution [16].

While the HLA type of an individual is mostly reported at the field 1 level (e.g., HLA-A*2), the beads in the SAB assay have HLA at the field 2 level (e.g., HLA-A*02:05, -A*02:10, etc.). If a donor type is HLA-A*2 and the recipient has antibody against A*02:05 but not against A*02:10 by SAB assay, then without knowing the field 2 type of donor HLA (is the donor A*02:05 or A*02:10?), it is not possible to determine whether the antibody against A*02:05 is a DSA or not.

Our approach in this patient was to proceed with transplantation. The MCS in the FCXM test was very low (our cutoff value to call an FCXM test positive is ≥40 MCS for T cells and ≥40 MCS for B cells), which provided reassurance regarding the negative FCXM. Our standard protocol for induction immunosuppression is three doses (each dose is 1.5 mg/kg body weight) of Thymoglobulin® with steroid-free maintenance immunosuppression with tacrolimus and mycophenolate. Because of the negative FCXM result, pre-transplant desensitization therapy was not provided for this patient. However, because there is a possibility of increased memory B cell response and antibody production after transplantation, we gave four doses of Thymoglobulin® with steroid-based maintenance immunosuppression.

Case Vignette Continued…

The transplant was uneventful and she achieved excellent kidney function. Her creatinine decreased to 0.8 mg/dl at discharge. At 3 months post-transplantation, her creatinine was 0.95 mg/dl, no proteinuria, and HLA-A2 DSA was 2600 MFI. Owing to the high immunologic risk, we monitored her plasma donor-derived cell-free DNA (dd-cfDNA); her dd-cfDNA at 3 months was 0.21%. At 6 months post-transplantation, her creatinine was 1.03 mg/dl; HLA-A2 DSA was 3500 MFI, and dd-cfDNA was 0.24%. She did not develop de novo *DSA. How should we interpret the dd-cfDNA results?*

Donor-Derived Cell-Free DNA Testing

Cell-free DNA (cfDNA) is a fraction of the total circulating nucleic acids and was first described in 1948. Most cfDNAs are double-stranded and contain approximately 100–200 base pairs of non-random fragments. In humans, approximately 50 billion cells undergo apoptosis every day, which remains the primary source of circulating cfDNA. Other mechanisms of cfDNA release include, but are not limited to, necrosis, pyroptosis, autophagy, phagocytosis, mitotic catastrophe, NETosis, and extracellular vesicles [17]. Such non-apoptotic sources usually contain fragments of approximately 10,000 bp. In healthy individuals, the normal concentration of cfDNA in plasma is 0–100 ng/mL, with a half-life of approximately 30–120 min [18]. Leukocytes are a major source of cfDNA in humans. Elevated levels of cfDNA can be seen in physiological states, such as exercise, or pathological states such as sepsis [19].

Polymorphisms are variations in DNA sequences (e.g., adenine nucleotide is replaced with a cytosine nucleotide at a specific location in the DNA sequence). For a variation to be called a polymorphism (versus a mutation), that variant should be present in ≥1% of the population. When polymorphisms occur in a single nucleotide, they are called single nucleotide polymorphisms (SNPs). SNPs occur in approximately 1 in 1000 nucleotides of the human genome. There are approximately four to five million SNPs in an individual. Most SNP's have no functional consequences. As recipients of organ transplants are also genome transplants, it is possible to monitor allograft injury by measuring dd-cfDNA in the plasma. Differentiation of recipient and donor cfDNA has evolved over the years. Initially, in kidney and liver transplant recipients, Y-chromosome-specific PCR was used to measure dd-cfDNA in female recipients of grafts from male donors [20]. Current commercial assays use droplet digital polymerase chain reaction or next-generation sequencing technology to measure cell-free DNA in plasma, and computational algorithms to determine the donor proportion of cfDNA based on the differences in SNPs. Importantly, donor proportions can be identified without the need for donor genotype information [21, 22].

dd-cfDNA is increasingly used to monitor kidney allograft recipients. After kidney transplantation, plasma dd-cfDNA levels are detectable on day 1 (~10% of total cfDNA) and exponentially decrease to a steady-state fraction of <0.5% by day 10 [23]. In stable transplant recipients, dd-cfDNA accounts for a small fraction (< 0.2%) of total circulating cfDNA. In the event of allograft injury, the dd-cfDNA fraction increase.

A recent study of about 2800 kidney transplant recipients from multiple centers in the USA and Europe found that dd-cfDNA (AlloSure or AlloSeq cfDNA, CareDx) had an odds ratio of 2.3 (95% confidence interval 1.9 to 2.7) for the diagnosis of rejection in the derivation cohort of 226 rejection biopsies among 1395 total biopsies, independent of estimated glomerular filtration rate, proteinuria, kidney graft instability, prior rejection, and DSA [24].

Despite being touted as a test for all types of rejection, dd-cfDNA may perform better in antibody-mediated (ABMR) than in T-cell-mediated rejection (TCMR) [25]. On a mechanistic basis, this is understandable given that endothelial cells are the targets of ABMR, and a blood test may detect endothelial injury more than an epithelial (tubulitis) injury in TCMR. As the donor-derived fraction increases due to any cause of allograft injury, dd-cfDNA is a marker of allograft injury rather than a specific type of rejection.

Another utility of dd-cfDNA is the longitudinal surveillance of allografts. In a registry study of 424 biopsies—26 ABMR, 62 TCMR, and 336 no rejection, dd-cfDNA fractions, but not serum creatinine, were significantly elevated before biopsy-confirmed rejection compared with no rejection biopsies. The median dd-cfDNA was >1% across months 1–6 before ABMR biopsies and 1 month before TCMR biopsies [26]. Although this registry study used a fixed 1% cutoff value, given that the dd-cfDNA fraction represents the total burden of allograft injury, the relative change value (RCV, sequential AlloSure results when the donor-derived fraction is above 0.20%) could be useful for clinical decision-making. According to

the manufacturer's recommendations for AlloSure, an RCV of ≤60% is due to biological variation, and a value of ≥149% represents a high likelihood of injury.

The three commercially available dd-cfDNA assays (AlloSure, Prospera, and TRAC) are all based on next-generation sequencing methodology and use different cutoff values to consider the test as positive: 1% for AlloSure and Prospera, and 0.7% for TRAC. No head-to-head comparisons of these three assays are available. However, a study that compared the paired sampling of AlloSure and Prospera found no statistically significant differences in their diagnostic characteristics [27]. The fourth commercial assay, VitaGraft, measures the absolute concentration of dd-cfDNA in addition to its fraction. Some studies have suggested that the absolute value of dd-cfDNA may be a better discriminant than the fraction for diagnosing acute rejection. The absolute value may be less influenced by clinical conditions characterized by minor fluctuations in leukocyte-derived cfDNA, such as sepsis, urinary tract infection, immunosuppressive therapy, etc. [28]. In addition, as the leukocyte counts decrease, the total cfDNA will decrease, and the dd-cfDNA fraction may increase. A combination of the dd-cfDNA fraction and absolute quantification may improve the diagnostic accuracy for kidney transplant rejection [29]. The dd-cfDNA assay has not yet been validated in patients who received multiple transplants from different donors.

An important consideration in interpreting dd-cfDNA results is the impact of the allograft biopsy procedure; the dd-cfDNA percentage increases following the biopsy procedure [30, 31]. Hence, the results should be interpreted with caution if obtained less than 24–48 h after a biopsy. A major advancement in the transport and processing of specimens for dd-cfDNA assay was the use of Cell-Free DNA BCT® tubes, also called Streck tubes (Streck, Inc.). A Streck tube is a direct-draw venous whole-blood collection device designed to stabilize nucleated blood cells and prevent the release of genomic DNA. These glass tubes maintain DNA stability for up to 14 days at room temperature [32].

In our patient, the dd-cfDNA results at 3 and 6 months were lower than the manufacturer-recommended cutoff value of 1%; thus, they were negative for tissue injury. Importantly, the value was stable between 3 and 6 months after transplantation. As the test had an NPV > 90%, we were reassured that there was no clinically significant allograft tissue injury. Nonetheless, we had similar patients with low dd-cfDNA levels, but whose biopsy showed C4d + ABMR. It is possible that low-level microvascular inflammation and the resulting allograft injury were not sufficient to increase circulating dd-cfDNA. We were cognizant of this possibility and continued to monitor our patient closely.

Case Vignette Continued…

Over the next six months, she developed multiple infectious episodes, including an episode of bacterial pneumonia that did not require hospitalization, two episodes of urinary tract infections, and low-level CMV viremia that was successfully treated. At 9 months after transplantation, BK virus was not detected in the urine, and at 12 months, it was 2.30E+07 IU/ml. Her serum BK level was < 200 IU/ml (detected, but below the lower limit of detection). She received tacrolimus (trough 7 ng/ml),

mycophenolate 1000 mg BID, and prednisone 5 mg daily. Her ImmuKnow ATP level was 55 ng/ml. How do we interpret this result?

Immune Cell Function Testing

In clinical practice, immunosuppression is typically assessed using therapeutic drug monitoring of calcineurin inhibitors. Importantly, drug dosing is guided by the trough levels. Initial studies evaluating the relationship between tacrolimus trough levels and rejection have shown contradictory results. However, more recent studies with lower target ranges and longer follow-up periods have shown a more consistent association between tacrolimus trough levels and graft outcomes and adverse effects [33]. Mycophenolate levels are not routinely measured due to significant between and within-subject variability and a second peak due to enterohepatic circulation, resulting in a poor correlation between trough levels and area under curve measurements [34].

ImmuKnow test is an immune cell function assay that detects cell-mediated immunity in immunosuppressed populations. Originally developed by Cylex Inc., ImmuKnow technology combines cell stimulation, cell selection, and quantification of metabolic markers (e.g., ATP) to measure the global cell-mediated immune response of CD4+ T lymphocytes.

This assay measures the ability of CD4+ T-helper lymphocytes to respond in vitro to mitogenic stimulation by phytohemagglutinin (PHA), a lectin isolated from red kidney beans, by quantifying intracellular ATP levels in CD4+ T cells. ATP is the energy source for all cells and is required for the effector functions of immune cells. In this assay, anticoagulated whole blood collected in sodium heparin tubes is incubated with PHA, and the CD4+ T cells are positively selected using magnetic particles coated with anti-human CD4 monoclonal antibodies. The cells are lysed to release intracellular ATP. The released ATP is detected using a luciferin-luciferase system and measured in a luminometer. ATP concentration (ng/ml) is calculated by comparing it with a calibration curve.

The manufacturer-recommended cutoff values are <225 ng/mL for low immune response, 225–524 ng/mL for moderate immune response, and ≥525 ng/mL for strong immune response. A low immune response in a transplant recipient can be envisioned as a quiescent CD4+ T cell state—reduced risk for rejection but increased risk of infection, and a strong immune response as an active/charged CD4+ T cell state—increased risk for rejection but reduced risk of infection. These cutoff levels were derived in an initial study of 155 healthy adults and 127 transplant recipients. For transplant recipients on calcineurin inhibitors, cellular ATP is an appropriate target, as these drugs inhibit mitochondrial respiration, which is a major source of intracellular ATP. Thus, reduced ATP production directly inhibits the cascade of events required for lymphocyte proliferation. Interestingly, this assay measures the absolute amount of intracellular ATP, without considering the total number of CD4+ T cells. In the initial study that established the cut-off value, there was no correlation between CD4 cell number and the level of ATP stimulation. Importantly, in this

study, whole blood tacrolimus levels did not correlate with the degree of biological immunosuppression, as measured by PHA-stimulated ATP levels. In addition, most patients who were between 4 and 12 months after transplantation had a moderate immune response zone [35]. Subsequent studies utilizing this test found it to be more effective in identifying over-immunosuppressed patients with a high risk of adverse events than in identifying those with a high risk of rejection [36].

Our patient's ATP level of 55 ng/ml indicated an over-immunosuppressed state, which partially explains the increased susceptibility to infections. Immune cell function assay is an important tool in our armamentarium to balance anti-allograft and anti-infective immunity. We reduced her tacrolimus dose to maintain a level of 6s. We also reduced the mycophenolate dose from 2000 mg/day to 1750 mg/day. Over the next *several months, her ATP level increased and stabilized between 200 and 250 ng/ml. Between 2.5 and 3 years after transplantation, her tacrolimus level remained stable at around 5.5 ng/ml; she received 750 mg twice a day of mycophenolate and 5 mg daily of prednisone. She did not experience any further episodes of pneumonia or urinary tract infection. Her blood CMV and urine BK virus levels were undetectable. Her most recent creatinine level was 1.22 mg/dl, with no proteinuria, and the dd-cfDNA level was 0.24%. She did not develop* de novo *DSA, and her HLA-A2 DSA was 1800 MFI.*

This case vignette provides an overview of the application of various molecular markers for the management of transplant recipients. In India, the cost and availability are major limitations for the universal application of such molecular monitoring tools. Currently, the SAB test costs Rs. 25,000 and the dd-cfDNA test costs Rs. 27,000. More importantly, commercial bead sets may not represent the HLA phenotypes in the Indian population. In one study, >50% of the alleles were underrepresented in SAB assays [37]. Nevertheless, promising studies on the molecular biomarkers in kidney transplant recipients are emerging. Two studies, one from *Stanley Medical College and Hospital, Chennai, and another from Sir Ganga Ram Hospital, New Delhi, India,* assessed dd-cfDNA using SNPs specific to the Indian population. These studies have established the feasibility of using dd-cfDNA assays to monitor kidney transplant recipients in India [38, 39].

Conclusion

Standard statistical rules for diagnostic tests are also applicable to molecular biomarkers. Thus, sensitivity and specificity represent the performance characteristics of the assay. They are inversely related and do not depend on the prevalence of the condition. Conversely, positive and negative predictive values represent the likelihood of an individual having or not having a condition, respectively, and depend on the prevalence of the condition. When attempting to diagnose a condition with a low prevalence, a positive test result is more likely to be false, even if the test is highly sensitive and specific. No single molecular biomarker is perfect, but the rational use of a combination of tests would allow us to effectively practice individualized patient-centric transplantation medicine.

References

1. Chaudhry D, Chaudhry A, Peracha J, Sharif A. Survival for waitlisted kidney failure patients receiving transplantation versus remaining on waiting list: systematic review and meta-analysis. BMJ. 2022;376:e068769. https://doi.org/10.1136/bmj-2021-068769.
2. Hariharan S, Israni AK, Danovitch G. Long-term survival after kidney transplantation. N Engl J Med. 2021;385(8):729–43. https://doi.org/10.1056/NEJMra2014530.
3. Huml AM, Wakam GK, Schold JD. Does the transplant system fail patients following kidney allograft loss? Am J Kidney Dis. 2025;85(3):267–9. https://doi.org/10.1053/j.ajkd.2024.12.001.
4. Josephson MA, Becker Y, Budde K, Kasiske BL, Kiberd BA, Loupy A, et al. Challenges in the management of the kidney allograft: from decline to failure: conclusions from a kidney disease: improving global outcomes (KDIGO) controversies conference. Kidney Int. 2023;104(6):1076–91. https://doi.org/10.1016/j.kint.2023.05.010.
5. Naesens M, Anglicheau D. Precision transplant medicine: biomarkers to the rescue. J Am Soc Nephrol. 2018;29(1):24–34. https://doi.org/10.1681/ASN.2017010004.
6. Loupy A, Mengel M, Haas M. Thirty years of the international Banff classification for allograft pathology: the past, present, and future of kidney transplant diagnostics. Kidney Int. 2022;101(4):678–91. https://doi.org/10.1016/j.kint.2021.11.028.
7. Anglicheau D, Suthanthiran M. Noninvasive prediction of organ graft rejection and outcome using gene expression patterns. Transplantation. 2008;86(2):192–9. https://doi.org/10.1097/TP.0b013e31817eef7b.
8. Bhaskaran MC, Heidt S, Muthukumar T. Principles of virtual Crossmatch testing for kidney transplantation. Kidney Int Rep. 2022;7(6):1179–88. https://doi.org/10.1016/j.ekir.2022.03.006.
9. Senev A, Emonds MP, Van Sandt V, Lerut E, Coemans M, Sprangers B, et al. Clinical importance of extended second field high-resolution HLA genotyping for kidney transplantation. Am J Transplant. 2020;20(12):3367–78. https://doi.org/10.1111/ajt.15938.
10. Gebel HM, Bray RA. HLA antibody detection with solid phase assays: great expectations or expectations too great? Am J Transplant. 2014;14(9):1964–75. https://doi.org/10.1111/ajt.12807.
11. Reed EF, Rao P, Zhang Z, Gebel H, Bray RA, Guleria I, et al. Comprehensive assessment and standardization of solid phase multiplex-bead arrays for the detection of antibodies to HLA. Am J Transplant. 2013;13(7):1859–70. https://doi.org/10.1111/ajt.12287.
12. Friedlander R, Putheti P, Diaz E, Menon A, Ponce B, Muthukumar T, et al. On the detection of anti-HLA antibodies using single antigen bead Luminex assay: lot-to-lot variations in MFI. Transplantation. 2013;96(4):e24–6. https://doi.org/10.1097/TP.0b013e31829c2481.
13. Robinson J, Barker DJ, Marsh SGE. 25 years of the IPD-IMGT/HLA database. HLA. 2024;103(6):e15549. https://doi.org/10.1111/tan.15549.
14. Gonzalez-Galarza FF, Mack SJ, Hollenbach J, Fernandez-Vina M, Setterholm M, Kempenich J, et al. 16(th) IHIW: extending the number of resources and bioinformatics analysis for the investigation of HLA rare alleles. Int J Immunogenet. 2013;40(1):60–5. https://doi.org/10.1111/iji.12030.
15. Patri P, Muthukumar T. Transplant Immunology. In: Molmenti EP, editor. Molmenti's kidney and pancreas transplantation: operative techniques and medical management. 2nd ed. McGraw Hill; 2023. p. 170–82.
16. Nunes E, Heslop H, Fernandez-Vina M, Taves C, Wagenknecht DR, Eisenbrey AB, et al. Definitions of histocompatibility typing terms. Blood. 2011;118(23):e180–3. https://doi.org/10.1182/blood-2011-05-353490.
17. Grabuschnig S, Bronkhorst AJ, Holdenrieder S, Rosales Rodriguez I, Schliep KP, Schwendenwein D, et al. Putative origins of cell-free DNA in humans: a review of active and passive nucleic acid release mechanisms. Int J Mol Sci. 2020;21(21):8062. https://doi.org/10.3390/ijms21218062.

18. Heitzer E, Auinger L, Speicher MR. Cell-free DNA and apoptosis: how dead cells inform about the living. Trends Mol Med. 2020;26(5):519–28. https://doi.org/10.1016/j.molmed.2020.01.012.
19. Kustanovich A, Schwartz R, Peretz T, Grinshpun A. Life and death of circulating cell-free DNA. Cancer Biol Ther. 2019;20(8):1057–67. https://doi.org/10.1080/15384047.2019.1598759.
20. Lo YM, Tein MS, Pang CC, Yeung CK, Tong KL, Hjelm NM. Presence of donor-specific DNA in plasma of kidney and liver-transplant recipients. Lancet. 1998;351(9112):1329–30. https://doi.org/10.1016/s0140-6736(05)79055-3.
21. Sharon E, Shi H, Kharbanda S, Koh W, Martin LR, Khush KK, et al. Quantification of transplant-derived circulating cell-free DNA in absence of a donor genotype. PLoS Comput Biol. 2017;13(8):e1005629. https://doi.org/10.1371/journal.pcbi.1005629.
22. Oellerich M, Christenson RH, Beck J, Schutz E, Sherwood K, Price CP, et al. Donor-derived cell-free DNA testing in solid organ transplantation: a value proposition. J Appl Lab Med. 2020;5(5):993–1004. https://doi.org/10.1093/jalm/jfaa062.
23. Gielis EM, Beirnaert C, Dendooven A, Meysman P, Laukens K, De Schrijver J, et al. Plasma donor-derived cell-free DNA kinetics after kidney transplantation using a single tube multiplex PCR assay. PLoS One. 2018;13(12):e0208207. https://doi.org/10.1371/journal.pone.0208207.
24. Aubert O, Ursule-Dufait C, Brousse R, Gueguen J, Racape M, Raynaud M, et al. Cell free DNA for the detection of kidney allograft rejection. Nat Med. 2024;30(8):2320–7. https://doi.org/10.1038/s41591-024-03087-3.
25. Wijtvliet VPWM, Plaeke P, Abrams S, Hens N, Gielis EM, Hellemans R, et al. Donor-derived cell-free DNA as a biomarker for rejection after kidney transplantation: a systematic review and meta-analysis. Transpl Int. 2020;33(12):1626–42. https://doi.org/10.1111/tri.13753.
26. Bromberg JS, Bunnapradist S, Samaniego-Picota M, Anand S, Stites E, Gauthier P, et al. Elevation of donor-derived cell-free DNA before biopsy-proven rejection in kidney transplant. Transplantation. 2024;108(9):1994–2004. https://doi.org/10.1097/TP.0000000000005007.
27. Melancon JK, Khalil A, Lerman MJ. Donor-derived cell free DNA: is it all the same? Kidney360. 2020;1(10):1118–23. https://doi.org/10.34067/KID.0003512020.
28. Cao C, Yuan L, Wang Y, Liu H, Cuello Garcia H, Huang H, et al. Analysis of the primary factors influencing donor derived cell-free DNA testing in kidney transplantation. Front Immunol. 2024;15:1435578. https://doi.org/10.3389/fimmu.2024.1435578.
29. Nie W, Wang Y, Fu Q, Wu C, Deng R, Yu X, et al. Integrating donor derived cell-free DNA fraction and absolute quantification for enhanced rejection diagnosis in kidney transplant recipients. Diagnostics (Basel). 2025;15(3):237. https://doi.org/10.3390/diagnostics15030237.
30. Kyeso Y, Bhalla A, Smith AP, Jia Y, Alakhdhair S, Ogir SC, et al. Donor-derived cell-free DNA kinetics Post-kidney transplant biopsy. Transplant Direct. 2021;7(6):e703. https://doi.org/10.1097/TXD.0000000000001149.
31. Li C, Wright S, Lamba P, Zhou M, Xu H, Sharma V, et al. Impact of ultrasound-guided percutaneous Core needle biopsy on biomarkers of human kidney allograft status. Transplantation. 2025;109(7):1209–21. https://doi.org/10.1097/TP.0000000000005355.
32. Medina Diaz I, Nocon A, Mehnert DH, Fredebohm J, Diehl F, Holtrup F. Performance of Streck cfDNA blood collection tubes for liquid biopsy testing. PLoS One. 2016;11(11):e0166354. https://doi.org/10.1371/journal.pone.0166354.
33. Marquet P. Getting tacrolimus dosing right. Ther Drug Monit. 2025;47(1):41–8. https://doi.org/10.1097/FTD.0000000000001266.
34. Staatz CE, Tett SE. Clinical pharmacokinetics and pharmacodynamics of mycophenolate in solid organ transplant recipients. Clin Pharmacokinet. 2007;46(1):13–58. https://doi.org/10.2165/00003088-200746010-00002.
35. Kowalski R, Post D, Schneider MC, Britz J, Thomas J, Deierhoi M, et al. Immune cell function testing: an adjunct to therapeutic drug monitoring in transplant patient management. Clin Transpl. 2003;17(2):77–88. https://doi.org/10.1034/j.1399-0012.2003.00013.x.
36. Millan O, Sanchez-Fueyo A, Rimola A, Guillen D, Hidalgo S, Benitez C, et al. Is the intracellular ATP concentration of CD4+ T-cells a predictive biomarker of immune status in stable

transplant recipients? Transplantation. 2009;88(3 Suppl):S78–84. https://doi.org/10.1097/TP.0b013e3181afeba6.

37. Kumar S, Doss SA, Stephen S, Pratheeba M, Jeyaseelan L, Daniel D. The challenge of using the virtual crossmatch as a singular tool for the detection of anti-HLA antibodies: a study from a tertiary care institute from South India. Transpl Immunol. 2021;65:101349. https://doi.org/10.1016/j.trim.2020.101349.
38. Fernando E, Sanathkumar HT, Ramanathan A, Srinivasaprasad ND, Thirumalvalavan K, Annadurai P, et al. Donor-derived cell-free DNA and active rejection in renal allografts. Indian J Nephrol. 2023;33(4):247–53. https://doi.org/10.4103/ijn.ijn_152_22.
39. Kumar N, Tandon A, Rana R, Rana DS, Bhalla AK, Gupta A, et al. Donor-derived cell-free DNA as a non-invasive biomarker for graft rejection in kidney transplant recipients: a prospective study among the Indian population. Diagnostics (Basel). 2023;13(23):3540. https://doi.org/10.3390/diagnostics13233540.

Approach to an Unsensitized Patient Awaiting First Renal Transplant: Algorithms of Choice

8

Vivek Kute, Manish R. Balwani, and Arpita Ray Chaudhury

Case Scenario

A 40-year-old male- Mr. A, diagnosed with end stage renal disease (ESRD) secondary to chronic glomerulonephritis due to IgA nephropathy, currently on maintenance hemodialysis via right radio cephalic fistula, presents to nephrology OPD for renal transplant evaluation—with spouse as a potential donor. He has not received any blood transfusion previously, nor does he have any major hospital admission history for infection. What should be the approach to immunologic evaluation for such a potentially non sensitized patient—awaiting first renal transplant with a living donor?

Risk assessment before kidney transplantation is done to differentiate between a high immunological risk recipient (such as those with history of recurrent transplants, blood transfusion and pregnancy) and standard immunological risk recipient who lack those sensitization histories. The patient described here fits the standard immunological risk criteria, and we will discuss the pre transplant workup of this patient, and the risk stratification will help the clinician to plan the most appropriate immunosuppression regime for the patient. A standard immunological risk patient is usually ABO-Compatible, may be haplo-identical when the living donor is one of the parents, siblings (25% siblings may be double haploidentical, 25% may be

V. Kute (✉)
Department of Nephrology, Institute of Kidney Diseases and Research Center Institute of Transplantation Sciences (IKDRC-ITS), Gujarat University of Transplantation Sciences (GUTS), Ahmedabad, Gujarat, India

M. R. Balwani
Saraswati Kidney Care Center, Nagpur, India

Department of Nephrology, Jawaharlal Nehru Medical College, Wardha, Maharashtra, India

A. R. Chaudhury
Department of Nephrology, North Bengal Medical College, Darjeeling, West Bengal, India

G. Abraham et al. (eds.), *Immunological Challenges in Organ Transplantation*,
https://doi.org/10.1007/978-981-95-1533-2_8

nonidentical), for spousal donation and unrelated living donation or with deceased donor, complete HLA Mismatch is often expected.

Pre-transplant workup can be done under four headings:

1. Assessing ABO compatibility.
2. Assessing HLA compatibility (In Indian context, for living relate donation, it is must in legal context to establish relationship, very helpful in identifying donor specific DSA, if DSA is present).
3. Looking for presence of Anti-HLA DSA (may be avoided in standard risk recipient/low resource center and presence of DSA is not a contraindication for transplant if cross match is negative).
4. Crossmatch tests (most important and indispensable).

Introduction

Globally, more than 50% of solid organ transplants are renal transplants, as the burden of patients with end stage kidney disease is also the highest, compared to all other solid organs [1], yet it is quite paradoxical that after more than five decades of experience in renal transplant, no uniform agreed upon guidelines exist to guide or provide an approach for immunological evaluation of any potential transplant recipient.

Each institute within a country may have its own policies for pre transplant assessment of recipients. Few countries like Australia and New Zealand have their own nationwide guidelines. KDIGO 2020 guidelines does confer some suggestions [2], yet a precise approach is lacking.

Methods of Immunologic Evaluation

Presence of ASnti HLADSA and Crossmatch nowadays can be done by two methodologies.

Broadly classified into:

1. *Cell Based Assay*
   (a) Complement Dependant Cell Cytotoxicity (CDC)
   (b) Panel Reactive Antibody (PRA)
   (c) Flow Cytometry based Cross Match (FCM/FCXM)
2. *Solid Phase Assay* (Fig. 8.1)
   (a) Flow Cytometry based Panel Reactive Antibody (Flow PRA)
   (b) Luminex Platform based Assay
      (i) Pooled Antigen panel
      (ii) Phenotypic Panel
      (iii) Single Antigen Bead Assay—and performing Virtual Cross Match
      (iv) Luminex Cross Match

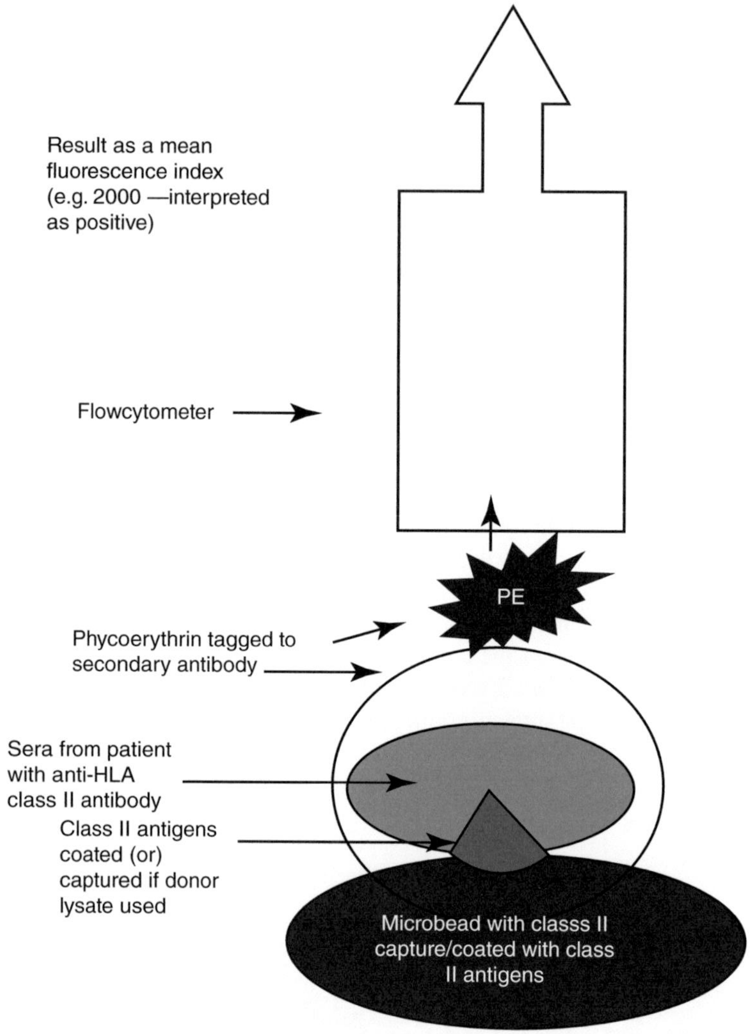

Fig. 8.1 Principle of the Luminex-based tests for antibody detection

Evolution of Immunological Evaluation of Kidney Transplant Recipient

In their seminal paper published in 1969, Patel and Terasaki demonstrated 24 out of 30 immediate graft failures in kidney transplants performed across a positive CDC crossmatch, compared to 8 out of 195 transplanted across a negative crossmatch [3]. Hence, CDC negativity came to be recognized as an important pre-requisite before proceeding for transplant. The flow cytometric crossmatch (FCXM), introduced in 1983, allowed for increased sensitivity in detecting IgG from recipient serum bound

to the surface of T and B lymphocytes through the use of a fluorescently labelled F(ab')2 anti-human IgG. Subsequent modifications in FCM with 96-well plate format allow for a rapid and more optimized protocol. The subsequent development of solid-phase HLA antibody screening techniques, including Luminex single antigen beads (SAB) assay, has greatly increased the sensitivity of antibody detection and, in conjunction with the donor HLA type, allows virtual crossmatch, that can identify anti HLA Donor Specific Antibodies (DSA), at levels that are below the limit of detection of FCXM. Hence, role of CDC cross match today remains less clear, due to availability of more sensitive methods and due to inherent limitations in CDC assay,

Complement-Dependent Cytotoxicity (CDC) Crossmatch (Figs. 8.2 and 8.3)

The CDC crossmatch is one of the oldest and most widely used methods to detect anti-HLA antibodies. In this assay, donor lymphocytes are incubated with the recipient's serum, which may contain anti-HLA antibodies. If the recipient has antibodies against the donor's HLA antigens, these antibodies will bind to the donor cells. The addition of complement, a group of proteins that enhances immune responses, will lead to the activation of the complement cascade and cause cell lysis (destruction). This cell lysis is indicative of a positive crossmatch.

In a positive CDC crossmatch, complement activation and cell death are observed, signaling the presence of anti-HLA antibodies in the recipient's serum. The result is generally considered a contraindication for transplant, as the presence of such antibodies increases the risk of AMR, a form of rejection mediated by the recipient's immune system.

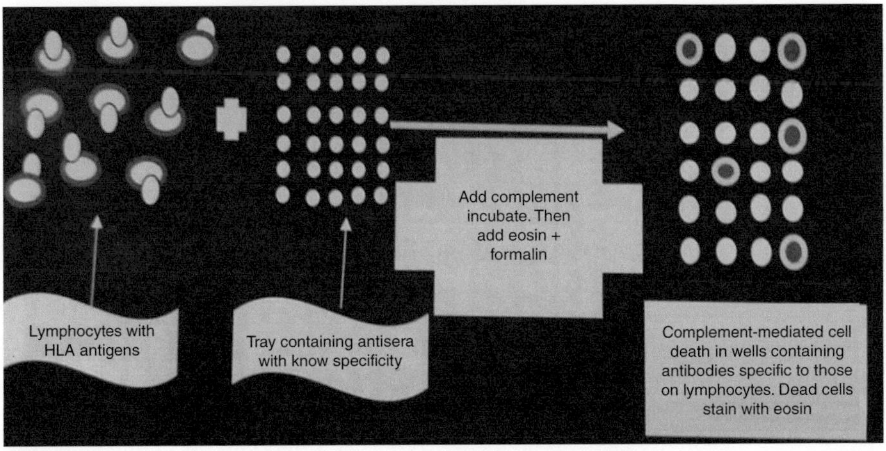

Fig. 8.2 Serological human leucocyte antigen typing based on the principle of complement dependent cytotoxicity

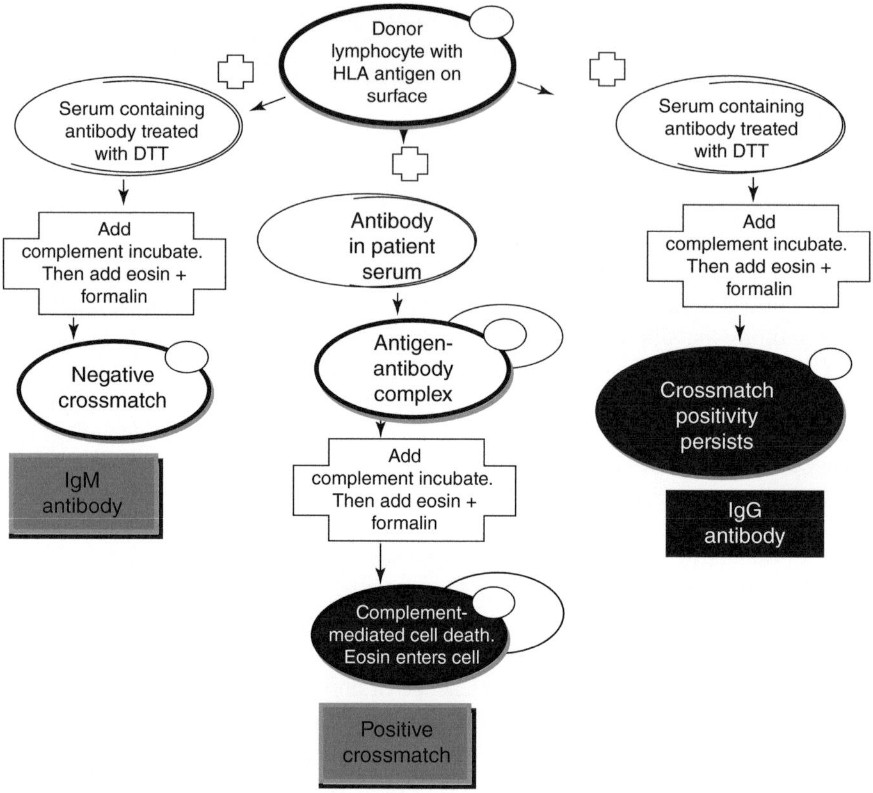

Fig. 8.3 The compliment dependent cytotoxicity cross match, with and without dithiothreitol, to differentiate the IgM and IgG antibody. *HLA* human leukocyte antigens, *CDC* complement dependent cytotoxicity, *DTT* dithiothreitol

The advantages of the CDC crossmatch include it is relatively low cost and simplicity. However, the assay has several limitations which includes:

(a) *Baseline viability of lymphocytes*—increased non-viable lymphocytes before cross match may potentially confound interpretation of cross match results.
(b) *Time consuming*—takes 6–8 h, versus about 4 h for FCM.
(c) *Cumbersome*—each step involves high degree of expertise and operator dependency.
(d) *Non-specific*—positive for non-specific IgM antibodies, post rituximab infusion for up to 6–8 months and non-HLA antibodies also—which, though an emerging area of interest, is generally not clinically relevant—as of now.
(e) *Less sensitive*—sufficiently high titers of antibody needed to activate complements to bring about a positive cross match result.
(f) *Subjective*—interpretation of final cross match result is highly subjective, hence for ambiguous positivity like 20–30%, high inter-observer variation present.

Despite these limitations, in many resource poor regions, due to current high costs associated with SAB assay or unavailability of expertise, infrastructure for FCM, many centers still rely on CDC cross match for transplant and this requires a paradigm shift.

Flow Cytometry Crossmatch (FCXM) (Table 8.1)

Flow cytometry crossmatch (FCXM) is a more modern and highly sensitive method for detecting anti-HLA antibodies. The FCXM test utilizes flow cytometry, a powerful technology that allows for the analysis of individual cells based on their physical and chemical characteristics. In this assay, donor cells lymphocytes are incubated with the recipient's serum. If the serum contains anti-HLA antibodies, they will bind to the donor cells. The cells are then incubated with fluorescently labelled secondary antibodies that recognize human antibodies.

FCM allows for the detection of antibody binding to donor cells with high precision and sensitivity. The fluorescence intensity of the labeled cells is measured using a flow cytometer, which analyzes the binding of antibodies on a cell-by-cell basis. The flow cytometer can detect low levels of antibodies that may not be detected by the CDC crossmatch, making it a more sensitive technique.

A positive FCXM result occurs when a proportion of donor cells are bound by recipient antibodies, suggesting the presence of anti-HLA antibodies. If the recipient's serum contains such antibodies, the result indicates a potential risk for Antibody Mediated Rejection (AMR), and further evaluation is needed before proceeding with transplantation. Overall, FCXM is a highly sensitive and reliable method for detecting anti-HLA IgG antibodies and is increasingly being used in transplant centers worldwide to predict transplant outcomes more accurately.

Table 8.1 Flow cytometry crossmatch (FCXM)

Strengths	Limitations
Increased sensitivity—more sensitive than CDC	**False positive**—highly sensitive, so may yield false positive results
No interference by IgM—as anti-human IgG antibodies used for detection	**Not a functional assay**—doesn't detect in-situ effect of antibody binding
Quantitative—amount of median channel shift is indicative of strength of antibody binding	**Effect of monoclonal antibodies**—like rituximab gives false positive result—Pronase treatment may help reduce this effect
No complement dependence—technique involves adding anti-human IgG antibodies, irrespective of whether or not—they fix complement	**Costly—**overall still an expensive test with respect to need of skilled personnel in performing the test, and cost of flow cytometer along with the whole assay used

Bead-Based Assays for Anti-HLA Antibody Detection

In recent years, bead-based assays have emerged as a more advanced and accurate method for detecting anti-HLA antibodies, specifically donor-specific antibodies (DSAs). These assays use microspheres or beads that are coated with a specific set of HLA antigens. The beads are then mixed with the recipient's serum, and if anti-HLA IgG antibodies are present, they will bind to the corresponding HLA antigen on the beads. The results can be analyzed using a variety of technologies, including flow cytometry or solid-phase immunoassays. Bead-based assays have the distinct advantage of enabling high-throughput screening and allowing for the detection of a broad range of antibodies with improved sensitivity.

Donor-Specific Antibody (DSA) Testing (Table 8.2)

Donor-specific antibody (DSA) testing is one of the most critical assays in modern kidney transplantation. DSAs are antibodies that specifically target HLA antigens

Table 8.2 Solid phase assay—Luminex platform based single antigen bead (SAB) assay

Strengths	Limitations
Highly sensitive—even more than FCM, the most sensitive assay available till date	**False positive**—highly sensitive assay, even antibodies with low avidity are detected
Correlation with FCM—though no single value above which a linear correlation with positive FCM seen, MFI values above 5000 are generally considered to be unacceptable for antibodies to HLA antigens—A/B/DR/DQ, as believed to be associated with positive FCM	**No uniformly accepted threshold**—no specific cut-off level of MFI agreed upon above which to consider antibodies as clinically significant
Reducing cold ischemia time (CIT)—in countries like Australia, they rely only on virtual cross match, such that physical cross match is not performed in deceased donor transplants, reducing CIT	**Costly**—up to 4 times the cost of LCM and 6 times the cost of CDC cross match
Semi quantitative—mean fluorescence intensity (MFI) gives a semi-quantitative estimate of strength of antibodies	**Neoepitopes**—denaturation of HLA antigens during bead fixation may expose neoepitopes and may give false positive result
	Prozone/hook effect—Detection of antibody by the assay, depends on optimal concentration of both—Antigen and antibody—Such that excess of any component may yield false negative result
	Peanut butter effect—For alleles shared by multiple HLA antigens, antibody gets distributed over many alleles; hence, resultant MFI may be underestimated
	Non physiologic allele representation—some antigenic alleles like those of HLA—C, DR/DQ, DRB3/4/5 which are normally expressed less on cell surface, are overrepresented on beads; hence, its physiologic relevance remains questionable

from the donor's tissue. These antibodies are a major risk factor for antibody-mediated rejection (AMR) and graft failure, making the detection and characterization of DSAs essential for transplant management.

In DSA testing, beads are coated with the donor's specific HLA antigens, and the recipient's serum is tested for antibodies that recognize those antigens. The beads are then analyzed using flow cytometry or a similar technology to measure the intensity of antibody binding. A positive DSA result indicates that the recipient has developed antibodies against the donor's HLA antigens, which may increase the risk of AMR and graft loss.

Panel Reactive Antibody (PRA) Testing

Panel reactive antibody (PRA) testing is another important bead-based assay used in organ transplantation. PRA testing measures the percentage of donor HLA antigens against which the recipient has antibodies. The test uses a broad panel of beads that are coated with various HLA antigens from a diverse group of donors. The recipient's serum is mixed with this panel, and the binding of antibodies is assessed to determine the overall level of sensitization.

A high PRA score indicates that the recipient has antibodies against a large percentage of potential donors, making it more difficult to find a compatible match. Conversely, a low PRA score suggests that the recipient has a relatively low level of sensitization and may have a broader pool of potential donors to choose from.

The Advantages of PRA Testing

Comprehensive assessment: PRA testing provides a broad overview of the recipient's sensitization status, allowing transplant centers to assess the likelihood of finding a compatible donor.

- **Widely available**: PRA testing is widely available and is often used as a standard pre-transplant evaluation in many transplant centers.

However, PRA testing has limitations:

- **Lack of specificity**: While PRA testing provides a measure of overall sensitization, it does not provide information on specific donor-recipient compatibility or the presence of DSAs.
- **Limited predictive value**: High PRA scores can indicate general sensitization but may not fully capture the risk of AMR or other complications associated with donor-specific antibodies.

The frequency of PRA testing varies depending on the transplant program's guidelines. In sensitized patients it may be done monthly. However, in non-sensitized patients, it may be done less often.
It may be advisable to do a PRA prior to transplantation and following transplantation to assess the risk of rejection.

Since 2007 UNOS has been using cPRA, which is a calculated PRA score. It is calculated based on the allelic frequency of different HLA groups within the population against which the recipient sera are expected to react.

To summarize we recommend the use of at least one cell based and one bead based assay before going ahead for transplant particularly in living donor transplants. In case of deceased donor transplant which are not sensitized, a simple virtual crossmatch to document the absence of unacceptable antigens may suffice, especially in the paucity of time.

Institutional Protocol at IKDRC Ahmedabad—Center with One of the Largest Number of Renal Transplant in Asia

Each potential living related renal transplant recipient (like Mr. A here) undergoes following sets of immunologic evaluation:

1. Single Antigen Screen test.
2. Single Antigen Bead assay—if screen test is positive.
   (a) Acceptable anti-HLA DSA MFI < 3000.
   (b) For anti-HLA DSA MFI 3000–5000—option of alternate donor versus desensitization versus Kidney Paired Donation (KPD) given—depending on patient preference and risk assessment.
   (c) For anti-HLA DSA MFI >5000—option of alternate donor versus KPD given.
3. Flow cross match if single antigen screen negative or acceptable MFI value in SAB assay—Median Channel Shift of up to 100 for B-cell and up to 50 for T-cell cross match—considered acceptable to proceed for transplant.
4. PRA is more used in deceased donor evaluation.

Immunologic Evaluation in Resource Intense Settings

A suggested algorithm in centers with sufficient expertise in performing different immunological tests includes the following: Flowcharts 8.1 and 8.2 shown as follows:

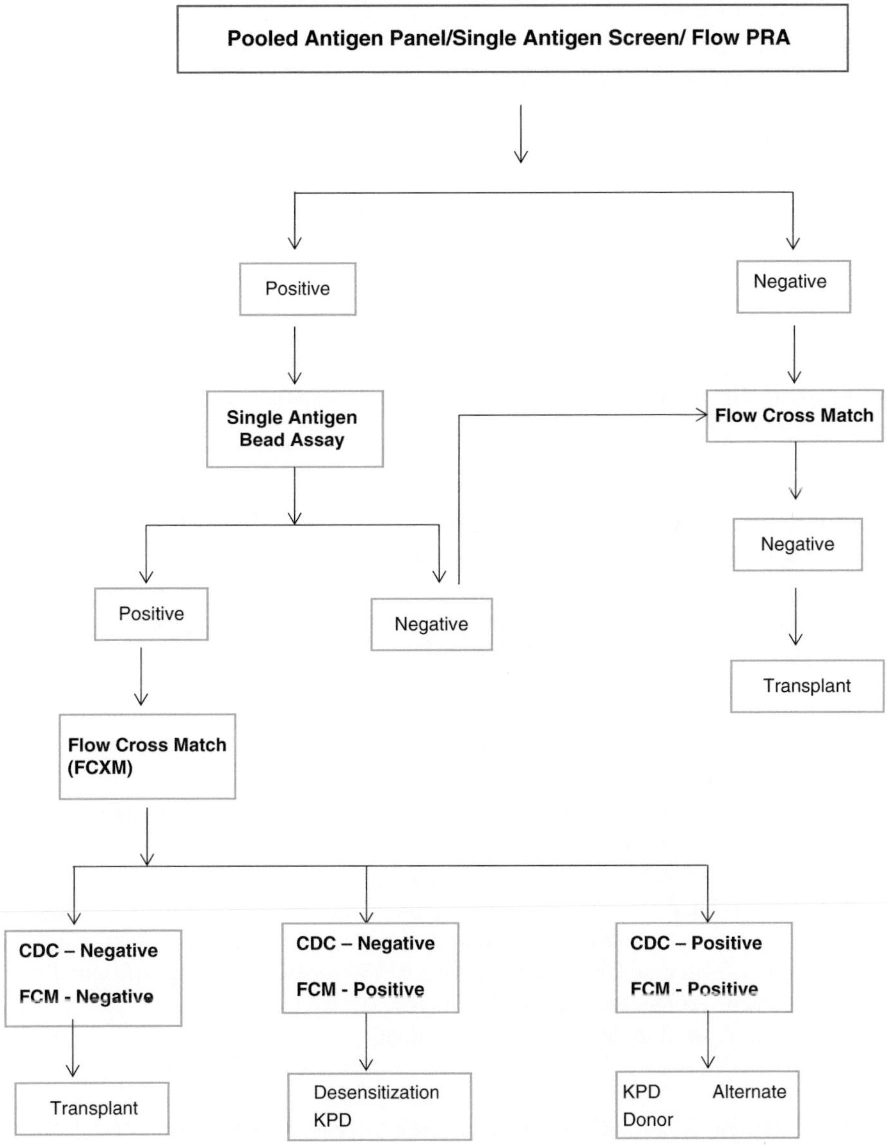

Flowchart 8.1 Showing immunological workup

Flowchart 8.2 Showing flow cross match and CDC cross match

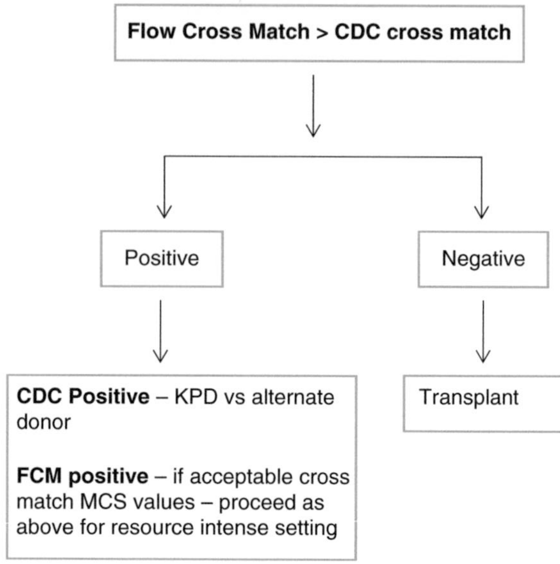

Table 8.3 Cost of individual assay in India at IKDRC

Assay	Cost
CDC cross match	₹ 5500
FCM	₹ 8000
PRA	₹ 10,000
SAB assay	₹ 33,000

Immunologic Evaluation in Limited Resource Settings

- For non-sensitized patients, can directly proceed with Floe cross match—which if negative ➔ proceed for renal transplant.
- For sensitized patients, will need assessment as above—with potential referral to tertiary care centers (Table 8.3).

References

1. IRODaT—International Registry on Organ Donation and Transplantation.
2. KDIGO-Txp-Candidate-GL-FINAL.pdf.
3. Mahowald GK. The CDC crossmatch in the era of flow cytometric cross-match and single antigen beads. J Bras Nefrol. 2021;43(3):299–300. https://doi.org/10.1590/2175-8239-JBN-2021-0110.

9 A Case Based Approach to the Sensitized Recipient with DSA on the SAB to the Proposed Donor

Dolly Daniel, Santosh Varughese, and V. G. David

Pre-transplant testing has evolved significantly over the last few decades. One of the key players that has evolved and found its place in numerous algorithms is the single antigen bead assay (SAB). With its high sensitivity, specificity, and being an assay that allows for identifying donor specific antibodies (DSA), deriving a panel reactive antibody status (PRA) and also analysing epitopes, the SAB has added much value.

Often referred to as the virtual crossmatch, the SAB is performed on the Luminex platform. The Luminex platform utilizes beads with attached HLA molecules. The unique fluorescent signature of each of these beads, with differential ratios of two fluorochromes, is the basis on which assays on this platform are performed with the final read out being on the Luminex fluorocytometer which has a dual laser flow. In the SAB assay, each bead is attached to a purified HLA molecule of a single specificity. When patient serum, containing anti HLA antibody is added, it binds the bead with the appropriate HLA molecule. A secondary antibody which is usually phycoerythrin (PE) labelled anti human globulin is then added. Following appropriate incubations, the test is read on the inbuilt dual laser flow and results expressed as a mean fluorescence intensity (MFI) (Fig. 9.1).

A positive result is based on cut offs laid down by laboratories. An MFI above 1000–1500 is often considered as positive by many laboratories.

D. Daniel (✉)
Transfusion Medicine and Immunohaematology,
Christian Medical College, Vellore, Tamil Nadu, India
e-mail: dollyd@cmcvellore.ac.in

S. Varughese · V. G. David
Department of Nephrology, Christian Medical College Vellore, Ranipet, Tamil Nadu, India
e-mail: vinoigd@cmcvellore.ac.in

G. Abraham et al. (eds.), *Immunological Challenges in Organ Transplantation*,
https://doi.org/10.1007/978-981-95-1533-2_9

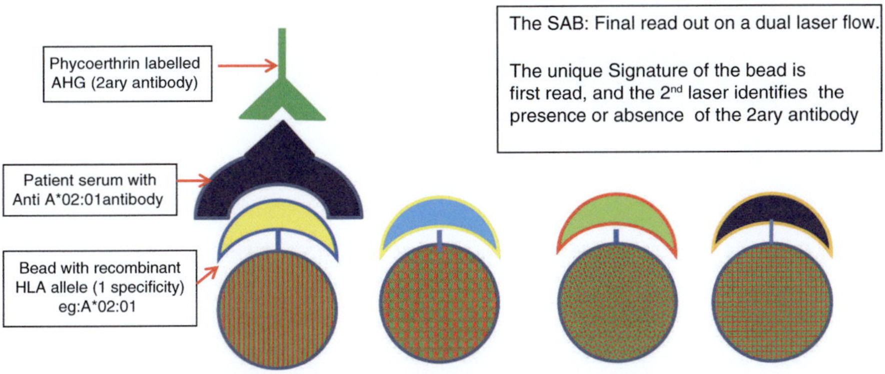

Fig. 9.1 The SAB: Each bead has it's unique fluorescent signature. Each bead has a recombinant HLA allele (single specificity) attached to it

Table 9.1 SAB: pros and cons

Pros	Cons
Highly sensitive	False positives due to antibody binding to denatured HLA
Highly specific	False negatives due to nonrepresentation
Can assess epitope reactivity	False negatives due to Prozone
Can use modifications to determine complement binding and IgM antibodies	Shared epitopes can decrease strength of positivity
Used as a semi quantitative assay though approved only as a qualitative assay by FDA	MFI does not account for variability of antigen expression and antibody affinity in vivo
Can be used to get a calculated PRA (cPRA)	Variability in antigen density on beads

It is critical that appropriate positive and negative controls are used in every run, and that results are carefully interpreted, keeping in mind the assays pros and cons (Table 9.1).

Clinical Relevance

The SAB is the gold standard for pre transplant evaluation, and post-transplant follow-up in many parts of the world. The clinical applicability has been demonstrated in many studies, particularly in the context of long term graft outcomes.

However most guidelines suggest a combination of assays to guide donor choice and decisions regarding desensitisation and modulation of immunosuppression [1].

This chapter will bring to you four different scenarios of SAB positivity, focussing on the importance of careful interpretation and the criticality of using multiple platforms to make the best possible clinical decisions for patients awaiting transplant.

Pre-transplant Immunologic Workup

Routine workup includes a high resolution 6 HLA typing, a CDC and flowcytometric crossmatch and a single antigen bead assay. Eleven locus HLA typing can be performed if available or is indicated based on the SAB results.

Case 1

A 28-year-old male Mr. XY with end stage renal disease (ESRD) was being evaluated for a renal transplant with his mother as a potential donor. There is a history of having received one red cell transfusion a year prior. HLA typing and CDC crossmatch results are as in Tables 9.2 and 9.3.

HLA typing: Haplomatch
CDC: Total (T and B) lymphocyte crossmatch
Autocrossmatch: Negative
Donor recipient crossmatch: Negative
Flowcytometric crossmatch results:
T cell crossmatch: Negative (Ratio < 1)
B cell crossmatch: Negative. (Ratio < 1.2)
Panel Reactive antibodies (PRA):
The PRA was 2% for class I antibodies and 1% for class II
Single antigen bead assay:
Class I:
Non DSA: present to the following specificities:

Table 9.2 HLA typing results: mismatched alleles are shaded yellow

Locus	A*	B *	C *	DRB1*	DQB1*	DPB1*	
Mr XY	01:01	27:05	04:01	11:01	03:03	09:01	
	24:02	13:01	07:01	15:01	03:02	09:01	
Mother donor	24:02	13:01	07:01	15:01	03:02	09:01	Haplomatch
	11:01	07:02	06:02	04:02	06:02	04:01	

Table 9.3 CDC Crossmatch results

Dilution	Auto	Auto with DTT	Donor recipient	Donor recipient with DTT	Inference
N	Neg	Neg	Neg	Neg	Negative auto and donor recipient crossmatch
N/2	Neg	Neg	Neg	Neg	
N/4	Neg	Neg	Neg	Neg	

A*02:01: MFI 1800, A* 68:01—MFI 2000
DSA: Absent
Class II:
Non DSA: Absent
DSA: Present
Antibody to DQB1*06:02, MFI 3500

Questions that arise:

1. Does the pattern of reactivity seen across the varied platforms used fit in?
2. How do we risk stratify this patient—which group does he fall into?

With the luminex platform and the SAB assay being the most sensitive of tests, an isolated positivity with DSA being present only on the SAB can occur when levels of antibodies are low. The European society of transplantation observe that MFIs in the 3000 to 4000 range generally do not reflect as a positive flow or CDC crossmatch. However, this antibody does heighten the risk for antibody mediated rejection. The ESOT risk stratification is as follows (Fig. 9.2) [2].

3. Is this donor an option that can be considered for this patient?
4. Is he a candidate for desensitisation prior to transplant? Or should other alternates such as paired kidney exchange be considered?
5. What would be the optimal management plan?

As per ESOT this patient fits into Category 3 (DSA Present and Negative Flow and CDC Cross Match at Day 0).

Patients in this category are clearly at a higher immunological risk. It is advisable to weigh risks and benefits, plan desensitization with the goal of minimizing ABMR, be prepared for an antibody rebound post-transplant, and assert great caution in terms of post-transplant monitoring and weaning of immunosuppression.

Case 2

A 35-year-old lady Mrs. SD with ESRD awaiting transplant comes for an evaluation with her husband being the prospective donor. She has had two children, but there is no history of transfusions in the past.

Results

HLA typing results 3/12 match as shown in Table 9.4.

CDC: (Total T and B) lymphocyte crossmatch
Autocrossmatch: Negative
Donor recipient crossmatch: Negative
Flowcytometric crossmatch results:
T cell crossmatch: Positive—ratio 3.0 (Negative <1.0)

Memory	Lab findings	Transplant	Risk	Risk level
Serological memory	Day zero DSA + Positive CDC	NOT possible		
	Day Zero DSA, Positive Flow and Negative CDC	Possible	Very high risk of acute ABMR and accelerated chronic AMR	
	Day zero DSA with negative flow	Possible	Risk for acute ABMR . Acceptable medium term graft survival	
Cellular memory	Absent day zero DSA with	Possible	Risk for ABMR increased	
	1. Probable cellular memory – due to a. historical DSA, b. challenged again by same antigens as in pregnancy or a previous transplant			
	2.Possible cellular memory - transfused with no information on blood donors			
Naive	No DSA and no cellular memory	Possible	Lower risk for ABMR. But risk for de novo DSAs persist	
	Patients with non DSA anti HLA antibodies – greater risk of developing de novo DSAs post transplant			

Fig. 9.2 Risk stratification (based on ESOT guidelines)

Table 9.4 HLA typing—the yellow highlighted alleles are those that are mismatched

Locus	A*	B*	C*	DRB1*	DQB1*	DPB1*	
Mrs SD	24:02	35:03	04:01	04:01	05:02	10:01	3/12 match
	11:01	27:04	06:01	04:04	06:01	09:01	
Husband donor	11:01	40:06	07:01	04:01	05:02	14:02	
	68:01	55:01	07:01	07:02	03:02	296:01	

B cell crossmatch: Positive—ratio 3.5 (Negative <1.2)

Panel Reactive antibodies:

The PRA was 7% for class I antibodies and 5% for class II

Single antigen bead assay:

Class I:

Non DSA: present to the following specificities:

A*02:01: MFI 1800, A*02:03—MFI 2500, A* 26:01—MFI 2000, B*40:01—MFI-7500, B*40:02—MFI-8000

DSA: Absent

Class II:

Non DSA present to the following specificities

DQB1*15:02, MFI 3500, DRB1*13:02—MFI 4000

DSA: Absent

Queries that need to be addressed:

1. How is the positive flow crossmatch explained in the absence of DSA on the SAB which is the more sensitive platform?
2. How do we risk stratify and plan management?

The single antigen bead assay is challenged by the number of specificities that can be represented on it [3]. In this patient donor combo, the flow crossmatch shows a clear positivity of both T and B cells, raising the possibility of it being either a class I antibody alone or a combination of both a Class I and II antibody.

When the mismatches between this patient donor pair are observed carefully, it is noted that the mismatched allele B*40:06 is not represented on the SAB kit being used. However the other B*40 family of alleles namely B*40:01 and B*40:02 are strongly positive with MFIs of 7500 and 8000.

Hence it is fair to deduce that this patient does have a DSA to the mismatched allele B*40:06, with an MFI intensity that is reflecting as a flow crossmatch positivity.

In addition, the patient has an antibody to alleles of the A*02 family. The mismatched allele A*68:01 belongs to the same CREG (cross reactive group) of alleles, and hence this too should be viewed with caution.

In the hierarchy of risk, he will therefore be a high risk—with positivity on SAB and flow, labelled by the ESOT as risk category 2.

Category 2 Patients (DSA Present with Positive Flow and Negative CDC Crossmatch at Day 0).

It would be prudent not to transplant across this immunologic barrier. A paired kidney exchange/alternate donor would be ideal. However in the absence of any other options, desensitization, with preparedness for rebound antibody causing ABMR and close post-transplant follow-up with continued immunosuppression will be crucial.

Case 3

A 45-year-old lady Ms. AXD with ESRD is being worked up for a renal transplant with an aunt as a potential donor. She has a history of three pregnancies. This is her first transplant and she has never received any transfusions prior. HLA typing and CDC results are shown in Tables 9.5 and 9.6 respectively.

HLA typing: 1/12 match
CDC: Total (T and B) lymphocyte crossmatch
Autocrossmatch: Negative
Donor recipient Crossmatch: Negative
Flowcytometric crossmatch results:
T cell crossmatch: Negative
B cell crossmatch: Negative
PRA:
The PRA was 12% for class I antibodies and 10% for class II
Single antigen bead assay:
Class I:
Non DSA: present
DSA: present to the following specificities:
B*35:02—MFI 11000
Class II:
Non DSA: present
DSA: present to the following specificities
DRB1*07:01—MFI 12000, DQB1*05:01—MFI 14000

Table 9.5 HLA typing: the yellow highlighted boxes indicate the mismatched alleles

Locus	A*	B*	C*	DRB1*	DQB1*	DPB1*	
Patient Ms. AXD	24:02	44:01	07:01	04:01	03:02	04:01	1/12 match
	29:01	27:02	07:01	15:02	03:01	02:01	
Donor Aunt	24:02	07:02	06:02	16:02	05:01	10:01	
	31:01	35:02	04:01	07:01	05:01	26:01	

Table 9.6 CDC Crossmatch results

Dilution	Auto	Auto with DTT	Donor recipient	Donor recipient with DTT	Inference
N	Neg	Neg	Neg	Neg	Negative auto and donor recipient crossmatch
N/2	Neg	Neg	Neg	Neg	
N/4	Neg	Neg	Neg	Neg	

Table 9.7 Repeat CDC

Dilution	Auto	Auto with DTT	Donor recipient	Donor recipient with DTT	Inference
Neat	Neg	Neg	Neg	Neg	Negative auto crossmatch
N/2	Neg	Neg	Neg	Neg	Positive donor recipient crossmatch in higher dilutions consistent with a prozone effect
N/4	Neg	Neg	Neg	Neg	
N/8	Neg	Neg	10% Pos	10% Pos	
N/16	Neg	Neg	20%	20%	
N/32	Neg	Neg	<5%	<5%	

Questions that arise:

1. Does the pattern of reactivity seen across the varied platforms used fit in?
2. How do we risk stratify this patient—which group does he fall into?

This patient has a very high level of both class I and class II anti HLA antibodies as detected on the SAB. The ESOT describes MFI strengths and how it correlates with differing platforms of testing [4]. While this may be variable, with the MFIs of DSA seen in this patient, one would expect it to reflect in a T and B cell flowcytometric crossmatch positivity. In fact with MFIs noted one should reconfirm reactivity on the CDC as well. Confirming the reactivity on these two platforms is absolutely critical as the risk category changes and decision making has to be within the context of an accurate risk category that the patient belongs to.

One of the most common causes this mismatch can be attributed to is the prozone phenomenon. This phenomenon occurs when high antibody levels lead to false negative results. Reactivity is picked up only when the optimal antigen antibody ratio is present. In this setting, diluting the serum is likely to yield more accurate results [5].

Repeat CDC crossmatch results are seen in Table 9.7:

CDC Crossmatch (Total T and B lymphocyte)
Auto crossmatch: Negative
Donor recipient crossmatch: Positive in greater dilutions (N/8, N/16)—consistent with a prozone phenomenon
Flowcytometric crossmatch using 1/10 and 1/50 dilutions 1
1/10 dilution
T cell crossmatch: Positive—ratio: 2.5

B cell crossmatch: Positive—ratio: 3.0

1/50 dilution

T cell crossmatch: Positive—ratio: 6.5

B cell crossmatch: Positive—ratio: 9.5

Discussion

The results reflect a strong DSA present on both the CDC and flowcytometric platforms, demonstrable when sera was diluted, confirming the prozone phenomenon. This patient therefore moves to a higher risk category than an isolated SAB positivity.

With a positive CDC, Flow, and SAB positivity, this donor should possibly NOT be considered. It is critical to understand that any platform can be prone to the prozone phenomenon. Many studies have demonstrated this phenomenon on luminex-based tests as well [6, 7].

On evaluation of a potential donor-recipient pair, it is important to consider the likelihood of a high immunological risk if any of the tests are positive. Even if there is no sensitizing event of a blood transfusion, pregnancy does confer an immunological risk and one must ensure that the transplant is safe to proceed. In this situation, one could have missed the high risk (even if flowcytometry was done) if DSAs were not looked for using SAB. This case highlights the importance of doing SAB even if two tests are normal, and also to ask for the tests in dilution if the SAB is unexpectedly high. With flowcytometry (in dilution) being highly indicative of preformed antibodies and the SAB showing high MFI, it would be prudent not to proceed with this prospective donor for transplantation.

Case 4

A 27-year-old female Ms. SDS with ESRD and a history of a mixed connective tissue disorder is being evaluated for a possible renal transplant with her mother as a prospective donor. She has a history of two pregnancies, but no transfusions.

HLA typing and CDC crossmatch results are shown in Tables 9.8 and 9.9 respectively.

HLA typing: Haplomatch.

Table 9.8 HLA typing results: yellow highlight indicates mismatched alleles

Locus	A*	B*	C*	DRB1*	DQB1*	DPB1*
Patient SDS	26:01	44:01	04:01	07:02	03:02	01:01
	24:01	52:02	04:04	04:02	05:02	296:01
Mother donor	26:01	44:01	04:01	07:02	03:02	01:01
	33:03	35:03	07:01	15:01	06:02	26:01

Table 9.9 CDC Crossmatch results: (total T and B) lymphocyte crossmatch

Dilution	Auto	Auto with DTT	Donor recipient	Donor recipient with DTT	Inference
Neat	10%	Neg	20%	Neg	Positive auto and donor recipient Crossmatch with features suggestive of an IgM antibody
N/2	5%	Neg	10%	Neg	
N/4	5%	Neg	10%	Neg	

Table 9.10 Repeat CDC Crossmatch results (Total T and B lymphocyte)

Dilution	Auto	Auto with DTT	Donor recipient crossmatch	Donor recipient with DTT	Inference
Neat	10%	Neg	20%	Neg	Positive auto and donor recipient crossmatch with features suggestive of an IgM antibody
N/2	5%	Neg	10%	Neg	
N/4	5%	Neg	10%	Neg	

Flowcytometric crossmatch results:

T Cell crossmatch: Negative

B cell crossmatch: Negative

PRA: Negative

SAB: Negative

The clinical significance of IgM antibodies in the context of renal transplants is debatable. However the probability of a class switch to an IgG should not be overlooked or ignored. Considering this risk, it is always advisable to repeat a crossmatch in 2–4 weeks on all three platforms to ensure that an IgG antibody is picked up if present.

Repeat testing in 3 weeks was performed and the results are shared in Table 9.10.

Flowcytometric crossmatch results:

T cell crossmatch: Positive—ratio 1.6

B cell cxmatch: Positive—ratio 2.3

Single antigen bead assay:

Class I:

DSA: present B*35:02- MFI-5000

Non DSA: Absent

Class II:

Non DSA: present

DSA: Absent

Discussion

Class switch of an IgM antibody to an IgG antibody is a well described phenomenon. This case highlights the caution one needs to have when an IgM antibody is identified on the donor recipient crossmatch. IgM antibodies can be detected on either the CDC or the flowcytometric crossmatch, the latter being possible if the

secondary antibody binding to IgM antibodies is used However this is often challenging as it is fraught with technical challenges, and the CDC continues to be a platform that detects both IgM and IgG antibodies and can alert us to this phenomenon. This case also highlights the importance of testing for DSA on more sensitive platforms. Despite the CDC continuing to show a pattern consistent with an IgM antibody, the previously negative flowcytometric crossmatch and virtual crossmatch have turned positive and low levels of IgG would have been completely gone undetected if only a repeat CDC was performed.

The results of this patient put her into a high risk category, which requires desensitization or consideration for a paired kidney exchange. The clinical significance of a persisting IgM antibody has been debated, but the detrimental effect of the same on graft outcome has been demonstrated in some studies [8–10].

Summary and Conclusion

The above four case scenarios summarize the evaluation of a patient for an alloimmune risk towards kidney transplantation and the importance of using the three platforms of testing to identify the Donor Specific Antibodies (DSA). The objectives of the importance of using the three platforms are twofold:

1. First, we assess HLA disparity between the recipient and donor to weigh the risk of the recipient developing a de novo alloimmune response.
2. Second, we identify any preformed circulating IgG antibodies against HLA in the recipient's serum.

Recent advancements in characterizing donor and recipient HLA disparities, along with the use of flow crossmatch and sensitive immunological tools to detect circulating IgG anti-HLA antibodies, have significantly transformed immune-risk profiling.

Approach to the Management of the High Risk Group Patients

Patients with HLA-DSA (Donor-Specific Antibodies) at the time of transplantation face significantly higher risks of acute antibody-mediated rejection (AMR) and poorer allograft outcomes [11–14]. The role of Mean Fluorescence Intensity (MFI) in Single Antigen Bead (SAB) testing for HLA-DSA remains contentious, as cut-offs vary by transplant centre [15]. In highly sensitized patients, assessing unacceptable mismatches based on SAB results must prioritize the heightened risk of AMR and the dangers of prolonged dialysis [16–19]. Current research shows no substantial benefits of DSA-negative transplants over DSA-positive ones regarding rejection, graft failure, or mortality [19]. The type of DSA critically influences outcomes. Patients with DSAs against HLA-A, HLA-B, and HLA-DR experience earlier AMR than those with DSA against HLA DQ. Additionally, 10-year graft survival is

markedly lower for patients with combined Classes I and II DSAs above 1000 MFI [20, 21]. Ultimately, transplantation despite the presence of DSA yields better survival rates than remaining on dialysis. A key goal in transplantation is to identify donor/recipient pairs with minimal incompatibility. Sensitized candidates have more options with a living donor, even if the donor is incompatible. The following can be the approach for a high risk group patient:

1. A trial of paired kidney exchange with a donor with avoidance of DSA is the best [22].
2. To consider ABO incompatible donor pairs will be an option because the ABO barrier is much easier to overcome than HLA incompatibility.
3. If unsuccessful, no compatible donors then proceed with transplantation with densitization with minimal post-transplant immunosuppression reduction. The risks of acute rejection, infections, long-term graft dysfunction, and regular monitoring of DSA in the post-transplant period need to be explained to the patient.
    (a) The current donor, the HLA incompatibility is surmountable (low levels of DSA, less than 1000 MFI) and is the only donor available to proceed with densitization.
    (b) High matched incompatibility with the current donor, then try with a KPD-matched donor with more amenable densitization.

Desensitization is a treatment that reduces preformed antibodies against a potential donor. The goal is to lower antibody levels so that the flow CDC and flow cross match is test is negative or below a specific cutoff. The optimal desensitization protocol before transplantation and the best maintenance immunosuppression regimen afterward are still under investigation. Different protocols have been used with varying success; however, most combine plasmapheresis, IVIG, rituximab, bortezomib, and early initiation of maintenance immunosuppression (Prednisolone, Tacrolimus, Mycophenolate) several weeks before the transplant. Low-level donor-specific antibodies (DSA) can have harmful long-term effects on the graft [23]. Desensitization allows for quicker transplants without a long wait on dialysis, but it comes with higher costs and increased risks of infections and complications, leading to poorer outcomes for both patients and grafts [24]. The cost-effectiveness of this therapy should be weighed against the savings and health benefits of a successful kidney transplant versus remaining on dialysis. Nonetheless, it may be the only option for some patients.

Enhancing graft longevity will always remain a priority and good immunologic monitoring and interventions based on the same, contribute greatly towards this long term goal. A combination of a comprehensive HLA assessment of patient and donor combined with a strategic combination of antibody testing tools can give valuable information to guide clinical decisions.

References

1. Tait BD, Süsal C, Gebel HM, Nickerson PW, Zachary AA, Claas FH, Reed EF, Bray RA, Campbell P, Chapman JR, Coates PT, Colvin RB, Cozzi E, Doxiadis II, Fuggle SV, Gill J, Glotz D, Lachmann N, Mohanakumar T, Suciu-Foca N, Sumitran-Holgersson S, Tanabe K, Taylor CJ, Tyan DB, Webster A, Zeevi A, Opelz G. Consensus guidelines on the testing and clinical management issues associated with HLA and non-HLA antibodies in transplantation. Transplantation. 2013;95(1):19–47. https://doi.org/10.1097/TP.0b013e31827a19cc.
2. Mamode N, Bestard O, Claas F, Furian L, Griffin S, Legendre C, Pengel L, Naesens M. European guideline for the Management of Kidney Transplant Patients with HLA antibodies: by the European society for organ transplantation working group. Transpl Int. 2022;35:10511. https://doi.org/10.3389/ti.2022.
3. Kumar S, Doss SA, Stephen S, Pratheeba M, Jeyaseelan L, Daniel D. The challenge of using the virtual crossmatch as a singular tool for the detection of anti-HLA antibodies: a study from a tertiary care institute from South India. Transpl Immunol. 2021;65:101349. https://doi.org/10.1016/j.trim.2020.101349.
4. European Guidelines for the Management of Kidney Transplant Patients with HLA antibodies: By the European Society for Organ Transplantation (ESOT) Working Group. Definition of sensitization. In: European guidelines for the management of kidney transplant patients with HLA antibodies. ESOT; 2022. p. 10–20.
5. Singh P, Tiwari AK, Mishra VC, Deshpande TV, Dorwal P, Bhardwaj AK, Kumari S, Raina V. Prozone phenomenon in pretransplant testing: an interesting conundrum involving solid-phase and cell-based assays. Asian J Transfus Sci. 2022;16(2):180–5. https://doi.org/10.4103/ajts.ajts_145_20.
6. Konvalinka A, Tinckam K. Utility of HLA antibody testing in kidney transplantation. J Am Soc Nephrol. 2015;26(7):1489–502. https://doi.org/10.1681/ASN.2014080837.
7. Rocha Y, Jaramillo A, Neumann J, Hacke K, Palou E, Torres J. Crossmatch assays in transplantation: physical or virtual? A review. Medicine (Baltimore). 2023;102(50):e36527. https://doi.org/10.1097/MD.0000000000036527.
8. McCalmon RT Jr, Tardif GN, Sheehan MA, Fitting K, Kortz W, Kam I. IgM antibodies in renal transplantation. Clin Transpl. 1997;11(6):558–64.
9. Babu A, Andreou A, Briggs D, Krishnan N, Higgins R, Mitchell D, Barber T, Daga S. Clinical relevance of donor-specific IgM antibodies in HLA incompatible renal transplantation: a retrospective single-center study. Clin Transpl. 2016;32:173–9.
10. Dodd P, Willicombe M, Brookes P, Roufosse C, Santos-Nunez E, Goodall D, Clarke C, Charif R, Galliford J, McLean A, Taube D. Significance of HLA IgM and IgM donor specific antibodies in renal transplantation: abstract# A123. Transplantation. 2014;98:435.
11. Amico P, Honger G, Mayr M, Steiger J, Hopfer H, Schaub S. Clinical relevance of pretransplant donor-specific HLA antibodies detected by single-antigen flow-beads. Transplantation. 2009;87:1681–8.
12. Orandi BJ, Garonzik-Wang JM, Massie AB, Zachary AA, Montgomery JR, et al. Quantifying the risk of incompatible kidney transplantation: a multicenter study. Am J Transplant. 2014;14:1573–80.
13. Gaston RS, Cecka JM, Kasiske BL, Fieberg AM, Leduc R, Cosio FC, et al. Evidence for antibody-mediated injury as a major determinant of late kidney allograft failure. Transplantation. 2010;90:68–74. https://doi.org/10.1097/tp.0b013e3181e065de.
14. Senev A, Coemans M, Lerut E, Van Sandt V, Daniëls L, Kuypers D, et al. Histological picture of antibody-mediated rejection without donor-specific anti-HLA antibodies: clinical presentation and implications for outcome. Am J Transpl. 2019;19:763–80. https://doi.org/10.1111/ajt.15074.

15. Bestard O, Thaunat O, Bellini MI, et al. Alloimmune risk stratification for kidney transplant rejection. Transpl Int. 2022;35:10138. https://doi.org/10.3389/ti.2022.10138.
16. Ling M, Marfo K, Masiakos P, et al. Pretransplant anti-HLA-Cw and anti HLA-DP antibodies in sensitized patients. Hum Immunol. 2012;73:879–83.
17. Morris GP, Phelan DL, Jendrisak MD, et al. Virtual crossmatch by identification of donor-specific anti-human leukocyte antigen antibodies by solid phase immunoassay: a 30-month analysis in living donor kidney transplantation. Hum Immunol. 2010;71:268–73.
18. Llorente S, Boix F, Eguia J, et al. C1q-fixing human leukocyte antigen assay in immunized renal patients: correlation between LuminexSAB-C1q and SAB-IgG. Transplant Proc. 2012;44:2535–7.
19. Buttigieg J, Ali H, Sharma A, Halawa A. Positive Luminex and negative flow cytometry in kidney transplantation: a systematic review and meta-analysis. Nephrol Dial Transplant. 2019;34:1950–60. https://doi.org/10.1093/ndt/gfy349.
20. Bachelet T, Martinez C, Del Bello A, et al. Deleterious impact of donor specific anti-HLA antibodies toward HLA-Cw and HLA-DP in kidney transplantation. Transplantation. 2016;100:159–66.
21. Otten HG, Verhaar MC, Borst HP, et al. Pre-transplant donor-specific HLA class-I and-II antibodies are associated with an increased risk for kidney graft failure. Am J Transplant. 2012;12:1618–23.
22. Segev DL, Gentry SE, Warren DS, Reeb B, Montgomery RA. Kidney paired donation and optimizing the use of live donor organs. JAMA. 2005;293:1883–90. https://doi.org/10.1001/jama.293.15.1883.
23. Keith DS, Vranic GM. Approach to the highly sensitized kidney transplant candidate. Clin J Am Soc Nephrol. 2016;11(4):684–93. https://doi.org/10.2215/CJN.05930615.
24. Kahwaji J, Sinha A, Toyoda M, et al. Infectious complications in kidney-transplant recipients desensitized with rituximab and intravenous immunoglobulin. Clin J Am Soc Nephrol. 2011;6(12):2894–900. https://doi.org/10.2215/CJN.03710411.

Non-HLA Antibodies in Kidney Transplantation

10

Surabhi Talwar, Hariharan Iyer, and Nithya Krishnan

Introduction

Immediate graft loss after renal transplantation is a rare occurrence today [1]. Usually attributed to severe acute or hyperacute rejection secondary to high titers of preformed circulating donor specific antibodies, it may also occur due to other reasons like vascular thrombosis (venous or arterial) and hemorrhage [1]. Hyperacute rejection, which is characterized by infarction and necrosis of the allograft, occurs within minutes to hours after revascularization during organ transplantation [2]. Plasmapheresis is usually attempted to eliminate these antibodies from the blood. Vascular thrombosis without any evidence of anastamotic/technical issues may be due to other modifiable or non-modifiable factors like dialysis modality, immunosuppression, infections, or native kidney disease. However, the absence of an obvious cause should prompt a thorough investigation for an underlying thrombophilic disorder and non-HLA antibodies [3, 4].

S. Talwar
Institute of Cardio Metabolic Medicine and The Centre for Healthcare and Communities, University Hospitals Coventry and Warwickshire NHS Trust, Coventry, UK
e-mail: surabhi.talwar2@uhcw.nhs.uk

H. Iyer
Division of Nephrology and Transplantation, Western University, London, ON, Canada
e-mail: hariharan.iyer@lhsc.on.ca

N. Krishnan (✉)
Institute of Cardio Metabolic Medicine and The Centre for Healthcare and Communities, University Hospitals Coventry and Warwickshire NHS Trust, Coventry, UK

The Centre for Healthcare and Communities, Coventry University, Coventry, UK
e-mail: Nithya.krishnan@uhcw.nhs.uk

G. Abraham et al. (eds.), *Immunological Challenges in Organ Transplantation*,
https://doi.org/10.1007/978-981-95-1533-2_10

In spite of a thorough and meticulous evaluation for the cause of allograft failure, sometimes no cause is established and hence, managing such patients remains a challenge in terms of both evaluation and management.

Case Report

A 28-year-old gentleman presented with nephrotic syndrome in 2002 and was diagnosed to have focal segmental glomerulosclerosis (FSGS) (NOS—not otherwise specified variant of FSGS)) on biopsy. He was treated with steroids, cyclosporine, and ACE-inhibitors. However, he did not respond well to treatment and progressed to end stage renal failure within 4 years. After a brief period of undergoing hemodialysis in 2006, he received his first renal transplant in 2007 at our center. It was a live related blood group compatible renal transplantation from his mother with a HLA mm of 1 (A),1 (B),1 (DR). The surgery was uneventful and he had good urine output immediately post-transplant with a urine protein creatininc ratio (PCR) of 2169 mg/mmol. However, his urine output dropped slowly over the next few days and he became anuric on the third day post-transplantation. An allograft biopsy was performed which revealed no glomerulitis or peritubular capillaritis. However, an arteriole revealed luminal occlusion due to severe intimal arteritis, indicative of rejection.

Despite treatment with high dose steroids and anti-thymocyte globulin (ATG), the graft could not be salvaged and he had to unfortunately undergo allograft nephrectomy on post-operative day 10. The kidney was grossly edematous and congested with thrombosed artery and vein. Histopathology confirmed complete renal infarction. Donor specific antibody (DSA) testing for anti-HLA antibodies was negative.

He was transitioned back on hemodialysis. However, due to significant proteinuria and hypoalbuminemia with residual urine output, he continued to have anasarca. Hence, he underwent bilateral native nephrectomy in 2008 and continued on hemodialysis.

In view of the poor outcome of first transplant, he was thoroughly evaluated for the cause of immediate graft loss before being considered for a second renal transplant. His thrombophilia workup, including lupus anticoagulant as well as anticardiolipin antibodies, factor V laden mutation and prothrombin gene mutation, were negative.

While being worked up for his second transplant from a friend, anti-HLA donor specific antibodies were not detectable. Therefore, tests were done for non-HLA antibodies against known antigenic targets. Anti-endothelial antibodies were negative. Anti-angiotensin receptor antibodies were equivocal at 14.84 μ/mL, and repeat sample revealed titers at 12.81 μ/mL (negative).

He underwent his second renal transplantation from a live unrelated donor in 2009, which was blood group compatible with an HLA mm of 2 (A),2 (B),2 (DR). He was given Basiliximab induction and maintained on triple immunosuppression with Tacrolimus, Azathioprine, and prednisolone. He also received double filtration

plasmapheresis and intravenous immunoglobulin (IVIg) pre and post-transplantation as a precautionary measure (given the adverse outcome with the first allograft). On the second postoperative day, he remained anuric and underwent re-exploration for peri-transplant hematoma and an allograft biopsy. Biopsy showed mild peri-tubular capillaritis suggestive of antibody mediated rejection. There were no vessels seen on this biopsy. Treatment was continued with high dose steroids, along with double filtration plasmapheresis and IVIg. However, he remained hemodialysis dependent and ultimately underwent allograft nephrectomy by post-operative day 10.

Histopathological examination of the kidney confirmed severe vascular rejection as there was widespread severe intimal arteritis and microcirculation inflammation. Anti-HLA donor specific antibody and anti-endothelial antibodies were negative. However, anti-angiotensin receptor type 1 antibody levels were positive at 17.76 and 18.44 μ/mL on repeat testing. He was extensively evaluated for atypical HUS, including ADAMTS13 activity, ADAMTS13 antibody levels, complement levels, complement factor H and I, anti-complement H antibodies, and C3 nephritic factor, all of which were negative. Genetic testing for atypical HUS showed a heterogenous variant of CFH and CFI, which was clinically insignificant.

With the above evaluation and outcomes, the patient was counselled regarding the high risk of early graft loss, should he pursue another transplant. He was placed on the deceased donor waiting list, while receiving maintenance home hemodialysis. He received his third transplant in December 2017, from a deceased donor. This transplant was blood group compatible with a HLA mm of 1 (A), 1 (B), 1 (DR).

He received pre-transplant Rituximab 1 g one off, double filtration plasmapheresis and IVIg 100 mg/kg following DFPP. Standard induction with Tacrolimus, Mycophenolate, and Prednisolone was used. (Can we clarify the induction and maintenance immunosuppression protocol?). Daily plasmapheresis and IVIg was continued post-transplant. Post-perfusion biopsy did not reveal any features of rejection. On post-operative day 3, he remained hemodialysis dependant. He underwent an allograft biopsy which revealed T-cell mediated rejection (Banff 2A). This was treated with IV Methylprednisolone and alemtuzumab. In view of positive ATR-1 antibodies. No DSAs were detected. He was started on Losartan immediately after transplant. His allograft started to function, and he came off hemodialysis.

However, he developed a peri-transplant hematoma, which was evacuated surgically 14 days after transplantation. His immunosuppression was reduced slowly as he developed candidemia with fluid collection in the paracolic gutter and peri-transplant area. He was treated with caspofungin. He was discharged from the hospital with a creatinine of 150 μmol/L and urine PCR of 154.6 mg/mmol. Plasmapheresis and IVIg were continued twice weekly at discharge and slowly reduced in frequency.

During the follow-up period, he had recurrent urinary tract infections and an enlarging peri-graft collection after ureteric stent removal along with moderate hydroureteronephrosis. A nephrostomy and nephrostogram revealed a urinary leak and an antegrade ureteric stent was inserted. His creatinine remained stable at 200–220 μmol/L. In view of recurrent infections and leucopenia, mycophenolate

was stopped. A repeat nephrostogram showed a long ureteric stricture, which was managed by upsizing of ureteric stent. Plasmapheresis and IVIg were slowly reduced to once every fortnight.

In June 2018, his creatinine increased to 280 μmol/L and urine PCR was 341 mg/mmol. He underwent an allograft biopsy which showed interstitial fibrosis and tubular atrophy (IFTA) involving 50% of the core and no evidence of acute rejection (see Image 10.1a). Electron microscopy showed diffuse podocyte effacement in 40% of glomeruli (see Image 10.1b). In view of the possibility of recurrence of FSGS, the frequency of plasmapheresis and IVIg was increased to weekly. He continued to have urinary infections and the nephrostomy was removed, leaving a ureteric stent in situ. With this treatment, his creatinine stabilised at 250–300 μmol/L with a urine PCR of 55–80 mg/mmol, and the frequency of plasmapheresis/IVIg was reduced to once every 2 weeks.

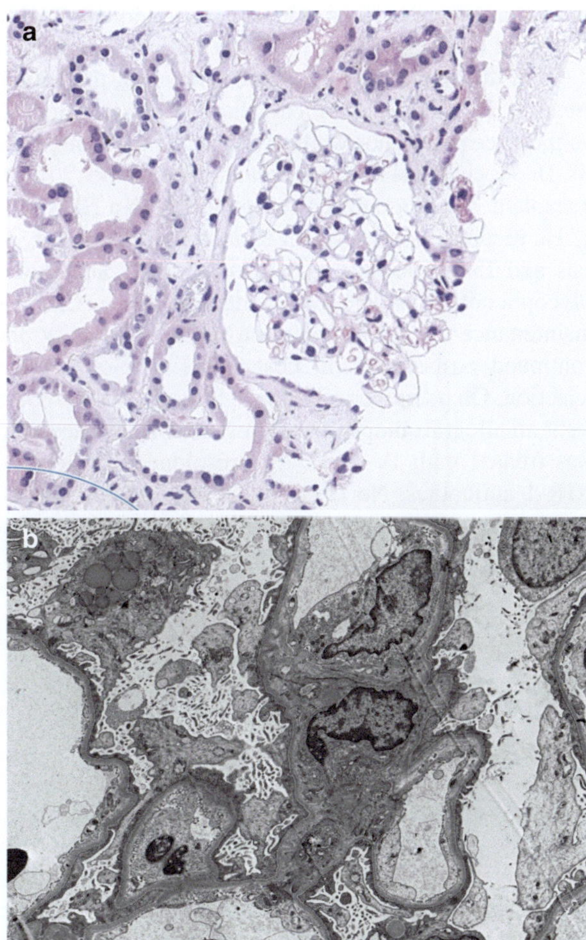

Image 10.1 Graft biopsy done in June 2018 with worsening of graft functions and proteinuria. (**a**) LM—Shows normal glomeruli. No evidence of rejection. (**b**) EM—Extensive podocyte effacement

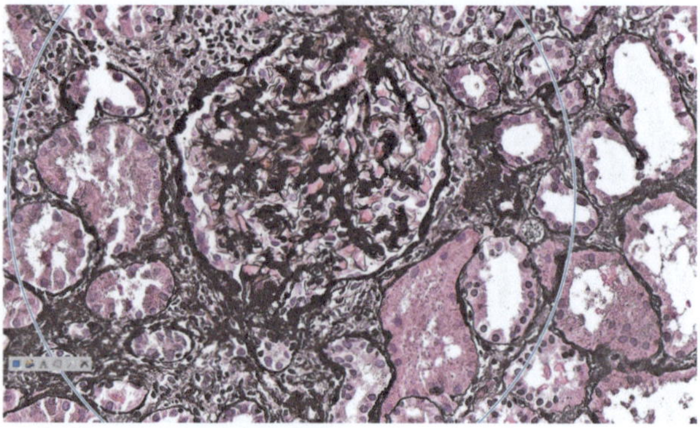

Image 10.2 Graft biopsy done in November 2018 with worsening graft function. (**a**) LM with Silver stain—T-Cell mediated rejection (Banff 1A). FSGS—tip lesion in glomerulus

He underwent a repeat transplant biopsy in November 2018 when his creatinine had risen to 350 μmol/L, and it showed T-cell mediated rejection (what Banff grade?). Light microscopy also showed two glomeruli with tip variant of FSGS lesions (see Image 10.2); however, overall podocyte effacement on electron microscopy had reduced. He was treated with high dose steroids and plasmapheresis and IVIg was tapered and stopped slowly over next few months. Due to persistent leucopenia, MMF was not restarted. His creatinine plateaued at 300–350 μmol/L.

In June 2019, he underwent distal transplant ureteric dilatation and ureteric reconstruction surgery at another center. Losartan was held postoperatively due an acute rise in creatinine. He developed another urinary infection that was treated appropriately. His creatinine rose to 600 μmol/L, and an allograft biopsy revealed T-cell mediated rejection (Banff t2, i1, v0) was treated with IV Methylprednisolone. Due to non-response to therapy, he was initiated on hemodialysis. Due to rapid worsening of creatinine and resistant hypertension, the possibility of a transplant renal artery stenosis was considered. This was confirmed by conventional angiogram and he underwent transplant renal artery angioplasty. Unfortunately, the allograft could not be salvaged and he remained haemodialysis dependent.

As he developed recurrent pain over the transplant site along with significant hematuria, allograft nephrectomy was performed in April 2020. Histopathology of the transplant kidney, revealed large areas of hemorrhagic necrosis, along with prominent peri-tubular capillaritis, tubulitis, and arteritis, all indicating an ongoing process of rejection (see Image 10.3).

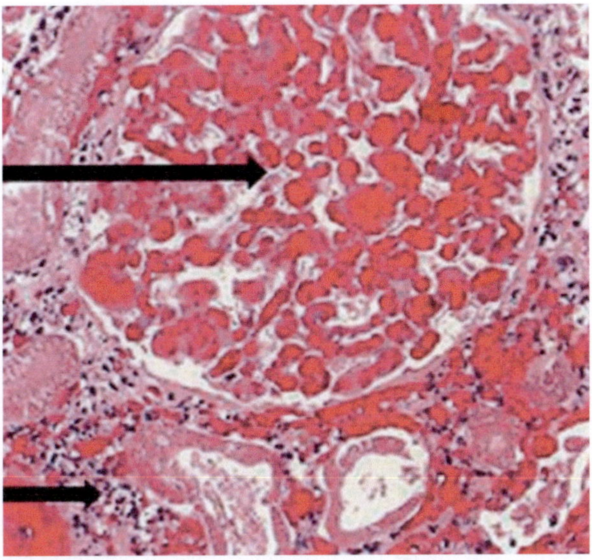

Image 10.3 Large areas of hemorrhagic necrosis, along with prominent peri-tubular capillaritis, tubulitis, and arteritis

Discussion

Renal transplant recipients not only have a better quality of life, but the mortality also falls after transplant from 12% in the first year of dialysis to 3% [5]. The above patient not only represents a challenge in terms of managing the graft, but also in decision making regarding offering a re-transplant as compared to maintenance hemodialysis.

Our patient lost his first graft with evidence of vascular rejection and thrombosis.

There are several studies suggesting an association between thrombophilia, vascular rejection, and allograft loss [4]. In a study of 97 patients by Heidenreich et al., the thrombophilic group was found to have higher rates of vascular rejection [6]. Works by Ekberg et al. and Fischereder et al. have linked vascular rejection with the presence of factor V laden mutation and G20210 mutation of the prothrombin gene respectively [7, 8].

In our patient, in view of negative anti-HLA DSAs and histopathology showing completely thrombosed vessels after allograft nephrectomy, a complete thrombophilia workup was carried out prior to second transplant, which was negative. The hematology service recommended routine heparin prophylaxis in the peri-operative period for his next transplant. However, he lost his second allograft as well with severe vascular rejection in the absence of anti-HLA DSAs. Antibody mediated rejection can occur in the absence of donor specific antibodies and has been explained by adsorption of DSAs within the allograft [9]. It may, however, be mediated by antibodies against various other cryptic antigenic targets [10] such as major-histocompatibility complex class I-related chain A (MIC-A), collagen V, killer immunoglobulin-like receptors (KIRs), human neutrophil antibody 3a (HNA-3a), and the activating antibodies targeting the angiotensin II type 1 receptor (AT_1R). Dragun et al. first demonstrated AT_1R antibodies (AT_1R -Ab) in a cohort of transplant recipients with severe

hypertension and vascular rejection [11].. They also suggested that AT_1R antibodies have structural similarities with anti-endothelial-cell antibodies. Our patient was tested for non-HLA antibodies- anti-endothelial antibodies and AT_1R -Ab.

Angiotensin II type 1 receptor antibody levels of >17 μ/mL (ELISA assay) have been associated with higher risk of rejection [12–14]. While our patient's anti-endothelial cell antibodies were negative, the AT_1R-Ab levels were positive before and after the second transplant. A correlation between AT_1R antibodies and recurrence of FSGS/de novo FSGS post-transplant has also been previously reported [15].

In our case, the differential diagnoses for immediate allograft failure were severe non-HLA antibody mediated vascular rejection and early recurrence of FSGS. Considering that histopathological analysis of the renal tissue on both occasions revealed features of vascular rejection with a persistently undetectable anti-HLA DSAs, the possibility of non-HLA antibody mediated rejection was strongly entertained. Nephrotic range proteinuria with hypoalbuminemia immediately after the first transplant could point towards recurrence of FSGS, which can occur in 30–50% of patients [16–18]. Podocyte effacement can be seen on electron microscopy even immediately after reperfusion [19]. A similar case with AT_1R-Ab, vascular rejection, and FSGS recurrence was reported by Alachkar et al. and was treated successfully with plasmapheresis, IVIg, Rituximab, and Losartan, along with the regular maintenance immunosuppression. Our patient was offered similar treatment at the time of his third transplant and was discharged with stable allograft function and mild proteinuria. However, his graft function worsened over time with multiple urological complications, recurrent UTIs and septicaemia, requiring reduction in immunosuppression. The patient developed worsening proteinuria with podocyte effacement, which later resolved on subsequent biopsies after treatment with plasmapheresis and IVIg. In light of significant IFTA, a creatinine of 350 μmol/L was likely to be his new baseline. This scarring was attributed to repeated and multiple insults to the allograft. This injury was further aggravated after stopping Losartan, which ultimately led to graft failure due to severe rejection (as evidenced by histopathologic examination of allograft nephrectomy).

Conclusion

Recurrence of FSGS and antibody mediated rejection are two possible causes of immediate graft loss after renal transplantation. The absence of anti-HLA DSAs should prompt a search for non-HLA antibodies such as activating antibodies against angiotensin II type I receptor (AT_1R-Ab). Co-existence of anti-HLA antibodies and AT_1R antibodies has also been reported (need a reference for this). Graft outcome and survival ultimately depends on the complex interplay between other factors associated with transplantation such as infection, septicemia, mechanical issues, and surgical complications.

Acknowledgements Our sincere thanks to Dr. Kishore Gopalakrishnan, Consultant Histopathologist, University Hospitals Coventry & Warwickshire, NHS Trust for providing the photographic images of the pathology slides.

References

1. Hamed MO, Chen Y, Pasea L, Watson CJ, Torpey N, Bradley JA, et al. Early graft loss after kidney transplantation: risk factors and consequences: early kidney graft loss. Am J Transplant. 2015;15(6):1632–43.
2. Puttarajappa C, Shapiro R, Tan HP. Antibody-mediated rejection in kidney transplantation: a review. J Transp Secur. 2012;2012:1–9.
3. Kazory A, Ducloux D. Acquired hypercoagulable state in renal transplant recipients. Thromb Haemost. 2004;91(4):646–54.
4. Irish A. Hypercoagulability in renal transplant recipients: identifying patients at risk of renal allograft thrombosis and evaluating strategies for prevention. Am J Cardiovasc Drugs. 2004;4(3):139–49.
5. McDonald SP. Australia and New Zealand dialysis and transplant registry. Kidney Int Suppl. 2015;5(1):39–44.
6. Heidenreich S, Dercken C, August C, Koch HG, Nowak-Göttl U. High rate of acute rejections in renal allograft recipients with thrombophilic risk factors. J Am Soc Nephrol. 1998;9(7):1309–13.
7. Ekberg H, Svensson PJ, Simanaitis M, Dahlbäck B. Factor V R506Q mutation (activated protein C resistance) is an additional risk factor for early renal graft loss associated with acute vascular rejection. Transplantation. 2000;69(8):1577–81.
8. Increased rate of renal transplant failure in patients with the G20210A mutation of the prothrombin gene. American Journal of Kidney Diseases. https://www.ajkd.org/article/S0272-6386(01)03877-X/abstract.
9. Detection of donor-specific anti-HLA antibodies with flow cytometry in eluates and sera from renal transplant recipients with chronic allograft nephropathy. https://pubmed.ncbi.nlm.nih.gov/12883199/.
10. Dragun D, Philippe A, Catar R. Role of non-HLA antibodies in organ transplantation. Curr Opin Organ Transplant. 2012;17(4):440–5.
11. Dragun D, Fritsche L, Kintscher U, Eckert D, Schönemann C, Neumayer H-H. Angiotensin II type 1–receptor activating antibodies in renal-allograft rejection. N Engl J Med. 2005;352:558–69.
12. Pearl MH, Zhang Q, Palma Diaz MF, Grotts J, Rossetti M, Elashoff D, et al. Angiotensin II type 1 receptor antibodies are associated with inflammatory cytokines and poor clinical outcomes in pediatric kidney transplantation. Kidney Int. 2018;93(1):260–9.
13. Pinelli DF, Friedewald JJ, Haarberg KMK, Radhakrishnan SL, Zitzner JR, Hanshew WE, et al. Assessing the potential of angiotensin II type 1 receptor and donor specific anti-endothelial cell antibodies to predict long-term kidney graft outcome. Hum Immunol. 2017;78(5–6):421–7.
14. Philogene MC, Bagnasco S, Kraus ES, Montgomery RA, Dragun D, Leffell MS, et al. Anti-angiotensin II type 1 receptor and anti-endothelial cell antibodies: a cross-sectional analysis of pathological findings in allograft biopsies. Transplantation. 2017;101(3):608–15.
15. Alachkar N, Gupta G, Montgomery RA. Angiotensin antibodies and focal segmental Glomerulosclerosis. N Engl J Med. 2013;368(10):971–3.
16. Rudnicki M. FSGS recurrence in adults after renal transplantation. Biomed Res Int. 2016;2016:1–7.
17. Sriprasad S. Ablative therapy in 2016 for small renal masses. Urol Nephrol Open Access J. 2017;4:6–10.
18. Agrawal V, Prasad N, Singh P. Primary focal segmental glomerulosclerosis recurring rapidly as collapsing glomerulopathy in a renal allograft recipient. Saudi J Kidney Dis Transplant. 2017;28(3):621.
19. Chang J, Pardo V, Sageshima J, Chen L, Tsai H-L, Reiser J, et al. Podocyte foot process effacement in post-reperfusion allograft biopsies correlates with early recurrence of proteinuria in focal segmental glomerulosclerosis. Transplantation. 2012;93(12):1238–44.

A Persistent CDC Positive Cross-Match in a Patient Awaiting Transplant: Variations of the CDC and Comparisons Across Platforms

11

Shruti Tapiawala, Dolly Daniel, Suchita Jogale, Edwin Fernado, and Sree Bhushan Raju

Case Capsule

A 45-year-old male with ESRD secondary to lupus nephritis underwent first renal transplant with mother being his donor. He lost the allograft to ABMR after 10 years of transplantation in the COVID era. A second transplant was contemplated. In view of absence of any living donor, he was listed for deceased donor kidney transplantation. Despite being listed for over 3 years, he had no suitable offers due to persistent CDC positivity against multiple potential donors.

Introduction

Testing for anti-human leukocyte antigen (anti-HLA) antibodies constitutes a cornerstone of the pre-transplant immunological evaluation. This critical step helps assess the immunological compatibility between the donor and recipient, thereby

S. Tapiawala (✉)
Institute of Renal Sciences, Gleneagles Hospitals, Mumbai, Maharashtra, India

D. Daniel
Transfusion Medicine and Immunohaematology, Christian Medical College, Vellore, Tamil Nadu, India
e-mail: dollyd@cmcvellore.ac.in

S. Jogale
ATDI Lab Private Limited, Mumbai, Maharashtra, India

E. Fernado
Stanley Hospital, Chennai, Tamil Nadu, India

S. B. Raju
Department of Nephrology, Nizam's Institute of Medical Sciences, Hyderabad, Telangana, India

G. Abraham et al. (eds.), *Immunological Challenges in Organ Transplantation*, https://doi.org/10.1007/978-981-95-1533-2_11

significantly influencing transplant outcomes. The introduction of the complement-dependent cytotoxicity (CDC) crossmatch by Patel and Terasaki in their seminal 1969 publication marked a pivotal moment in the history of transplantation medicine. Their ground-breaking work established a standardized method for detecting donor-specific antibodies in recipients, fundamentally re-shaping the approach to graft selection and improving graft survival rates [1].

Over the ensuing decades, advancements in immunology and laboratory technology have driven significant evolution in this field. While the CDC crossmatch remains historically important, it has been largely supplemented—and in some cases, replaced—by more sophisticated, highly sensitive, and specific assays. Techniques such as flow cytometry crossmatch, solid-phase assays including enzyme-linked immunosorbent assays (ELISA), and Luminex-based single antigen bead assays have enabled more precise detection and characterization of anti-HLA antibodies. These innovations have not only improved diagnostic accuracy but also enhanced risk stratification, allowing for tailored immunosuppressive regimens and better prediction of potential antibody-mediated rejection. As a result, modern transplant immunology now offers a more nuanced and comprehensive assessment of alloimmune risk, ultimately contributing to improved graft longevity and patient survival. This chapter shall provide detailed description of the CDC crossmatch and comparisons with other technologies available to us to evaluate and troubleshoot a deceased or a living donor transplant [2, 3].

The CDC Crossmatch Principle

The Complement-Dependent Cytotoxicity (CDC) assay represents one of the earliest and most foundational techniques used in transplant immunology for assessing donor-specific antibodies (DSAs) and determining the compatibility between donor and recipient. This assay operates on the principle of complement-mediated cell lysis, a natural immune mechanism wherein antigen-antibody complexes on the surface of target cells activate the classical complement pathway, ultimately leading to the formation of a membrane attack complex (MAC) and subsequent cell death [4, 5].

When employed in the context of a crossmatch assay, the CDC test is typically performed using donor-derived lymphocytes, which are incubated with recipient serum on a Terasaki tray, a specialized U-bottom microplate designed for such reactions. The lymphocyte population may include unseparated total lymphocytes (T and B cells) or may be separated into isolated T lymphocytes, which predominantly express HLA class I antigens, and B lymphocytes, which express both class I and class II antigens. This separation allows for more nuanced detection of class-specific antibodies [6].

Following the initial incubation of cells and serum, rabbit-derived complement is added to the reaction wells. This step is crucial, as it facilitates the formation of the MAC in the presence of pre-existing antibodies. After the addition of complement, eosin dye and formalin are introduced. Eosin is a vital dye that penetrates only those

cells whose membranes have been compromised by complement-mediated lysis. Viable cells exclude the dye and appear refractile under an inverted phase-contrast microscope, while non-viable (lysed) cells take up the dye and appear darkly stained. The result is interpreted by assessing the percentage of dead cells, with readings typically performed manually under microscopy using the principle of dye exclusion.

Other methods have utilized fluorescent dyes as well for this purpose. However, the traditional CDC assay—often referred to by various names including the Standard Crossmatch, NIH Crossmatch, Basic Crossmatch, or No-Wash Crossmatch—remains widely used in many laboratories due to its ability to detect both IgG and IgM anti-HLA antibodies, particularly those that are complement-fixing and thus clinically relevant in hyperacute rejection [6].

Despite its clinical utility, the CDC crossmatch has certain limitations. It is relatively less sensitive compared to modern solid-phase assays, and results can be influenced by non-HLA antibodies, serum quality, and the viability of donor cells. Nevertheless, it remains a valuable tool for identifying high-risk donor-recipient pairs and is still considered a gold standard in many transplant centers when evaluating the risk of immediate graft loss due to antibody-mediated injury [7]. See Fig. 11.1 for CDXM procedure.

Figure 11.2 illustrates the morphological differences between live and dead lymphocytes under phase-contrast microscopy.

To enhance the sensitivity of this method, many modifications and alterations were done to this method.

The CDC crossmatch detect preformed donor-specific antibodies (DSAs) in a transplant recipient's serum. These antibodies may be of the immunoglobulin M (IgM) or immunoglobulin G (IgG) isotype. Of these, IgG antibodies are considered clinically significant, as they are associated with antibody-mediated rejection and graft loss [8]. In contrast, IgM antibodies, although occasionally detected, are typically transient and not strongly implicated in transplant rejection [9].

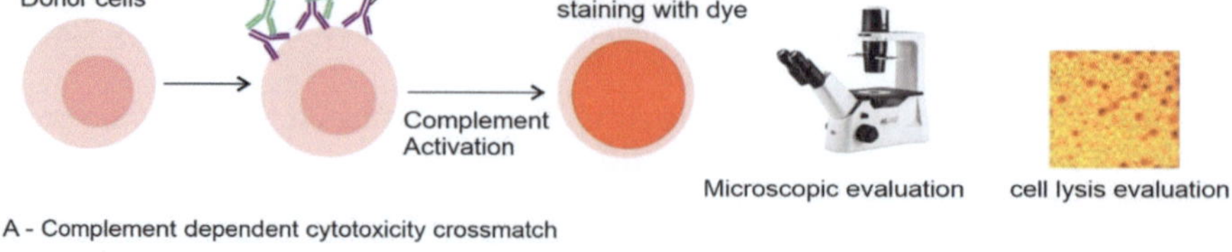

Fig. 11.1 Illustration of CDC crossmatch method

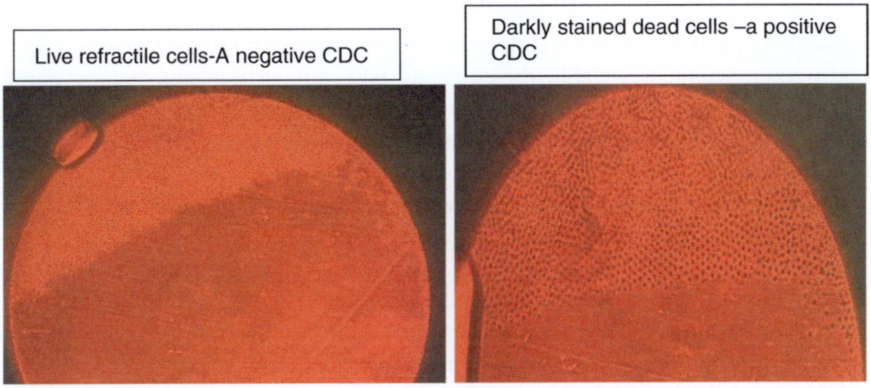

Fig. 11.2 Morphological differences between live and dead lymphocytes under phase-contrast microscopy

DTT Treatment to Differentiate Antibody Isotypes

To distinguish between IgM and IgG activity, recipient serum may be treated with 1,4-dithiothreitol (DTT), a reducing agent that disrupts the disulfide bonds critical to IgM's pentameric structure [10]. Following DTT treatment, IgM antibodies are rendered non-functional, whereas IgG antibodies remain intact. Thus, the loss of cytotoxicity in the DTT-treated serum suggests the presence of IgM antibodies, which are generally not considered clinically actionable. Conversely, persistent cytotoxicity after DTT treatment implicates IgG antibodies, necessitating further evaluation.

T and B Cell Crossmatch: Identifying HLA Specificity

A refined interpretation of CDC crossmatch results involves separating donor lymphocytes into T and B cell subsets:

- T cells express HLA class I antigens.
- B cells express both HLA class I and class II antigens.

By analyzing the pattern of positivity:

- Positive T and B cell crossmatch indicates the presence of anti-HLA class I antibodies.
- Isolated B cell positivity, with a negative T cell crossmatch, suggests the presence of anti-HLA class II antibodies [6].

This differential pattern provides critical information for assessing immunologic risk prior to transplantation.

Avoiding the Prozone Effect: Importance of Serum Dilutions

To mitigate the risk of a false-negative CDC result, the test should be performed with serial serum dilutions. This strategy helps overcome the prozone phenomenon, where high antibody concentrations can paradoxically inhibit complement activation due to antibody excess [11]. In such cases, diluting the serum restores optimal antigen-antibody ratios, revealing true cytotoxic potential.

Auto-Crossmatch: Detecting Autoantibodies

An auto-crossmatch, wherein the recipient's serum is tested against their own lymphocytes, serves as an essential internal control. A positive auto-crossmatch suggests the presence of autoantibodies or non-HLA antibodies that could interfere with donor-specific crossmatch interpretation [12]. Identifying such interference is crucial in patients with autoimmune disorders or those sensitized through transfusions, prior transplants, or pregnancies.

Enhancements to the CDC Crossmatch: Improving Sensitivity and Diagnostic Yield

The CDC crossmatch, while historically foundational in transplant immunology, is recognized as the least sensitive among the various antibody detection platforms currently in use. Despite its specificity and clinical utility, particularly in detecting complement-fixing donor-specific antibodies (DSAs), its limitations have prompted the development of several modifications aimed at enhancing its sensitivity. These enhancements help to reduce the incidence of false-negative results and improve the detection of clinically significant antibodies that might otherwise go unnoticed with standard protocols.

Several modifications have been widely adopted in clinical laboratories to increase the sensitivity of the CDC assay. These enhancements target different aspects of the crossmatch procedure, including incubation time, temperature, antibody removal, and detection of non-complement-fixing antibodies.

Extended Incubation Protocol

One of the most straightforward and effective modifications involves prolonging the incubation period between recipient serum and donor lymphocytes. In the standard CDC protocol, the cells and serum are typically incubated together for 30 min at room temperature, followed by the addition of complement and a further 60-min incubation. In the extended incubation variation, the initial incubation period is doubled to 60 min, and after the addition of complement, the mixture is incubated for an additional 2 h.

This prolonged contact time allows for more robust antibody-antigen interactions, especially in cases of low-affinity antibodies or low antibody titers that may not produce detectable cytotoxicity in shorter incubation windows. This enhancement is particularly beneficial in sensitized patients where a heightened immune surveillance is critical [13].

Incubation at Physiological Temperature (37 °C)

Another well-documented modification involves performing the serum-cell incubation at 37 °C, which is the physiological body temperature. This adjustment mirrors in vivo conditions more closely and may improve the binding kinetics of antibodies to their corresponding HLA targets. Some low-affinity or temperature-sensitive antibodies, especially those of the IgG subclass, exhibit improved reactivity under physiological thermal conditions.

Studies have shown that this temperature modification enhances assay sensitivity, particularly for detecting antibodies that may be weak or transiently reactive under standard room temperature protocols [14].

Triple Wash Technique

Prior to the addition of complement, a triple wash step can be employed to remove unbound serum antibodies and anti-complementary factors that could otherwise inhibit the CDC reaction. This technique involves:

- Flooding the wells containing lymphocytes and serum with phosphate-buffered saline (PBS) or similar washing buffer.
- Allowing the cells to settle either by gravity or by a gentle centrifugation step ("soft spin").
- "Flicking" the liquid out of the wells, a process by which the buffer is swiftly expelled to avoid disturbing the settled cells.

This wash cycle is repeated three times, hence the name "triple wash." By removing unbound immunoglobulins and complement-interfering substances, the method significantly increases the specificity and clarity of the assay results. However, it is important to note that the "flicking" step introduces potential variability into the assay. Differences in technique between technologists—such as the force or angle of flicking—can influence the residual buffer volume and, consequently, the assay outcome. As a result, this variation is recognized as a common source of inter-laboratory and intra-laboratory variability in CDC crossmatch results [15, 16].

Use of Anti-Human Globulin (AHG)

A further enhancement involves the addition of Anti-Human Globulin (AHG), also known as Coombs reagent, to the crossmatch assay. AHG functions by bridging low levels of bound IgG antibodies that may not activate complement efficiently. This makes it possible to detect non-complement-binding IgG antibodies, which can still be clinically relevant despite their inability to initiate the CDC reaction on their own.

The AHG-augmented CDC crossmatch is especially useful in cases where the patient may harbor low-affinity or non-complement-fixing antibodies that could escape detection by the conventional CDC method. By increasing the assay's ability to detect a broader spectrum of antibody specificities, this enhancement improves overall sensitivity and has been particularly valuable in highly sensitized transplant candidates [6, 17].

These modifications—extended incubation, physiological temperature incubation, triple washing, and AHG enhancement—collectively serve to increase the sensitivity and diagnostic accuracy of the CDC crossmatch. Although the CDC assay has been largely supplanted by flow cytometry and solid-phase assays in many high-throughput laboratories, it remains a vital tool in resource-limited settings and in specific clinical scenarios. Understanding these enhancements ensures more precise immunologic risk assessment and improves transplant outcomes.

Crossmatch with Current and Peak Sera

According to the American Society for Histocompatibility and Immunogenetics (ASHI) guidelines, CDC crossmatch is typically performed using serum dilutions ranging from 1:2 up to 1:16. Dilution helps to reduce non-specific reactivity and unmask high-affinity antibodies that may be obscured in the presence of high serum immunoglobulin concentrations.

A well-documented immunological artefact encountered in CDC-XM is the Prozone phenomenon. This occurs when high concentrations of antibodies saturate available antigen-binding sites without allowing optimal lattice formation required for complement activation. As a result, a strongly sensitized patient may show a false-negative result in neat (undiluted) serum, whereas a diluted serum reveals strong cytotoxicity.

Historic and Current Sera

To fully understand a recipient's immunological risk, it is important to perform CDC crossmatch using multiple serum time points: peak, historic, and current.

Peak Sera

Peak sera refers to the serum sample containing the highest recorded titer of DSAs prior to transplant. This represents the maximum immunological threat and is

invaluable for predicting possible post-transplant complications even if current antibody levels are lower due to immunosuppressive treatment or natural decline [7].

Historic Sera

Historic sera are samples collected at various time points prior to transplant, which can help document the trajectory of immune sensitization. Even if DSAs are undetectable in the current serum, prior sensitization can result in anamnestic (memory) immune responses, which may culminate in antibody-mediated rejection (ABMR) [18].

Current Sera

Current serum reflects the immunologic status at the time of transplantation. A positive CDC crossmatch using current serum indicates circulating DSAs that pose a risk for hyperacute or accelerated rejection, necessitating the reconsideration of the donor-recipient pairing.

Crossmatching with all three types of sera enables a comprehensive risk assessment. This multi-sera approach is especially vital for highly sensitized patients, and for those previously exposed to alloantigens through transfusions, pregnancies, or previous transplants.

Crossmatch with Titers

Titration involves serially diluting patient serum and determining the highest dilution at which cytotoxicity is still detectable. This dilution correlates with the strength of the DSAs and offers an objective measure of antibody burden. This method can be used to monitor the strength of DSAs during desensitization process [18].

Clinical Application in Desensitization Protocols

In the setting of desensitization therapy, DSA titration serves as a quantitative biomarker for therapeutic efficacy. A reduction in DSA titer, culminating in a negative CDC crossmatch, suggests a successful desensitization, allowing the transplant team to proceed with the transplantation with reduced risk of acute rejection.

Scoring of CDC

The scoring is based on the percentage of dead cells seen in a well of a Terasaki tray (see Table 11.1). The ASHI grading system is as below [6].

The CDC crossmatch though set aside in some centers still retains its relevance in many parts of the world and a positive crossmatch indicating the presence of IgG anti HLA antibodies remains a contraindication for transplant. It is always prudent to not interpret or use a CDC in isolation, but to use it in combination with a flow-cytometric/solid phase antibody testing alongside.

Table 11.1 Scoring of dead cells and interpretation in CDC crossmatch

Score	% dead cells	Interpretation
1	0–10	Negative
2	11–20	Doubtful negative
4	21–50	Weak positive
6	51–80	Positive
8	81–100	Strongly positive
0		Not interpretable

Fallacies of a CDC Crossmatch

Despite its historical significance in transplant immunology, the CDC crossmatch is limited by several well-recognized fallacies that compromise its reliability and interpretability. One of the primary limitations is its relatively low sensitivity, particularly in detecting low-level donor-specific antibodies (DSA) that may still be clinically relevant. Unlike solid-phase assays, such as Luminex single-antigen bead testing, CDC crossmatch requires a threshold level of antibody to activate the classical complement cascade and produce observable cell lysis. As a result, weak or non-complement-binding antibodies—some of which may still pose a risk for antibody-mediated rejection (AMR)—can go undetected [19]. Furthermore, CDC crossmatch is vulnerable to false positives, often due to the presence of non-HLA antibodies, autoantibodies, or high levels of IgM, especially in patients with autoimmune diseases or infections. These antibodies may bind to donor lymphocytes and activate complement without being clinically relevant to graft outcomes [20]. Additionally, inter-laboratory variability in techniques such as cell viability, complement source, and incubation conditions can further confound results, making standardization challenging [21]. The use of pronase treatment to remove Fc receptors and reduce background reactivity has been employed in some centers, yet this too introduces variability and potential cell damage. Finally, the CDC crossmatch does not provide antigen specificity, which limits its utility in guiding desensitization protocols or monitoring alloimmune responses over time [22]. Given these fallacies, while the CDC crossmatch remains a useful tool in certain clinical scenarios—particularly in highly sensitized patients or when used with titrations—its role has largely been supplanted by more sensitive and specific assays in modern transplant immunology.

In contemporary transplant immunology, most laboratories have adopted a multifaceted approach to assess immunologic compatibility between donors and recipients. This involves the integration of both cellular and solid-phase assays, which together aid in the selection of appropriate donors and the stratification of immunologic risk in patients. The use of multiple assay types has become essential due to the complexity and variability of individual immune responses, as well as the need for increasingly precise evaluations in transplant settings.

Among the most widely utilized platforms for HLA antibody testing are the complement-dependent cytotoxicity (CDC) assay, enzyme-linked immunosorbent assay (ELISA), flow cytometry, and the Luminex platform. Each of these methods

has its own unique advantages, limitations, and level of sensitivity. The CDC assay, a traditional method, is considered the least sensitive and primarily detects complement-fixing antibodies. Despite its limitations, it remains in use for its ability to provide functional assessment of antibody-mediated cytotoxicity. ELISA and flow cytometry offer intermediate levels of sensitivity, with the latter allowing for enhanced resolution in detecting both complement and non-complement binding antibodies. The most sensitive platform currently available is the Luminex-based assay, which employs bead-based multiplex technology to detect HLA antibodies with high specificity and minimal sample volume (see Fig. 11.2).

Regardless of the testing platform employed, the assay formats themselves may vary significantly. Common formats include the antibody screening test, which provides a general assessment for the presence of HLA antibodies in the recipient; the panel reactive antibody (PRA) test, which estimates the percentage of the general donor pool to which a patient may react immunologically; HLA antibody identification called Single Antigen Bead assay to determine specificity of HLA antibodies present in serum and the crossmatch test, which directly evaluates compatibility between a specific donor and recipient. Crossmatching can be further categorized into physical crossmatching—such as CDC or flow-cytometric crossmatch—and virtual crossmatching, which is a predictive analysis based on known antibody and donor antigen profiles.

The nuances and specific information conveyed by each of these testing methods and formats are crucial for clinical decision-making. A comparative overview of these tests and what they reveal about donor-recipient compatibility is provided in Fig. 11.3.

Comparisons with Alternative Crossmatch and Antibody Detection Platforms

The CDC crossmatch has long served as the gold standard for detecting preformed DSAs in renal transplantation, primarily due to its historical role in reducing hyperacute rejection. However, advances in immunological assays have brought about more sensitive and specific methods that have challenged the pre-eminence of

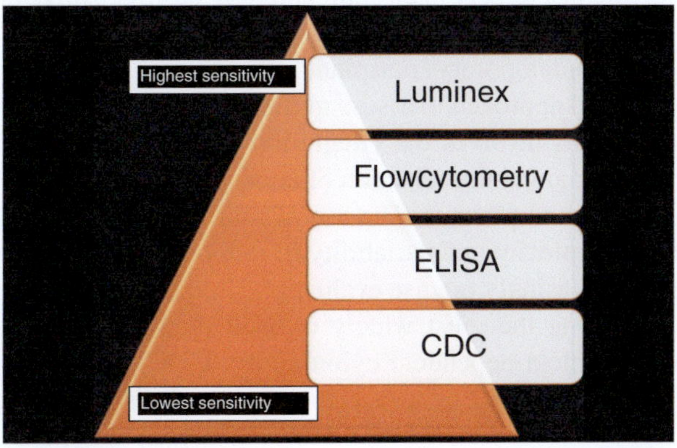

Fig. 11.3 HLA antibody testing platforms and their sensitivity

CDC. While CDC remains a valuable screening tool, particularly for detecting high-titer complement-fixing antibodies, it lacks sensitivity in detecting low-level or non-complement-fixing DSAs, which may still be clinically significant. This aligns with prior studies demonstrating that CDC crossmatch can fail to detect class II antibodies or low-avidity antibodies that may contribute to chronic antibody-mediated rejection (AMR) [19].

Flow cytometric crossmatch provides a significantly higher sensitivity compared to CDC due to its ability to detect both IgG and non-complement-fixing antibodies. It is particularly advantageous in identifying antibodies against HLA class I and class II antigens, thereby enabling better stratification of immunologic risk. A study by Tambur et al. found that positive FCXM in the presence of a negative CDC crossmatch was associated with increased incidence of early AMR, underscoring the importance of FCXM in pre-transplant immunologic risk assessment [23].

Solid-phase assays, particularly SAB assays, offer the highest sensitivity and specificity, capable of detecting DSAs at very low levels and providing antibody specificity. This facilitates the identification of unacceptable antigens and helps guide donor selection in sensitized patients. However, these assays do not assess the functional capacity of the antibodies, such as complement activation, which can be clinically relevant. To address this, the C1q and C3d binding assays have been introduced to assess the complement-fixing ability of DSAs, providing additional prognostic information [24]. (Fig. 11.4).

Virtual crossmatching, which relies on the integration of recipient antibody profiles and donor HLA typing data without a physical assay, allows for rapid organ allocation and has shown comparable predictive value to traditional crossmatches in low-risk patients [25]. However, its accuracy depends on the quality and completeness of both antibody and HLA typing data.

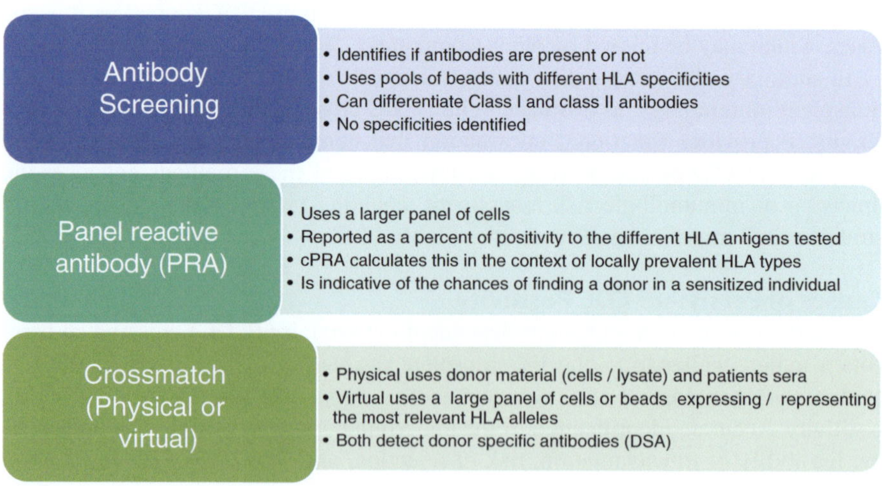

Fig. 11.4 Details of each assay

There are many studies that have compared CDC crossmatch with flow cytometric crossmatch (FCXM), solid-phase assays (SPA) including single antigen bead (SAB) testing, and Luminex-based virtual crossmatching to evaluate their clinical utility, sensitivity, specificity, and impact on transplant outcomes.

Study done by J.P. Peräsaari et al., studied about 83 crossmatches performed against DSA, the best performance was observed with FCXM (83%). With LXM and CDCXM, the DSA detection rates were 71% and 66%, respectively. FCXM, the best accuracy was achieved with both 1000 and 5000 MFI and for CDCXM with 5000 MFI [26].

A study by Nuchjarnun Choktaweesak et al. revealed that, the crossmatch results of FCXM were compared with the standard CDC-AHG-XM. The sensitivity and specificity of FCXM in T and B cells were 100% and 100% and 98.86% and 99.41%, respectively [27].

In a retrospective study done by Prashant Pandey et al., the correlation between donor-specific antibody (DSA) strength measured by solid-phase assay and crossmatch outcomes was evaluated in 102 prospective renal transplant recipients. Stratification by L-SAB-DSA-MFI values revealed that IgG-T-cell-FC-XM positivity increased with higher MFI levels, while T-cell-CDC-AHG-XM remained largely negative until MFI values exceeded 5000. Notably, all cases with L-SAB-DSA-MFI >5000 were positive in both FCXM and CDCXM, underscoring a strong association between high antibody strength and cytotoxic crossmatch positivity. Statistically significant correlations were found for L-SAB-DSA-MFI >2215 with FCXM positivity and >4689 with CDCXM positivity ($p < 0.001$), highlighting the predictive value of MFI thresholds in assessing the risk of positive crossmatch outcomes [28].

Despite the superior sensitivity of newer assays, the CDC crossmatch remains relevant in certain clinical scenarios. It is particularly important in identifying highly sensitized recipients with high-titer cytotoxic antibodies, in whom a positive CDC crossmatch remains a strong contraindication to transplantation in many centers. Additionally, CDC detects IgM antibodies and non-HLA antibodies in some cases, which may be missed by other techniques [29].

In summary, the comparative analysis supports a multimodal approach to pre-transplant immunologic assessment. While CDC crossmatch has limitations in sensitivity, it provides functional information that complements the high-resolution specificity of FCXM and SAB assays. Integration of these methods enhances the precision of immunologic risk assessment, enabling tailored immunosuppression strategies and improved graft outcomes (Table 11.2).

Causes of Persistent CDC Positivity

A persistent positive complement-dependent cytotoxicity (CDC) crossmatch presents a major challenge in the pre-transplant evaluation of sensitized candidates. Multiple immunological and technical factors can contribute to sustained CDC positivity. Accurate identification of the underlying cause is crucial, as it influences the feasibility of transplantation, risk of rejection, and selection of desensitization strategies.

Table 11.2 Comparison of all HLA crossmatch methods

Feature	CDC	Flow Cytometric Crossmatch (FCXM)	Solid Phase Assays (e.g., Luminex Single Antigen Bead—SAB)
Principle	Complement-mediated cytotoxicity	Detection of antibody binding to donor cells via fluorescence	Antibody binding to HLA-coated beads, detected by fluorescence
Sensitivity	Low	Moderate to high	High to very high
Specificity	Moderate	High	Very high (with well-characterized beads)
Detection of non-complement binding abs	No	Yes	Yes
Quantification	Semi-quantitative	Quantitative	Quantitative (mean fluorescence intensity—MFI)
Detection of non-HLA antibodies	No	Limited (if cell-based expression)	Yes (when included on bead panels)
Clinical utility	Historical gold standard, useful for high-risk or sensitized patients	Detects low-level DSAs missed by CDC	Offers detailed antibody profiling; essential for risk stratification
Limitations	Subjective interpretation, low sensitivity, false positives (esp. B-cell CDC)	Requires viable donor cells, more technical	Potential for false positives, especially with low-level antibodies or denatured antigens
Complement activation information	Yes	No (unless modified)	Only if C1q/C3d assays are used
Time to result	Rapid	Moderate	Rapid (depending on platform)

Alloantibodies

Sensitization Events

The most common cause of persistent CDC positivity is the presence of high-titer, complement-fixing anti-human leukocyte antigen (HLA) alloantibodies. Sensitization events such as prior organ transplantation, multiple pregnancies, or blood transfusions can expose the recipient's immune system to non-self HLA antigens, triggering an alloimmune response [30]. Over time, repeated exposures lead to the development of a broad antibody repertoire, often characterized by high-avidity, complement-fixing IgG antibodies capable of activating the classical complement pathway and causing cytotoxicity in CDC assays.

Persistent Memory B Cell Activation

In sensitized individuals, long-lived plasma cells, maintained by continuous antigenic stimulation, can perpetuate the production of high-avidity alloantibodies even in the absence of recent immunological events [31]. These memory B cells and their progeny produce antibodies that not only bind HLA antigens with high affinity but also readily fix complement, maintaining persistent crossmatch positivity over extended periods.

Autoantibodies

Autoimmune Conditions

Systemic autoimmune diseases, such as systemic lupus erythematosus (SLE) or mixed connective tissue disease, can induce the formation of non-HLA-directed autoantibodies, including anti-endothelial cell antibodies (AECAs) and other autoreactive IgG species [32]. These autoantibodies may nonspecifically bind to donor lymphocytes, leading to complement activation and cytotoxicity in CDC assays, thereby mimicking HLA-specific sensitization.

Non-specific Positivity

Drugs used for desensitization treatment such asrituximab or other monoclonal antibodies (e.g., alemtuzumab) targeting lymphocyte antigens can result in CDC positivity. These antibodies may be present in serum and cause complement activation against donor B cells [33].

Technical or Non-specific Reactivity

Pan-Reactive and IgM Antibodies

Non-specific pan-reactivity, particularly mediated by polyreactive IgM antibodies, is another cause of persistent CDC positivity. Infections, immune dysregulation, or systemic inflammatory states may transiently expand populations of polyreactive B cells producing natural IgM antibodies capable of binding to multiple antigens with low specificity [34].

Artifactual Binding

Technical issues in donor cell preparation, such as the exposure of cryptic or denatured antigens, can also lead to artifactual binding by recipient antibodies. Improper handling or prolonged storage of cells can alter membrane integrity, unmasking epitopes that elicit non-specific antibody binding and false-positive CDC reactions [35].

Workup Required for Persistently Positive CDC Crossmatch

A systematic and comprehensive workup is essential to determine the cause of persistent CDC positivity and guide subsequent clinical management.

Confirm Antibody Specificity

- **Solid-Phase Assays:** Solid-phase immunoassays, particularly Luminex-based single antigen bead (SAB) platforms, are critical tools for identifying the presence and specificity of anti-HLA antibodies [19]. Detection of donor-specific antibodies (DSAs) confirms alloimmuneetiology and correlates with increased risk of hyperacute rejection.

Evaluate Crossmatch Type

- **T cell vs. B cell Reactivity:** T cell crossmatch positivity predominantly indicates anti-HLA class I antibodies, while isolated B cell positivity may suggest class II antibodies or non-specific binding.
- **DTT Treatment:** Dithiothreitol (DTT) treatment of sera can help differentiate IgM from IgG antibodies, as DTT disrupts IgM pentamers. Persistence of crossmatch positivity after DTT treatment points toward IgG-mediated reactivity.

Exclude Non-specificity

- Non-specific positivity in the complement-dependent cytotoxicity crossmatch (CDC-XM) should be further evaluated using more sensitive and specific assays, such as flow cytometriccrossmatch (FCXM) and solid-phase immunoassays, particularly the single-antigen bead (SAB) assay. If these supplemental tests do not identify donor-specific anti-HLA antibodies (DSAs), the CDC-XM positivity is considered non-specific and is typically not clinically relevant in the context of transplant eligibility. In such cases, the observed CDC reactivity does not reflect alloimmune sensitization and should not preclude proceeding with transplantation [33]

Autoimmune Screen

- **Autoantibody Panels:** Comprehensive autoimmune serologies, including antinuclear antibody (ANA) and extractable nuclear antigen (ENA) panels, should be performed. Detection of autoantibodies may support a diagnosis of an underlying autoimmune process contributing to crossmatch positivity.

Clinical Implications

Persistent CDC positivity, especially when mediated by donor-specific anti-HLA antibodies, portends a significantly increased risk of hyperacute and acute antibody-mediated rejection (AMR) [36]. Even in the absence of demonstrable DSAs, persistent CDC reactivity complicates candidate management and necessitates careful risk stratification.

Key Clinical Considerations

- **Waitlist Priority:** Candidates with persistent CDC positivity may face significant barriers to transplantation from standard donors. In highly sensitized populations, achieving a negative crossmatch without desensitization may be unattainable.
- **Desensitization Protocols:** Desensitization therapies, including plasmapheresis, intravenous immunoglobulin (IVIG), rituximab (anti-CD20 monoclonal antibody), bortezomib (proteasome inhibitor targeting plasma cells), or complement inhibition with agents such as eculizumab, may be employed to reduce antibody titers or their pathogenicity [37].
- **Acceptable Mismatch Strategies:** Strategies such as permissive mismatch allocation, based on detailed HLA epitope matching or sensitization history, can expand donor options for highly sensitized patients and facilitate successful transplantation despite persistent CDC positivity [38].

Case Capsule Discussion

His Luminex testing revealed broad class I and II sensitization, with multiple high-MFI DSAs. Treatment with plasmapheresis and IVIG reduced CDC reactivity, but positivity recurred within weeks.

Persistent CDC positivity in transplant candidates represents a complex immunological barrier, often due to high-risk DSAs, non-HLA antibodies, or autoantibodies. A comprehensive diagnostic and therapeutic approach is essential to navigate transplant eligibility and outcomes (see Figs. 11.5 and 11.6). Advances in immunogenetics, antibody characterization, desensitization with addition of newer biological (Imlifidase, clazakizumab) may expand options of treatment for these highly sensitized patients [39, 40].

We describe below three cases where the CDC has impacted on clinical decision making.

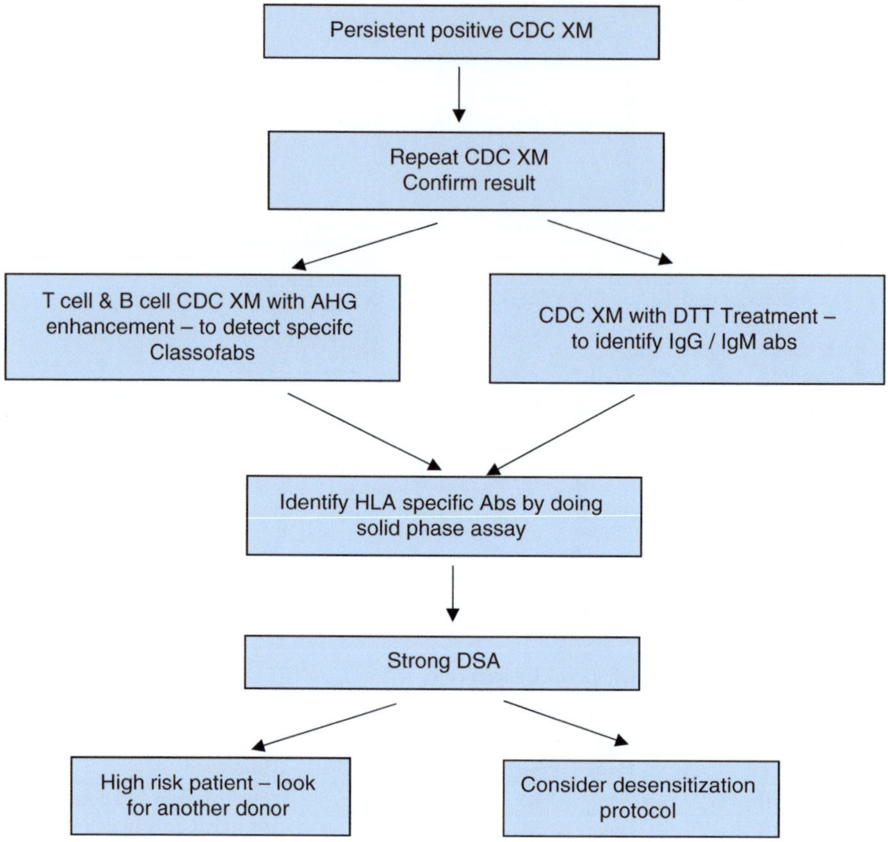

Fig. 11.5 Approach to navigate transplant eligibility and outcomes

Case 1

A 28-year-old gentleman with end stage renal disease (ESRD) was being worked up for a renal transplant with his mother as a potential donor. There was a history of two red cell transfusions 1 month prior. Pre-transplant immunological workup includes six locus high resolution HLA typing of patient and donor, a CDC and flow-cytometriccrossmatch, and a single antigen bead assay.

Immunology workup:

HLA typing (six locus high resolution): Haplomatch (Table 11.3)

Flow-cytometric crossmatch:

T cell crossmatch: Negative

B cell crossmatch: Negative

Single antigen bead assay: Negative (all mismatched alleles represented)

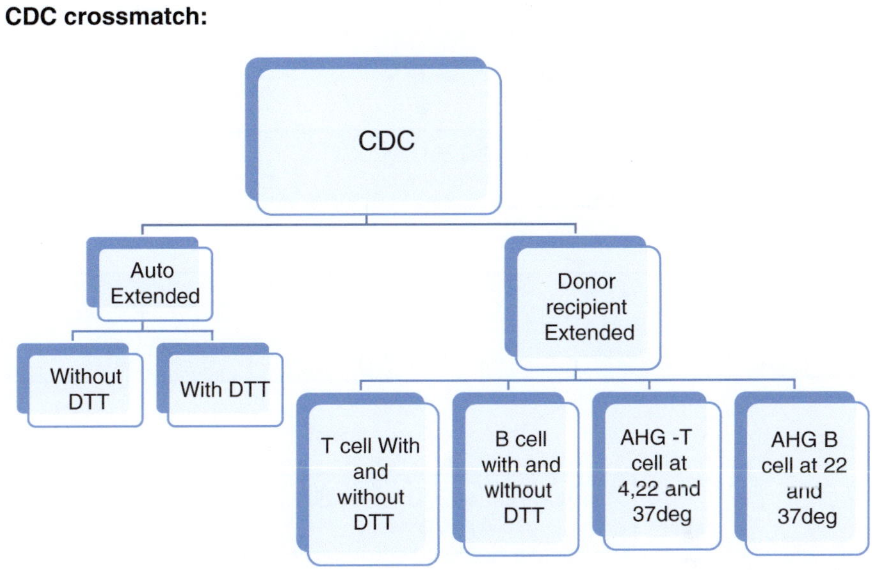

Fig. 11.6 Work-up done with CDC XM

Table 11.3 CDC Crossmatch (Total T and B cell)—Case 1

Sera dilutions	Auto crossmatch (%)	Auto with DTT	Donor recipient XM (%)	Donor recipient with DTT	Inference
Neat	10	Neg	40	Neg	Positive auto and donor recipient crossmatch with features suggestive of an IgM antibody
N/2	10	Neg	20	Neg	
N/4	10	Neg	10	Neg	

Question: Can this patient be taken up for transplant? Can the CDC positivity be labelled as non-specific?
How do we proceed with this patient's management?

A patient with a positive CDC crossmatch reflecting a pattern consistent with an IgM antibody should be advised to have a repeat crossmatch in 2 weeks to assess if class switch to IgG has occurred and whether a clinically significant anti HLA antibody has developed (see Table 11.4). As per protocol the repeat antibody profiling was performed on all three platforms.

Flow Cytometric Crossmatch:
T cell crossmatch: Positive—Ratio3.5 (>1.0—positive)
B cell crossmatch: Positive—Ratio 4.0 (>1.2—positive)

SAB assay:
Class I: DSA present

Table 11.4 Repeat CDC XM after 2 weeks results

Repeat testing	Auto crossmatch (%)	Auto with DTT	Donor recipient cx match (%)	Donor recipient with DTT (%)	Inference
Neat	10	Neg	40	20	Positive autocrossmatch
N/2	10	Neg	20	5	With features suggestive of an IgM antibody
N/4	10	Neg	10	5	Positive donor recipient crossmatch with features suggestive of an IgG antibody

Class II: DSA present
Antibody to mismatched allele B*20:02 present: MFI-7500
Antibody to mismatched allele DQB1*05:02 present: MFI-11000

Class switch of an IgM antibody to an IgG antibody is a well described phenomenon. This case highlights the caution one needs to have when an IgM antibody is identified on the donor recipient crossmatch. IgM antibodies can be detected on either the CDC or the flow cytometric crossmatch, the latter being possible if the secondary antibody binding to IgM antibodies is used. However this is often challenging as it is fraught with technical challenges, and the CDC continues to be a platform that detects both IgM and IgG antibodies and can alert us to this phenomenon.

This case highlights the importance of repeat testing after a minimum of 2 weeks. A positive CDC, flow cytometric crossmatch and DSA on the SAB indicates the presence of strong anti HLA antibodies and makes this donor unsuitable for this patient. Transplant across this barrier will almost certainly result in a hyperacute rejection.

The results of this patient put her into a high risk category, which requires a new donor or consideration for a paired kidney exchange.

Case 2

A 55-year-old lady with ESRD Ms. KLM, on dialysis, is being evaluated for a renal transplant with her husband Mr. SDA as a potential donor. She has had two pregnancies 28 and 25 years prior, but not received any transfusions. Her pre-transplant immunologic workup was as below (Table 11.5).

CDC Crossmatch: Results
Auto crossmatch: Negative
Donor recipient crossmatch (total T and B cell): Negative
Donor recipient T cell crossmatch: Negative

Table 11.5 High-resolution HLA typing results of recipient and donor

	A*	B*	C*	DRB1*	DQB1*	DPB1*
Ms. KLM (recipient)	01:01	35:03	04:01	04:01	01:01	01:01
	68:01	07:02	06:01	04:02	03:01	09:01
Mr. SDA (donor)	02:01	51:01	O6:01	15:01	05:01	26:01
	24:01	08:02	07:01	07:04	02:01	296:01

Donor recipient B cell crossmatch: Negative
AHG crossmatch:
T cell crossmatch: Positive—20% dead cells
B cell crossmatch: Positive—20% dead cells
Flow cytometric crossmatch:
T cell crossmatch: Positive—Ratio 2.3
B cell crossmatch: Positive—Ratio 2.5
Single antigen bead assay:
Class I: Positive.
DSA Present:
A*24:01—MFI 6000
B*08:02—MFI 4000
Class II: Negative
Questions:
How does one explain the variability in CDC positivity in its varied formats?
How do we proceed with management of this patient's potential transplant?

This patient has an anti HLA antibody that is donor specific, detected on all three platforms—CDC, Flow cytometry, and the Luminex SAB. However on the CDC it shows variable results with the extended incubation format being negative, but the AHG format being positive. It is well known that AHG eliminates low-affinity antibodies, amplifies the cytotoxicity reaction promoted by low titer complement-fixing antibodies, and detects non complement fixing antibodies as well, thus increasing the sensitivity of the CDC crossmatch.

As expected, the results of this patient conform to that found in literature. Previous studies have shown that there is a significant difference in antibody detection between the CDC with extended incubation and the solid phase assay, while this difference was not demonstrable when an AHG CDC was compared to the same solid phase assay (Saito et al.). As early as 1982, it was deciphered that the AHGG-CDC technique converts cytotoxicity-negative absorption-positive (CYNAP) HLA alloantibody to direct cytotoxic antibody. This is a consistent phenomenon, that enhances accuracy of antibody detection and specificity.

Clinically this again translates to a high risk transplant. While a hyperacute rejection is unlikely in this setting, acute rejection and compromise of graft function are highly likely. The alloimmunizing event is probably her two pregnancies. With her husband being the prospective donor, even post desensitization, she is likely to rebound with a strong anamnestic response. The best options for her may be an

alternate donor/paired kidney exchange or to be added to a deceased donor waiting list.

Case 3

A 35-year-old male, Mr. DAN, with ESRD is referred for a pre transplant immunological work up, as he had been found to be crossmatch positive elsewhere, with his mother Ms. FS who is the prospective donor. There is no history of any prior sensitising events such as transfusions and this is his first transplant. There is no history of any drugs.

HLA typing:

Six locus high resolution typing: Patient is a haplomatch with his mother

CDC crossmatch:

Auto crossmatch: Positive. After DTT treatment of serum—Negative (Table 11.6)

Flow cytometric crossmatch:

T cell: Negative

B cell: Negative

SAB assay:

Class I and Class II: Negative.

(In both the SAB and the flow cytometric crossmatch, the secondary antibody is an IgG labelled with Phycoerythrin)

As per guidelines and protocol, the tests were repeated after a period of 2–4 weeks.

Repeat testing after 3 weeks:

The results were identical on the CDC, Flow cytometric crossmatch, and the SAB

Questions:

Can this patient proceed for transplant with the same donor?

Does it increase his risk of rejection?

Table 11.6 Comprehensive CDC XM results

	Total T&B cell cxmatch (%)	T&B cxmatch with DTT	T cell Cxmatch 22 & 37 deg (%)	T cell cxmatch with DTT	B cell cxmatch 22 & 37 deg (%)	B cell Cxmatch with DTT	Inference
Neat serum	40	Neg	40	Neg	40	Neg	Positive donor recipient crossmatch with features suggestive of an IgM antibody
N/2	30	Neg	40	Neg	40	Neg	
N/4	20	Neg	30	Neg	40	Neg	

This patient has a persisting, fairly strong IgM antibody, demonstrated on the CDC platform. IgM antibodies have generally been considered clinically insignificant if they do not class switch. Initial studies in the 90s have shown no difference in outcome post-transplant with regards to both rejections and graft function.

However more recently there is data to suggest that post-transplant IgM antibodies contribute to graft failure, while some other studies point to the relevance of pre transplant IgM antibodies in patients with poor graft outcome. The latter emphasizes the need to document and categorize these antibodies as IgM DSA as opposed to nonspecific antibodies. The ability to identify IgM antibodies on assays such as the flow cytometer and the SAB has made this a realistic possibility.

However, as per most current guidelines, an IgM antibody detected on the CDC crossmatch is not a contraindication for transplant. It would be prudent to monitor these patients more closely in the post-transplant period and also while weaning immunosuppression.

Conclusion

As noted, the CDC assay, while often superseded by more sensitive and specific methodologies such as flow cytometric crossmatch (FCXM) and solid-phase assays (e.g., Luminex-based single-antigen bead testing), it continues to hold clinical relevance by offering unique functional insight into the complement-fixing potential of alloantibodies. This aspect is particularly valuable in assessing the pathogenicity of donor-specific antibodies (DSAs), as the ability to activate complement has been associated with a higher risk of antibody-mediated rejection (AMR). Understanding its variations and interpreting results in conjunction with newer platforms allows for a more comprehensive assessment of immunological risk in transplantation. Consequently, the CDC crossmatch remains a useful adjunctive tool in pre-transplant immunological assessment.

References

1. Terasaki PI. Humoral theory of transplantation. Am J Transplant. 2003;3:665.
2. Tait BD. Detection of HLA antibodies in organ transplant recipients—methodologies, controversies, and clinical significance. Hum Immunol. 2016;77(11):1030–8.
3. Cai J, Terasaki PI. Humoral theory of transplantation: mechanism, prevention, and treatment. Hum Immunol. 2005;66:334.
4. Patel R, Terasaki PI. Significance of the positive crossmatch test in kidney transplantation. N Engl J Med. 1969;280(14):735–9.
5. Terasaki PI, McClelland JD. Microdroplet assay of human serum cytotoxins. Nature. 1964;204:998.
6. Hamrick CW, Lebeck LK. Flow cytometric T and B cell crossmatching. In: Hahn AB, Land GA, Strothman RM, editors. ASHI laboratory manual. 4th ed. American Society for Histocompatibility and Immunogenetics; 2000. p. VI.B.4.1–5.
7. Mulley WR, Kanellis J. Understanding crossmatch testing in organ transplantation: a case-based guide for the general nephrologist. Nephrology (Carlton). 2011;16(2):125–33.

8. Kracker S, Radbruch A. Immunoglobulin class switching: in vitro induction and analysis. Methods Mol Biol. 2004;271:149–59.
9. McCalmon RT Jr, Tardif GN, Sheehan MA, Fitting K, Kortz W, Kam I. IgM antibodies in renal transplantation. Clin Transpl. 1997;11(6):558–64.
10. Okuno T, Kondelis N. Evaluation of dithiothreitol (DTT) for inactivation of IgM antibodies. J Clin Pathol. 1978;31(12):1152–5. https://doi.org/10.1136/jcp.31.12.1152.
11. Tambur AR, Schinstock C. Clinical utility of serial serum dilutions for HLA antibody interpretation. HLA. 2022;100(5):457–68.
12. Bryan CF, Martinez J, Muruve N, et al. IgM antibodies identified by a DTT-ameliorated positive crossmatch do not influence renal graft outcome but the strength of the IgM lymphocytotoxicity is associated with DR phenotype. Clin Transpl. 2001;15(Suppl 6):28.
13. Miller J, Ozturk G, Terasaki PI. Use of extended incubation crossmatch for kidney transplantation. Transplant Proc. 1984;16(6):1678–9.
14. Park H, Lee BH, Lim YM, Han B, Song EY, Park MH. False-positive T-cell cytotoxicity crossmatch results due to autoantibodies in Korean network for organ sharing crossmatch tests. Clin Transplant Res. 2017;31:150–5.
15. Amos DB, Bashir H, Terasaki PI. The use of the Amos wash technique for reducing false-positive crossmatch results. Transplant Proc. 1977;9(2):147–51.
16. Ozturk OG, Terasaki PI, Miller J, Bentz WS. Practical methods for improving B cell crossmatching. Transplant Proc. 1999;31(1–2):1868–9.
17. Terasaki PI, Perkins H, McClelland JD. Antihuman globulin: an effective addition to the crossmatch test for detecting lymphocytotoxic antibodies. Am J Clin Pathol. 1978;69(1):103–10.
18. Guerin C, Pomier G, Fleuru H, Laverne S, Le Petit JC, Berthoux FC. Renal transplantation with positive allocrossmatch (B/T; historical/current) but current T negative. Transplant Proc. 1991;23(1 Pt 1):417–8.
19. Tait BD, Süsal C, Gebel HM, Nickerson PW, Zachary AA, Claas FH, et al. Consensus guidelines on the testing and clinical management issues associated with HLA and non-HLA antibodies in transplantation. Transplantation. 2013;95(1):19–47.
20. Bamoulid J, Courivaud C, Déschamps M, Mercier P, Ferrand C, Penfornis A, Ducloux D. Subclinical rejection in kidney transplant recipients: diagnostic value of DSA and histopathology. Transplantation. 2011;91(10):1116–22.
21. Worthington JE, Martin S, Al-Husseini DM, Dyer PA. Interlaboratory variation in crossmatch techniques: the need for standardization. Transplant Proc. 2010;42(9):3501–4.
22. Tambur AR, Campbell P, Claas FH, Feng S, Gebel HM, Jackson AM, et al. Sensitization in transplantation: assessment of risk (STAR) 2017 working group meeting report. Am J Transplant. 2018;18(7):1604–14.
23. Tambur AR, Leventhal JR, Zitzner JR, Kaufman DB, Friedewald JJ, Abecassis MM. The presence of donor-specific anti-HLA antibodies after positive flow cytometrycrossmatch kidney transplantation is associated with poor allograft outcome. Transplantation. 2000;70(7):1098–101.
24. Loupy A, Lefaucheur C, Vernerey D, Prugger C, Van Huyen JPD, Mooney N, et al. Complement-binding anti-HLA antibodies and kidney-allograft survival. N Engl J Med. 2013;369(13):1215–26.
25. Ziemann M, Pfitzner T, Blasczyk R. Virtual crossmatching: a tool for enhancing the safety of transplantation. Transfus Med Hemother. 2012;39(1):20–7.
26. Peräsaari JP, Jaatinen T, Merenmies J. Donor-specific HLA antibodies in predicting crossmatch outcome: comparison of three different laboratory techniques. Transpl Immunol. 2018;46:23–8.
27. Choktaweesak N, Jayavesa P, Phiancharoen S, Na-thalang O, Sroyson S, Dhitavat V. Comparison between complement-dependent cytotoxicity and flow cytometry crossmatches in deceased donor kidney transplantation. Hematol Transfus Med. 2023;33:179–87.
28. Pandey P, Pande A, Kumar Devra A, Kumar Sinha V, Prasad Bhatt A. Comparative analysis of complement-dependent lymphocytotoxicity crossmatch and flow cytometry crossmatch results versus Luminex single-antigen bead-based donor-specific IgG class I antibody MFI values

in live related renal transplant cases; a retrospective observation in 102 cases. J Immunoass Immunochem. 2020;42(3):300–13.
29. El-Awar NR, Akaza T, Terasaki PI, Nikaein A. Human leukocyte antigen class I epitopes: update to 103 total epitopes, including the C locus. Transplantation. 2011;91(9):1052–7.
30. Leffell MS, Zachary AA. The problem of transplanting the sensitized patient. Clin Lab Med. 2002;22(2):265–80.
31. Halliley JL, Tipton CM, Liesveld J, et al. Long-lived plasma cells are contained within the $CD19^{-}CD38hiCD138^{+}$ subset in human bone marrow. Immunity. 2015;43(1):132–45.
32. Pereira LM, Ng W-F, Utz PJ. Human autoantibodies: growing in the medicine garden. J Clin Invest. 2018;128(5):1691–8.
33. Reindl-Schwaighofer R, Oberbauer R. False-positive CDC x-match after rituximab. Transpl Int. 2014;27(12):124–5.
34. Zhou ZH, Zhang Y, Hu YF, et al. The broad immunoregulatory role of natural IgM. Cell MolImmunol. 2007;4(1):27–36.
35. Morris PJ, Knight SC. Cytotoxic antibodies and transplantation. Transplant Rev. 1990;4(2):66–87.
36. Kerman RH, Süsal C, Wiesner RH, et al. Importance of preformed cytotoxic antibodies in chronic renal allograft rejection. Transplantation. 1991;51(1):113–8.
37. Stegall MD, Gloor J, Winters JL, et al. A comparison of plasmapheresis versus high-dose IVIG desensitization in renal transplant candidates with high levels of donor-specific antibody. Am J Transplant. 2011;11(6):1138–45.
38. Duquesnoy RJ. A structurally based approach to determine HLA compatibility at the humoral immune level. Hum Immunol. 2016;77(5):405–13.
39. Vo AA, Huang E, Ammerman N, Toyoda M, Ge S, Haas M, Zhang X, Peng A, Najjar R, Williamson S, Myers C, Sethi S, Lim K, Choi J, Gillespie M, Tang J, Jordan SC. Clazakizumab for desensitization in highly sensitized patients awaiting transplantation. Am J Transplant. 2022;22(4):1133–44.
40. Jordan SC, Legendre C, Desai NM, Lorant T, Bengtsson M, Lonze BE, Vo AA, Runström A, Laxmyr L, Sjöholm K, Schiött Å, Sonesson E, Wood K, Winstedt L, Kjellman C, Montgomery RA. Imlifidase desensitization in crossmatch-positive, highly sensitized kidney transplant recipients: results of an international phase 2 trial (Highdes). Transplantation. 2021;105(8):1808–17.

Kidney Transplant Rejection: T-Cell and Antibody-Mediated Rejection

12

Mariam Priya Alexander, Nalumackal Vijayan Seethalekshmy, Radhika Krishna Patil, and Carrie Schinstock

Case-Based Discussion

A 28-year-old man with end-stage renal disease secondary to IgA nephropathy underwent a living unrelated kidney transplantation. This was a negative T-cell and B-cell flow crossmatch. Donor-specific antibodies were not identified prior to transplant. Following basiliximab induction, he was maintained on triple immunosuppression (pred/MMF/tacrolimus). Three months post-transplant, he had a rise in serum.

In evaluating acute kidney injury post-transplant, the clinical differential diagnosis includes the following:

1. Acute cellular rejection (ACR)
2. Acute antibody-mediated rejection (ABMR)
3. Acute tubular injury (ATI) from nephrotoxic agents or ischemic injury
4. Urinary tract obstruction/infection (imaging and cultures needed)
5. Vascular complications included renal artery thrombosis and stenosis

M. P. Alexander (✉)
Department of Laboratory Medicine and Pathology, Mayo Clinic, Rochester, MN, USA
e-mail: alexander.mariam@mayo.edu

N. V. Seethalekshmy
Department of Pathology, Amrita Institute of Medical Sciences, Kochi, Kerala, India
e-mail: seethalekshmy@aims.amrita.edu

R. K. Patil
Department of Nephropathology, Shri Balaji Kidney Care Pvt. Ltd., Madhapur, Hyderabad, Telangana, India

C. Schinstock
Department of Nephrology and Hypertension, Mayo Clinic, Rochester, MN, USA
e-mail: Schinstock.carrie@mayo.edu

G. Abraham et al. (eds.), *Immunological Challenges in Organ Transplantation*, https://doi.org/10.1007/978-981-95-1533-2_12

Pertinent investigations included a serum creatinine of 1.7 mg/dL (increased from a baseline of 1 mg/dL). Tacrolimus level was 4.71 ng/mL. Urinalysis showed trace protein. Urine microscopy revealed 15–20 WBCs/HPF, and 10–15 RBCs/HPF. Urine culture showed no growth. Imaging studies revealed a bulky allograft kidney (11.5 × 5.7 cm) with normal echotexture. Hydronephrosis/ hydroureter/ perinephric collection/thrombi were not identified. A renal allograft biopsy was performed.

Renal Allograft Biopsy Light Microscopy

The allograft biopsy was adequate with two cortical cores and included three interlobular arteries. At least 20 non-sclerosed glomeruli were identified. There was no evidence of glomerulitis, glomerular thrombi, or glomerular basement membrane remodeling. The tubulointerstitial compartment was characterized by tubular injury. There was epithelial cell swelling and the presence of necrotic cell debris within the tubular lumen. Lymphocytic tubulitis was also noted with infiltration by 7–8 mononuclear cells per tubular cross section in the most affected tubule (t2). There was diffuse interstitial infiltrate, involving over 75% of the non-scarred cortex (i3). The inflammatory infiltrate was composed of lymphocytes, plasma cells, and an occasional eosinophil. Subendothelial mononuclear infiltration was noted in one interlobular artery with focal fibrinoid necrosis (v3). C4d immunofluorescence studies were performed and were negative in the peritubular capillaries of the cortex (Fig. 12.1).

Allograft Biopsy Diagnosis

Acute T Cell-Mediated Rejection, Banff 2022 Grade III (g0, t2, i3, v3, ptc0, C4d0)

The patient was treated with a high-dose pulse of corticosteroids and anti-thymocyte globulin with vigilant follow-up and reassessment of renal function. Subsequent monitoring revealed an improvement in serum creatinine levels, indicating a favourable response to therapy. At 9 months follow-up, graft function was stable with no further rejection episodes.

In this chapter, we review the most recent guidelines and concepts relating to the two broad categories of rejection, namely **T cell-mediated and antibody-mediated rejection** as discussed in The Banff 2022 Kidney Meeting Report [1].

T-Cell-Mediated Rejection

T cell-mediated rejection (TCMR) is characterized by the infiltration of lymphocytes that affects the tubules, interstitium, and sometimes the arterial intima. It occurs when T cells respond to donor antigens that are expressed within the tubules, interstitium, blood vessels, and glomeruli of the transplanted organ. These are most

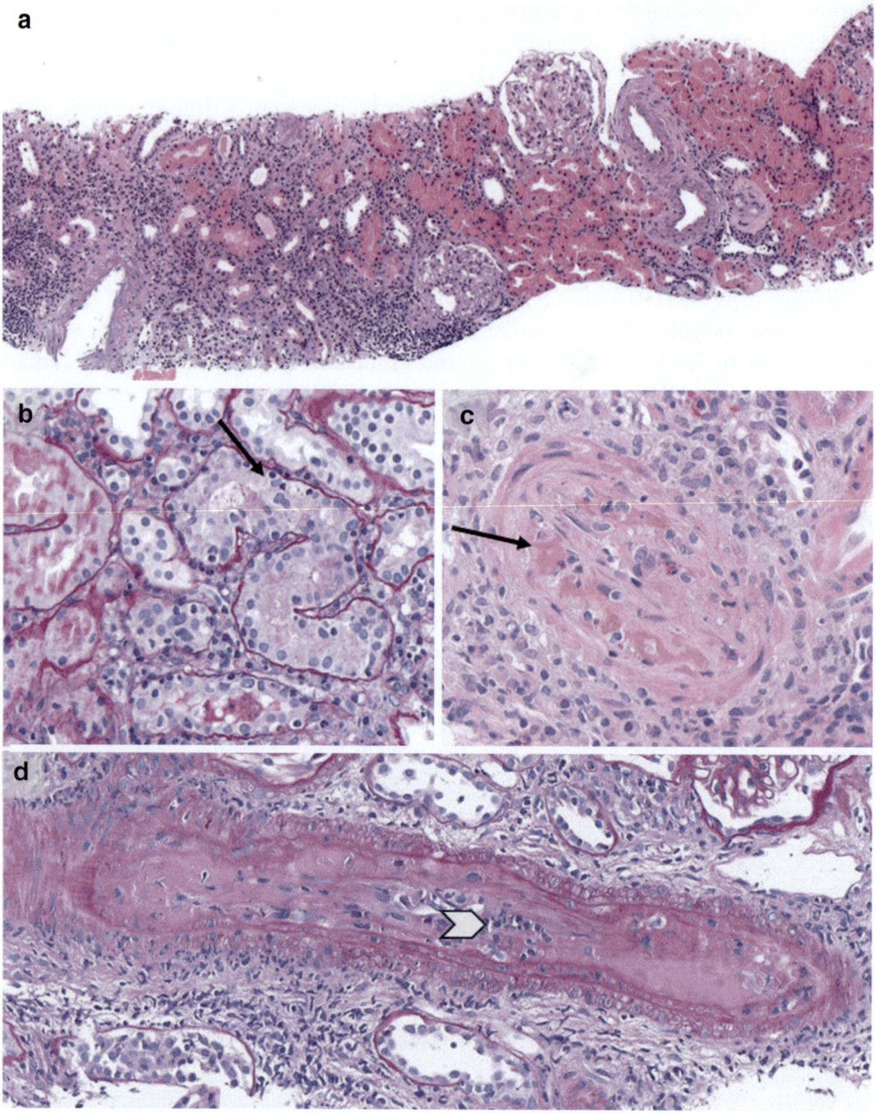

Fig. 12.1 Acute T cell mediated rejection. (**a**) There is diffuse interstitial inflammation involving over 50% of the non-scarred cortex. (**b**) Multifocal moderate tubulitis (arrow) is noted, involving non-atrophic tubules. (**c, d**) Focal fibrinoid necrosis (arrow) in one artery, and another artery shows subendothelial mononuclear inflammation (endothelialitis) (arrowhead)

commonly directed against histocompatibility antigens. Other immune cells, including natural killer cells and macrophages, may also contribute to this rejection process.

The spectrum of T-cell-mediated rejection includes borderline rejection, acute T-cell-mediated rejection, and chronic active T-cell-mediated rejection.

Borderline Rejection in Renal Transplant Recipients

Borderline rejection (BL-R) is a pathological diagnosis in renal transplant recipients that suggests possible T cell–mediated rejection (TCMR). The diagnostic criteria are outlined in Table 12.1. However, the clinical implications of BL-R remain uncertain. Nankivell, in a single-center cross-sectional cohort study of 551 renal transplant recipients (146 cases of borderline rejection, 55 cases of acute T-cell mediated rejection, and 826 normal controls) compared the functional and histological outcomes between these three groups [2]. The study found that BL-R was associated with renal dysfunction, acute as well as progressive tubular injury, and increased risk of subsequent acute rejection, allograft failure, and patient mortality. Patients with BL-R identified on biopsies performed for clinical renal dysfunction had suboptimal functional recovery, persistent inflammation, and high rates of acute rejection despite treatment. Additionally, BL-R on longer follow-up was associated with new-onset microvascular inflammation, C4d staining, transplant

Table 12.1 Defining criteria for borderline and T-cell mediated, by Banff 2022 classification of renal allograft pathology

Pattern of rejection	Banff diagnostic guidelines
Borderline rejection	Less than or equal to 25% interstitial inflammation (i1) with any tubulitis (t1, t2, or t3) or >25% interstitial inflammation (i2 or i3) with mild tubulitis (t1)
Acute T cell mediated rejection	
Type IA	More than 25% interstitial inflammation (i2 or i3) with moderate tubulitis (t2)
Type IB	More than 25% interstitial inflammation (i2 or i3) with severe tubulitis (t3)
Type IIB	Severe intimal arteritis (v2) with or without interstitial inflammation and tubulitis
Type III	Transmural arteritis and/or arterial fibrinoid change and necrosis of medial smooth muscle cells with accompanying lymphocytic inflammation (v3)
Chronic T cell mediated rejection	
Grade IA	*ti* ≥ 2, and i-IFTA ≥ 2, other known causes of i-IFTA (e.g., pyelonephritis, BK-virus nephritis, etc.) ruled out, and t2
Grade IB	*ti* ≥ 2 and i-IFTA ≥ 2, other known causes of i-IFTA ruled out, and t3
Grade II	Arterial intimal fibrosis with mononuclear cell inflammation in fibrosis and formation of neointima

glomerulopathy, and de novo donor-specific antibodies. In contrast, BL-R inflammation in protocol biopsies resolved in 62% of untreated cases. These findings highlight the heterogeneous nature of BL-R, with some cases having mild inflammation while others have significant consequences leading to inferior functional and histological outcomes.

Acute T Cell-Mediated Rejection

T cell-mediated rejection (TCMR) continues to be a significant concern in kidney transplant recipients, even with current standard immunosuppressive therapy using tacrolimus and mycophenolate. The diagnostic criteria are outlined in Table 12.1. Type 1 accounts for approximately 65% of acute T-cell mediated rejection, Type 2 for approximately 30%, and Type 3 for less than 5%. The chief diagnostic features, namely tubulitis, interstitial inflammation and endothelialitis are depicted in Fig. 12.1.

An observational study of 775 kidney transplant recipients found that 30% experienced initial TCMR (includes BL-R, and acute TCMR) as detected on 17% for-cause and 13% surveillance biopsies; 64% had either persistent or recurrent episodes on follow-up. Greater frequency and severity of TCMR were associated with higher HLA-DR/DQ molecular mismatch, and both initial and recurrent TCMR significantly increased the risk of graft loss, highlighting under-immunosuppression in many patients and the need for improved therapies [3]. Complementing these findings, another study of 586 patients who underwent surveillance biopsies within the first year post-transplant showed that early subclinical inflammation (SCI), particularly with tubulitis (SCI-T), was a strong predictor of future clinical biopsy proven acute rejection (C-BPAR) and graft loss over 5 years. Patients with SCI-T had nearly four times the risk of subsequent rejection and twice the risk of graft loss compared to those without significant inflammation [4]. Together, these studies underscore the critical role of early detection and more effective immunosuppressive strategies to prevent long-term immune-mediated graft injury. The differential diagnosis on histology includes acute antibody mediated rejection, BK polyoma virus interstitial nephritis, pyelonephritis, and acute allergic tubulointerstitial nephritis.

Chronic Active T-Cell-Mediated Rejection (cATCMR)

Chronic active T-cell-mediated rejection was initially introduced into the Banff classification in 2017. This is the expression of a persistent or recurrent T-cell mediated rejection that leads to chronic allograft pathology that includes transplant arteriopathy, inflammation in areas of interstitial fibrosis, and tubular atrophy. The diagnostic histologic criteria are outlined in Table 12.1. It was recognized as an important entity due to the growing evidence that i-IFTA was associated with graft loss. cATCMR was associated with under-immunosuppression and prior episodes

of active T-cell-mediated rejection. The greatest challenge of this diagnostic entity is the lack of specificity given that i-IFTA lesions are nonspecific and are recognized in the setting of previous episodes of T-cell-mediated rejection, as well as following episodes of pyelonephritis, polyomavirus-associated nephropathy, recurrent or de novo glomerulonephritis, and antibody-mediated rejection.

Nakagawa et al. assessed the predictive value of the updated Banff 2017 criteria for chronic active T cell-mediated rejection (CA-TCMR) in kidney transplant patients [5]. Among 406 recipients who underwent a 1-year protocol biopsy, 8% were diagnosed with CA-TCMR under the new criteria, including some previously classified as normal or having other forms of rejection. CA-TCMR was linked to poorer graft outcomes—specifically, a higher risk of doubling serum creatinine or graft loss. Factors associated with CA-TCMR included cyclosporine use, prior acute rejection, and BK virus nephropathy. The diagnosis showed strong interobserver agreement and was a significant independent predictor of poor prognosis, second only to antibody-mediated rejection. These findings support the clinical relevance of the Banff 2017 criteria in identifying patients at higher risk for graft failure.

Hegelson et al. evaluated inflammation in areas of fibrosis (i-IFTA) and chronic active T cell-mediated rejection (CA TCMR) in kidney transplant biopsies from two cohorts: a prospective group ($n = 585$) and a cross-sectional group ($n = 458$) [6]. The study found that i-IFTA, particularly when combined with other histologic features ($t \geq 2$, $ti \geq 2$), was associated with both a cellular and humoral activity. Notably, most patients with i-IFTA did not have a history of acute rejection; additionally, the study also highlighted that i-IFTA ≥2, along with $t \geq 2$ and $ti \geq 2$, could occur without prior TCMR. The findings of this study raise questions about the current diagnostic criteria for CA TCMR.

Hyperacute Rejection

Hyperacute rejection occurs immediately upon implantation and perfusion of the graft. It is due to pre-existing donor reactive HLA or blood group antibodies at the time of implantation [7]. This form of rejection is uncommon now due to pretransplant antibody testing.

This pattern of rejection is secondary to the presence of preexisting antibodies to the donor endothelium. The antibodies can be against HLA (class 1 or class 2) or ABO blood group antigens. The antibody titers must be sufficiently high to cause immediate rejection. There is often complement activation, endothelial cell activation, and concurrent platelet activation.

The most common presentation is of a primary non-functioning graft or a non-functioning graft within hours after implantation.

On gross examination, the graft is cyanotic within minutes to hours post-implantation. Microscopically, there is evidence of thrombi within the glomeruli and arterioles. Neutrophil-rich margination is seen within the glomerular capillary loops and peritubular capillaries. Within a span of 24 h, widespread thrombosis

involving both glomeruli and arteries is noted. Fibrinoid necrosis of arteries with incipient or frank cortical and medullary necrosis may also be seen. Immunofluorescence studies highlight linear staining of cortical peritubular capillaries with C4d (Fig. 12.2a, b). The absence of this finding does not argue against the diagnosis of hyperacute rejection.

Antibody-Mediated Rejection

Antibody-mediated rejection (AMR) is the most common cause of late allograft loss after kidney transplantation. It was first identified as a diagnostic entity in 2001 and was incorporated into the Banff Classification in 2003. The clinicopathological spectrum of AMR is highly varied. It can range from a hyperacute form of rejection (with primary graft nonfunction) to early acute ABMR that is noted in positive crossmatch kidney transplants, to gradually progressive graft dysfunction occurring years post-transplant. Antibody-mediated rejection (ABMR) can occur in the setting of preformed donor-specific antibody (DSA) or in patients who develop de Novo DSA. Antibody-mediated rejection (ABMR) may also be concurrent with T-cell-mediated rejection [8–10].

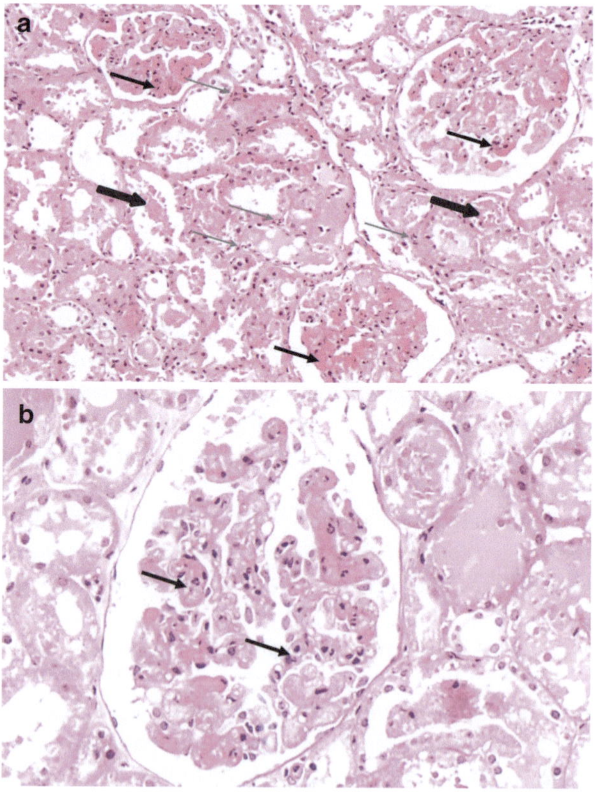

Fig. 12.2 (**a**) H&E Stain: 10×: Glomeruli with congested capillary lumina and showing glomerulitis (black arrows) along with tubular injury with loss of the brush border and proteinaceous and necrotic debris in tubules (broad black arrows) and mild peritubular capillaritis (grey arrows). (**b**) Hematoxylin & Eosin Stain: 20×: Glomeruli with congested capillary lumina and glomerulitis with neutrophils (black arrows)

The diagnostic categories of AMR, although constantly evolving, were originally based on fulfilling three criteria: (1) histological evidence of tissue injury, (2) evidence of current or recent antibody interaction with the endothelium, and (3) serological evidence of donor-specific antibody to either HLA or non-HLA antibodies. Over the years, several iterations of this classification with modifications of the criteria have been made. In this chapter, we focus on the phenotypes/subcategories of the Banff 2022 classification of antibody-mediated rejection and microvascular inflammation/injury, before which we review the most recent criteria [1].

Histological Features of AMR and/or Microvascular Inflammation (Criterion 1)

Criterion 1 for Active lesions of ABMR includes:

- Microvascular inflammation (glomerulitis and/or peritubular capillaritis: ($g > 0$ and/or ptc > 0), in the absence of recurrent or de novo glomerulonephritis, although in the presence of acute TCMR, borderline infiltrate, or infection, ptc ≥ 1 alone is not sufficient, and g must be ≥1.
- Intimal or transmural arteritis ($v > 0$).
- Acute thrombotic microangiopathy, in the absence of any other cause.

Criterion 1 for Chronic lesions of ABMR includes:

- Transplant glomerulopathy (cg > 0) if no evidence of chronic TMA or chronic recurrent/de novo glomerulonephritis; includes changes evident by electron microscopy (EM) alone (cg1a).
- Severe peritubular capillary basement membrane multilayering (ptcml1; requires EM).

Other lesions that can be noted in antibody-mediated rejection, lending strength to the diagnosis but in themselves are not diagnostic include Arterial intimal fibrosis (CV) of new onset excluding other causes; leukocyte within the sclerotic intima (favors chronic antibody mediated rejection if there is no prior history of T-cell mediated rejection), and acute tubular injury, in the absence of any other apparent cause.

Criterion 2 to diagnose AMR/MVI was previously termed as *evidence of current/recent antibody interaction with vascular endothelium*, but is now revised in the Banff 2022 to read as, "*at least 1 or more of the following,*" namely:

- Linear C4d staining in peritubular capillaries or medullary vasa recta (C4d2 or C4d3 by IF on frozen sections, or C4d > 0 by IHC on paraffin sections).
- Moderate microvascular inflammation ([g + ptc] ≥ 2) in the absence of recurrent or de novo glomerulonephritis, although in the presence of acute TCMR, borderline infiltrate, or infection, ptc ≥ 2 alone is not sufficient, and g must be ≥1.

- Biopsy-based transcript diagnostics for AMR/MVI above a defined threshold, if thoroughly validated for use as a substitute for MVI, and available.

Criterion 3 to diagnose antibody-mediated rejection was previously termed, "*Serological evidence of circulating donor-specific antibodies (DSA to HLA or other antigens).*" The Banff 2022 now uses the terminology for criterion 3 to read as, "*Evidence of circulating donor-specific antibodies (DSA to HLA or other antigens).*" The nuances of this category include:

- Thorough testing for DSA should be done following the STAR guidelines.
- The detection of non-HLA antibodies may also be used as serological Banff criteria to diagnose AMR if testing protocols are standardized and clinically validated.
- C4d staining may serve as a surrogate for DSA.

The key features of AMR are illustrated in Fig. 12.3.

Phenotype/Subcategories of Banff 2022 Category 2: Antibody-Mediated Rejection and Microvascular Inflammation/Injury (AMR/MVI)

Seven Phenotypes:

1. Active antibody-mediated rejection
2. Chronic active antibody-mediated rejection
3. Chronic antibody-mediated rejection
4. C4d staining without evidence of rejection
5. Microvascular inflammation, DSA negative, C4d negative
6. Probable antibody-mediated rejection
7. C4d staining with acute tubular injury

This incorporates **active, chronic active, and chronic AMR**. Once the diagnostic criteria for AMR are identified on the biopsy, the next step is to ascertain if *the MVI threshold* ($g + ptc \geq 2$) and or biopsy-based transcripts for AMR/MVI are met. When *MVI is at or above the threshold*, the presence of positive C4d and positive DSA, or either DSA or C4d being positive, is diagnostic of antibody-mediated rejection. If the diagnostic features of AMR/MVI lesions are present, but *below the MVI threshold*, a positive C4d, independent of DSA status, is diagnostic of antibody-mediated rejection.

Once the diagnosis of AMR is made, it is called active AMR if only active features (cg = 0; ptcml = 0), including C4d positivity, are present. It is called chronic active AMR if there are both active and chronic features (cg > 0 and/or severe ptcml), and it is called chronic AMR if only chronic features are present. In the case of chronic AMR, a prior documented diagnosis of active or chronic active AMR, or prior evidence of DSA, counts as DSA positivity.

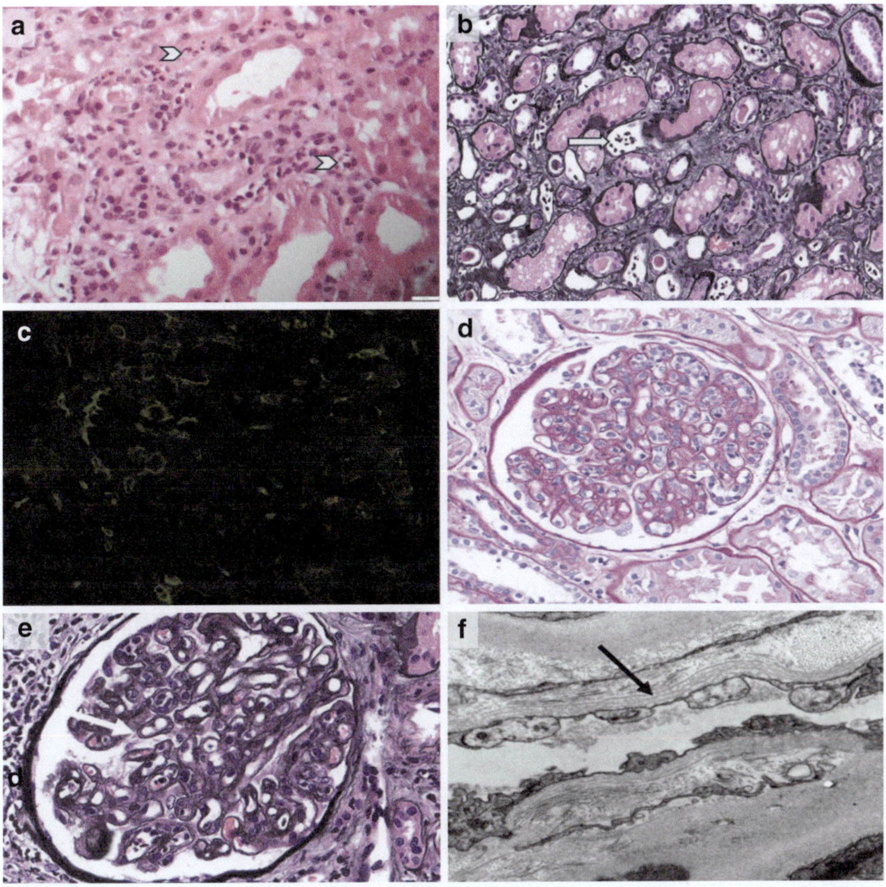

Fig. 12.3 Antibody-mediated rejection. (**a**) Acute tubular injury is seen with notable neutrophilic peritubular capillaritis (arrowhead). Hematoxylin and eosin 20×. (**b**) Moderate mononuclear peritubular capillaritis is identified in the cortical peritubular capillaries (gray arrow), Jones methenamine silver stain, 20×. (**c**) C4d immunofluorescence stain highlights diffuse linear peritubular capillaries staining (C4d3). (**d**) Glomerulitis characterized by endothelial swelling and luminal occlusion of 75% of capillary lumina with marginated mononuclear cells. (**e**) Notable capillary wall remodeling of at least 30% of the visualized glomerular capillary loops. Marginated mononuclear cells are seen within their lumina. Jones methenamine silver, 40×. (**f**) Peritubular capillaries identified on ultrastructural evaluation highlighting circumferential basement membrane multilayering

MVI, DSA-negative, and C4d-negative: MVI observed above the histologic threshold, in the absence of circulating DSA and with negative C4d staining in the peritubular capillaries, may be seen in the setting of both normal and abnormal kidney function. The underlying cause for this specific phenotype is poorly understood. Further research is needed to determine the prevalence, causes, and related biological processes, as well as the optimal treatment approaches for this pattern. These cases may indicate undetected HLA-DSA, alloreactive T cell-mediated responses,

autoreactive or alloreactive non-HLA antibodies, primary NK cell activation due to missing self, viral infections, other mechanisms of innate immune activation, ischemia-reperfusion injury, among other factors.

Probable AMR is diagnosed when DSA is detected and MVI is present, but below the histological threshold, and C4d deposition along peritubular capillaries is absent. This phenotype, which can be seen in patients with normal or abnormal kidney function, likely indicates antibody-mediated rejection.

Removal of ATI as a Key Histologic Feature of AMR

Acute tubular injury, on its own, is too frequent and nonspecific to be considered a histological feature of antibody-mediated rejection. This was confirmed by a Banff 2022 survey. Additionally, it was further emphasized that there are no standardized criteria for the definition of acute tubular injury, nor the qualification of the severity of acute tubular injury. In light of these limitations, acute tubular injury on its own is removed from the diagnostic features of AMR/MVI.

C4d Staining with Acute Tubular Injury

There are three important clinical scenarios in which acute tubular injury and C4d positivity are noted in the absence of other histological features of AMR/MVI) namely:

1. In an ABO incompatible transplant, reflects **accommodation**.
2. Early posttransplant in a DSA-sensitized (positive crossmatch) patient, reflects **probable antibody-mediated rejection.**
3. Conventional risk transplant, DSA negative, reflects **no antibody-mediated rejection.**

Arterial Intimal Fibrosis of New Onset

This lesion is therefore now eliminated from among the key diagnostic lesions. Recognition that this lesion can be the expression of both ongoing antibody-mediated rejection and chronic active T-cell-mediated rejection is critical.

Clinical Presentation of AMR

This varies based on whether the DSA is preformed or de novo. Preformed DSA occurs pre-transplant, while de novo DSA occurs later in the course of the transplant, in patients with combined AMR and T-cell mediated rejection, and or under immunosuppression (iatrogenic or due to nonadherence). In the setting of preformed DSA, the diagnosis of active AMR encompasses a wide spectrum of clinicopathological features. These include a more aggressive form of active AMR, namely early acute antibody-mediated rejection in positive crossmatch kidney transplant, and a more indolent form of active AMR.

Early acute antibody-mediated rejection (AMR) occurs within the first month (<30 days) after transplantation. This pattern of injury is caused by high levels of

preformed DSA, which increase from memory B-cell response following antigen stimulation. This type of ABMR is characterized by very high serum donor-specific antibody (DSA) levels, as measured during biopsy, as well as C4d deposition in the peritubular capillaries. It often leads to acute tissue injury in the graft, including thrombi and acute tubular injury. Histologically, the biopsy shows acute tubular injury, thrombi, and mild, occasionally moderate microvascular inflammation (most commonly neutrophils) of the glomerular and peritubular capillaries. Almost all cases will show diffuse staining of the cortical peritubular capillaries with C4d, the latter corresponding to the serum levels of donor-specific antibody. Clinically, early acute ABMR is uncommon but can be challenging to treat. If not recognized and addressed promptly, there is a high risk of graft loss. The current standard of care for managing this specific form of acute ABMR typically involves plasmapheresis, intravenous immunoglobulin, and corticosteroids. Additional therapies, such as complement inhibitors (eculizumab), rituximab, and splenectomy, may also be considered.

Active late AMR is a more indolent form of antibody-mediated rejection, which occurs after 30 days post-transplant, usually between 3 months and 5 years post-transplant, and patients commonly have normal graft function without proteinuria. The pathophysiology in this subset is the formation of de novo DSA or an increase in preformed DSA in the setting of suboptimal immunosuppression and/or concomitant cellular rejection. The biopsy shows mild to moderate microvascular inflammation with C4d staining of peritubular capillaries ranging from negative to focal, and less commonly, diffuse C4d staining.

Chronic antibody-mediated rejection may be active or inactive. The histological hallmarks of chronicity include transplant glomerulopathy and peritubular capillary basement membrane multilayering (peritubular capillaropathy). Transplant glomerulopathy is present in approximately 4% of surveillance biopsies taken 1 year after transplantation. Even when diagnosed early, TG is progressive [11] and frequently leads to allograft failure. A study from the Mayo Clinic [12] found that in patients with a negative T-cell crossmatch, elevated levels of anti-HLA-II antibodies were associated with a higher risk of developing transplant glomerulopathy (TG). Additionally, higher antibody levels correlated with the presence of C4d in the peritubular capillaries of TG biopsies. Moreover, the presence of C4d in TG is linked to reduced graft survival.

Donor Specific Antibodies and Antibody-Mediated Rejection

Donor-specific antibodies (DSA) are an independent risk factor for active antibody-mediated rejection (AMR) and graft loss. Preformed DSA is identified pre transplant and is less of a concern due to improved donor selection criteria. The most common type of AMR is associated with de novo DSA (dnDSA). DnDSA develops

in 15–25% of kidney transplant recipients within the first 5 years after transplantation, with an annual incidence rate of 2% in compliant patients. Nonadherence to medication, reduced immunosuppression following physician direction, a higher eplet mismatch, younger age, and a history of T-cell-mediated rejection (TCMR) are risk factors for dnDSA formation. Specific IgG subclasses of dnDSA, higher antibody titers, and the ability to bind complement are associated with worse outcomes [8].

Activity and Chronicity Indices

The activity and chronicity indices are developed based on kidney transplant biopsy findings to evaluate their association with graft survival in patients with active or chronic active antibody-mediated rejection (AMR). The activity index is based on the summation of scores for each of the following indices: i, t, v, c4d, to a maximum of 12. The chronicity index is determined by the summation of ci, ct, cv, and cg. The score for cg is doubled. The maximum chronicity score is 15. The chronicity index was found to be significantly associated with graft loss, with a threshold value of 4 or greater being predictive of graft loss. This association was independent of other factors such as glomerular filtration rate, type of AMR, and donor-specific antibody status. The findings were confirmed in an independent cohort of patients, suggesting that the chronicity index may be a valuable tool for decision-making in patients with AMR [13].

Ancillary Diagnostic Tools

Biopsy-Based Transcript Diagnostics

Since its introduction in 2013, there have been significant advancements in the exploration of renal allograft pathology through biopsy-based transcript platforms [14–16]. The two currently available platforms are the Molecular Microscopic Diagnostic System (MMdx) and the nCounter B-HOT panel.

The MMdx Kidney platform is licensed for commercial use as a send-out test in a CLIA-approved laboratory and functions as a laboratory-developed test (LDT), which means it does not require FDA approval for clinical use in the USA. The clinical availability of this platform is gradually increasing.

The consensus B-HOT panel can be used to analyze allograft biopsies, but experience with it is still limited. Many technical questions remain unresolved, and the clinical validity of the platform needs to be demonstrated. While the Nanostring nCounter platform is FDA-approved in the USA, the classifiers based on the B-HOT panel have not yet been validated.

Single-Cell Spatial Technology

Novel tools such as imaging mass cytometry and spatial transcriptomics are being increasingly used to study allograft biopsies [17]. They provide novel insights into commonly encountered patterns of injury.

References

1. Naesens M, Roufosse C, Haas M, Lefaucheur C, et al. The Banff 2022 kidney meeting report: reappraisal of microvascular inflammation and the role of biopsy-based transcript diagnostics. Am J Transplant. 2024;24(3):338–49. https://doi.org/10.1016/j.ajt.2023.10.016.
2. Nankivell BJ, Agrawal N, Sharma A, Taverniti A, P'Ng CH, Shingde M, Wong G, Chapman JR. The clinical and pathological significance of borderline T cell-mediated rejection. Am J Transplant. 2019;19(5):1452–63. https://doi.org/10.1111/ajt.15197.
3. Rampersad C, et al. The negative impact of T cell-mediated rejection on renal allograft survival in the modern era. Am J Transplant. 2022;22(3):761–71.
4. Mehta RB, Melgarejo I, Viswanathan V, Zhang X, Pittappilly M, Randhawa P, Puttarajappa C, Sood P, Wu C, Sharma A, Molinari M, Hariharan S. Long-term immunological outcomes of early subclinical inflammation on surveillance kidney allograft biopsies. Kidney Int. 2022;102(6):1371–81.
5. Nakagawa K, Tsuchimoto A, Ueki K, Matsukuma Y, Okabe Y, Masutani K, Unagami K, Kakuta Y, Okumi M, Nakamura M, Nakano T, Tanabe K, Kitazono T, Japan Academic Consortium of Kidney Transplantation Investigators. Significance of revised criteria for chronic active T cell-mediated rejection in the 2017 Banff classification: Surveillance by 1-year protocol biopsies for kidney transplantation. Am J Transplant. 2021;21(1):174–85. https://doi.org/10.1111/ajt.16093.
6. Helgeson ES, Mannon R, Grande J, Gaston RS, Cecka MJ, Kasiske BL, Rush D, Gourishankar S, Cosio F, Hunsicker L, Connett J, Matas AJ. i-IFTA and chronic active T cell-mediated rejection: a tale of 2 (DeKAF) cohorts. Am J Transplant. 2021;21(5):1866–77. https://doi.org/10.1111/ajt.16352.
7. Tittelbach-Helmrich D, Bausch D, Drognitz O, Goebel H, Schulz-Huotari C, Kramer-Zucker A, Hopt UT, Pisarski P. Hyperacute rejection of a living unrelated kidney graft. Case Rep Med. 2014;2014:613641. https://doi.org/10.1155/2014/613641.
8. Schinstock CA, Mannon RB, Budde K, Chong AS, Haas M, Knechtle S, Lefaucheur C, Montgomery RA, Nickerson P, Tullius SG, Ahn C, Askar M, Crespo M, Chadban SJ, Feng S, Jordan SC, Man K, Mengel M, Morris RE, O'Doherty I, Ozdemir BH, Seron D, Tambur AR, Tanabe K, Taupin JL, O'Connell PJ. Recommended treatment for antibody-mediated rejection after kidney transplantation: the 2019 Expert Consensus from the Transplantion Society Working Group. Transplantation. 2020;104(5):911–22. https://doi.org/10.1097/TP.0000000000003095.
9. Cornell LD. Histopathologic features of antibody mediated rejection: the Banff classification and beyond. Front Immunol. 2021;12:718122. https://doi.org/10.3389/fimmu.2021.718122.
10. Schinstock C, Stegall MD. Acute antibody-mediated rejection in renal transplantation: current clinical management. Curr Transplant Rep. 2014;1(2):78–85.
11. Gloor JM, Sethi S, Stegall MD, Park WD, Moore SB, DeGoey S, Griffin MD, Larson TS, Cosio FG. Transplant glomerulopathy: subclinical incidence and association with alloantibody. Am J Transplant. 2007;7(9):2124–32.
12. Issa N, Cosio FG, Gloor JM, Sethi S, Dean PG, Moore SB, DeGoey S, Stegall MD. Transplant glomerulopathy: risk and prognosis related to anti-human leukocyte antigen class II antibody levels. Transplantation. 2008;86(5):681–5.

13. Haas M, Mirocha J, Huang E, Najjar R, Peng A, Sethi S, Vo A, Anglicheau D, Jordan SC, Rabant M. A Banff-based histologic chronicity index is associated with graft loss in patients with a kidney transplant and antibody-mediated rejection. Kidney Int. 2023;103(1):187–95. https://doi.org/10.1016/j.kint.2022.09.030.
14. Short S, Lewik G, Issa F. An immune atlas of T cells in transplant rejection: pathways and therapeutic opportunities. Transplantation. 2023;107(11):2341–52. https://doi.org/10.1097/TP.0000000000004572.
15. Lazarou C, Moysidou E, Christodoulou M, Lioulios G, Sampani E, Dimitriadis C, Fylaktou A, Stangou M. Non-invasive biomarkers for early diagnosis of kidney allograft dysfunction: current and future applications in the era of precision medicine. Medicina (Kaunas). 2025;61(2):262. https://doi.org/10.3390/medicina61020262.
16. Chancharoenthana W, Traitanon O, Leelahavanichkul A, Tasanarong A. Molecular immune monitoring in kidney transplant rejection: a state-of-the-art review. Front Immunol. 2023;14:1206929. https://doi.org/10.3389/fimmu.2023.1206929.
17. Alexander MP, Zaidi M, Larson N, Mullan A, Pavelko KD, Stegall MD, Bentall A, Wouters BG, McKee T, Taner T. Exploring the single-cell immune landscape of kidney allograft inflammation using imaging mass cytometry. Am J Transplant. 2024;24(4):549–63. https://doi.org/10.1016/j.ajt.2023.11.008.

Immunology of ABO Incompatible Transplantation

13

Narayan Prasad, Mythri Shankar, and K. S. Nayak

Case Summary of a Successful ABOI KT

A 39-year-old female, blood group O positive, developed ESKD due to IgA nephropathy. She has three pregnancies, one stillbirth, and she has been on maintenance dialysis for the last 4 years and received 5 units of blood during dialysis vintage. She had a single-parent mother, wanted to donate a kidney, but her blood group was B positive. She waited for a deceased donor kidney for 4 years. The transplant unit decided for ABO incompatible transplantation, and her workup revealed an anti-A IgG titer of 1:250. She had donor-specific antibodies against HLA-A*02:01 (MFI 2200) and HLA-DR*11:02 (MFI 1600), possibly due to sensitization from pregnancy and blood transfusion. She was explained the ABOi KT and the protocols of desensitization. Her Flow cytometric cross-match was negative, and the CDC also was negative. She was desensitized using Rituximab 14 days before KT, and Tacrolimus and mycophenolate mofetil (MMF) were initiated on day 10 before KT. Five sessions of PP followed by low-dose IVIG were given for desensitization. A titer of 1:8 and a repeat anti-HLA antibody level of less than 700 were achieved before KT. Patient had successful KT with immediate graft function on day 3 with serum creatinine of 1.2 mg/dL, but on day 5, she had reduced output and serum creatinine increased to 1.8 mg/dL with anti A titer was 1: 64. Three session of PP was given and her urine output impoved and serum creatinine settled to

N. Prasad (✉)
Department of Nephrology, Sanjay Gandhi Postgraduate Institute of Medical Sciences, Lucknow, Uttar Pradesh, India

M. Shankar
Department of Nephrology, Institute of Nephro-Urology, Bengaluru, Karnataka, India

K. S. Nayak
Virinchi Hospital, Hyderabad, Telangana, India

G. Abraham et al. (eds.), *Immunological Challenges in Organ Transplantation*, https://doi.org/10.1007/978-981-95-1533-2_13

0.9 mg/dL. The protocol biopsy was done at 3 months and showed no other evidence of rejection, except C4d positivity.

History of ABO-Incompatible Transplantation

Until the mid-1990s, the poor viability of allografts in ABO-incompatible (ABOI) transplantation, with a 1-year graft survival rate of approximately 4%, limited the feasibility of ABOI kidney transplantation (KT) in routine clinical practice [1, 2]. Chung and colleagues [3] made the initial effort for ABOI-KT in 1955 and revealed that eight of ten kidney transplants failed within a few days after transplant surgery. In 1987, a breakthrough observation revealed that 12 out of 20 transplants from A2 blood group donors into blood group O recipients remained functional for an extended period, primarily due to the weaker expression of the red blood cell A antigen in the A2 donor compared to that of the A1 donor [4]. Moreover, the prevalence of low A2 donors limited the utility to small populations.

Subsequently, it became evident that an effective desensitization protocol using plasma exchange/plasmapheresis (PP), surgical removal of the spleen, donor's thrombocytes/immature T cells transfusion, together with potent immunosuppressants, can achieve favourable outcomes in living donor ABO-KT. This approach also reduced the titers of anti-A or -B antibodies, use of anti-thymocyte globulin (ATG), and other immunosuppressant drugs, such as calcineurin inhibitors, antimetabolites, corticosteroids, and most importantly, concomitant splenectomy, being practised in many parts of the world, an invasive procedure in desensitization [5]. At 1 year, 75% allograft survival and 88% patient survival were observed in the 23 recipients on follow-up [6]. The success story subsequently led to the broader utilization of ABOI-KT, in Japan in the late 1980s, in the USA from the mid-1990s, in Europe from the early 2000s, and subsequently in India and other parts of the world. Still, KT is best performed with ABO compatibility.

ABOI-KT partially fills the gap in reduced supply and high demand for donors, thereby addressing the high burden of ESKD patients. A majority of ABOI-KT since 1989 has been performed in Japan [7], reaching up to 14% of all living donor KTs [4], and presently, up to 20% at our institute. With a notable success story from Japan utilizing novel desensitization techniques and newer immunosuppressive therapies, ABOI-KT also emerged with hopes in Europe and the USA in the early 2000s [8], and lately, in South Asia [9–11]. The success of the ABOI-KT surpassed ABO antibody barriers, reducing waiting time and increasing transplantation rates by up to 30% across the ABOI barrier. A modern desensitization protocol can reduce the need for 90% of patients with an ABO incompatibility with an available living donor.

ABO Antigen Systems

The ABO blood group antigen system consists of oligosaccharides that are predominantly expressed on RBCs, endothelial cells, tubules, and glomeruli of the kidneys, making it necessary to match the ABO antigen for successful KT. The antigen density varies significantly across different blood groups. Blood group A2 recipients, comprising 20% of all Caucasian individuals and only 1.1% in South Indians, [12–14] have a poor exhibition of blood group antigen (30–50%) on the RBC surface, that entrust lower immunogenicity of organs from blood group A2 donors, compared to that of the Blood group A1 and B [13, 14]. Transplantation from A2 to O, A2 to B, or A2B to B, has relatively better outcomes, as compared to those who had a significant A1 barrier [15, 16]. It occurs mainly due to the low expression of A antigen on the endothelial surface of kidneys with A2 kidneys [15, 17]. ABOI-KT with A2 organs has been performed using standard immunosuppressive therapy without extra applications [6]. Moreover, Blood type O recipients produce more anti-A/anti-B immunoglobulin-G (IgG) than A or B blood group recipients [18–21]. Moreover, IgG is primarily associated with both acute antibody-mediated rejection and chronic rejection [22, 23]. Anti-A/B antibodies are formed upon exposure to gut bacteria during early infancy. Inherent anti-A/B antibodies, particularly in individuals with blood group type O, are predominantly of the IgM class, but also include IgG and IgA to some extent [24, 25]. These antibodies are naturally occurring and are not induced by prior exposure to A or B blood group antigens. The presence of IgG anti-A/B antibodies in individuals with blood type O is notable because these antibodies can cross the placenta [24, 25]. While the pathogenic importance of anti-A/B antibodies in solid organ transplantation is established, the relative contribution of the immunoglobulin isotypes and their subclasses to organ rejection remains to be elucidated [24]. The detection of anti-A and anti-B antibodies before transplantation remains a crucial step and methods need to standardized for successful transplantation [26].

Blood Group Antigens as a Barrier, Their Antigenicity and Pathogenicity

Figure 13.1 illustrates the mechanisms of complications and barriers associated with ABOI transplantation Anti-A/B can bind to A/B antigen expressed on the endothelial cells of the glomerulus and stimulate the complement system, leading to the formation of the membrane attack complex and the release of C3a and C5a, which recruit neutrophils to the site of reaction [27]. The activated endothelial cells also secrete cytokines, contributing to inflammation, and neutrophils release lytic enzymes, which cause the lysis of endothelial and other cells. The activated endothelial cells also activate platelets and secrete von Willebrand factors, leading to the formation of thrombi in tiny capillaries, finally manifesting as hyperacute rejection and severe acute antibody-mediated rejection [28].

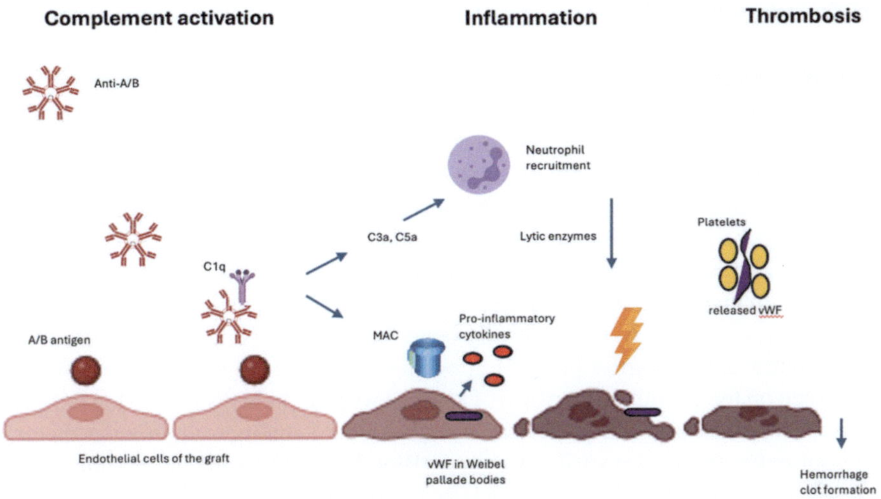

Fig. 13.1 Mechanisms of injury of the allograft with anti-A/B antibodies

Methods for Detecting Anti-A/B Before ABOI-KT

The methods to detect anti-A/B titers and desensitization protocols vary widely across the centers around the globe; however, the basic principles are consistently the same to decrease the anti-A/B titer to an admissible level from baseline and prevent the further formation of antibodies for 10–14 days till accommodation. There are three commonly used different methods for anti-A/B typing (Table 13.1), [26]: 1. Test tube methods, 2. Gel card/column agglutination method, and 3. Flow cytometric methods [29, 30]. There are some automated methods, such as the

Table 13.1 Comparison of the different procedures for detecting anti-A and anti-B antibodies

Characteristics	Tube method	Column agglutination/ gel card method	Flow cytometry
A column component	No requirement	Sephadex gel or glass bead	No requirement
Red blood cells use	Y	Y	Y
AHG (coombs reagent)	Y	Y	No
Secondary antibody	No	No	Y
Immunoglobulin-M (auto-antibody) expunction	DTT or 2ME	DTT or 2ME	Not needed
Elucidation and result	Agglutination, titer	Agglutination, titer	Fluorescence detection, or titer
Cost effectiveness	Low	Intermediate	Relatively high
Duration of test	30–60 min	30–60 min	1–2 h

AHG antihuman globulin, *Y* yes, *DTT* dithiothreitol, *ME* 2-mercaptoethanol, *MFI* mean fluorescence intensity

Galileo NEO, that can also be used for ABO antibody titration; however, they are not routinely used [31, 32].

The Tube Test (TT) method involves serially diluting the patient's serum and reacting it with known A and B red blood cells in tubes. The maximum dilution that causes visible agglutination (clumping) indicates the antibody titer. The Column Agglutination Technique (CAT), also known as the Gel Card Method, utilizes a column or gel card to perform agglutination reactions, which can be more sensitive and easier than the tube method and is typically recommended for ABOi transplantation. Blood group antibodies are commonly considered as titers using the twofold dilution method, which involves serial dilution of the serum to determine the maximum dilution at which the antibody is still detectable. The titer is considered as the reciprocal of the highest dilution that produces a positive reaction. Dithiothreitol (DTT) treatment can be used to inactivate and dissolve IgM antibodies, which helps to differentiate IgG and IgM antibodies. A titer of <64 is generally considered a low titer, and more than 64 is regarded as a high titer, in general practice. However, the titer threshold for ABOI transplantation may vary from center to center depending on the methods used and the center's experience. For all practical purposes, a threshold of 1:8 of the IgG titer of anti-A/B is considered safe to proceed for ABOI KT. Moreover, there are reports of successful ABOI KT at higher titers. Even though the anti-A/B antibodies recur at high levels after about 2 weeks of transplantation without damaging the organs, an event defined as accommodation. Flow cytometric analysis is rarely used to monitor titers in ABOI transplantation, mainly due to the cost and complexity of the technique. The anti-A/B antibodies (isoagglutinins) titer is usually reduced using desensitization protocols and maintained below a limit considered safe for the initial 2 weeks after transplantation, e.g., <1:32 in the tube technique and mostly 1:8 in column agglutination/gel card techniques.

Accommodation After Transplantation

Accommodation is defined as a condition in which an allograft tissue exhibits resistance to immunological injury and graft dysfunction, even in the presence of antibodies. Accommodation may be induced by manipulating changes in the potency and expression of antigens or antibodies, complement action, and acquired resistance to injury [26, 33].

Accommodation in ABOI Transplantation

The mechanisms of accommodation involve changes in the levels of glycosyltransferases, a product of ABO blood group genes in the graft, that regulate the ABO antigens' expression on the allograft's capillary endothelium [34, 35]. The process of accommodation, which typically occurs within 2 weeks of transplantation, prevents organ and tissue damage from a rebound of high anti-A/B antibodies. The possible mechanisms are considered to be increased local expression of CD45,

CD55, and CD59 complement regulatory proteins, which occurs due to inactivation of the ERK1/2 signalling pathway, that depends on anti-A/B antibodies titer [36]. The reduced antigenicity of the ABO antigens prevents or delays the antigen-antibody reaction. Vascular endothelial injury may occur, potentially leading to the production of inhibitors and antibodies against glycosyltransferases. Besides the upregulation of protective or regulatory genes, other possible mechanisms for accommodation include the replacement of the donor endothelium by host endothelium and the intrinsic resistance of cells and tissues to complement-mediated injury [33, 37]. Host-related factors also play important roles, such as changing immune phenotypes from T helper cells 1 (Th1) to T helper cells 2 (Th2), and the availability of accommodated antibodies, such as IgG2b, which have been less effective in activating complement during accommodation [38–40].

Differences in the Accommodation Process in ABOI and HLAI Transplantation

By inducing cytoprotective genes and regulating apoptotic molecules, including the BCL-2-associated death promoter (BAD), the phosphoinositide 3-kinase/Protein Kinase B (PI3K/AKT) signalling pathway, which is responsible for altering cellular function, plays a key role in defending cells against external stress and apoptosis [41]. Experimental evidence suggests that even low levels of anti-HLA class I antibodies can activate AKT and achieve a shielding effect by inducing antiapoptotic genes, such as A20, Bcl-2, and Bcl-xl, as well as cytoprotective genes, including heme oxygenase 1 (HO-1) and ferritin H [42].

In contrast, ERK1/2 inactivation after anti-A and anti-B antibody bonding exhibits resistance to complement-mediated injury by inducing CD55 and CD59, a crucial condition for accommodation in ABOI transplantation [27]. This complement inhibition typically results in the fact that C4d deposition does not lead to allograft damage in ABOI transplantation; however, C4d deposition is considered a significant factor in antibody-mediated rejection in HLAI transplantation [40]. Finally, it is assumed that accommodation happened when C4d deposition was detected in patients with normal function and structure of the graft in ABOI [43].

Accommodation in HLA Incompatibility

A study showed that preincubation with a higher level of anti-A/B antibody is required to increase the expression of complement regulatory proteins, suggesting that accommodation could be acquired more readily in ABOI transplantation than in HLAI transplantation [33, 44], but ABOI showed a poor prognosis compared with HLAI in liver transplantation, whereas ABOI has a better outcome than HLAI in kidney transplantation [45, 46]. However, the differences in clinical course between graft organs have yet to be elucidated.

After HLA antibody binding to endothelial cells, a higher HLA antibodies titer is required to activate ERK and mammalian target of rapamycin complex- 1 signaling pathway, resulting in cell proliferation [41]; in contrast, low titers of anti-HLA antibodies activate Rho, FAK, and AKT in the mammalian target of rapamycin complex 2 pathway, leading to cell survival through antiapoptotic gene expression. Anti-A/B antibodies do not affect the PI3K/AKT pathway but significantly inactivate the ERK1/2 pathway, increasing complement regulatory proteins without significant changes in cytoprotective proteins such as HO-1 or ferritin H. There are no changes in protective gene expression, such as HO-1, A-20, or Bcl-2 [42]. Experimental evidence also suggests that antibodies to alpha-Gal (carbohydrate epitopes similar to the blood type A/B structure) induce upregulation of CD59, leading to endothelial cell protection, emphasizing the consideration of multiple factors at both the recipient and donor sides [47]. As for graft-related factors, the expression of complement regulatory, antiapoptotic, or antioxidant proteins that can exert a protective function is reportedly involved in accommodation [48].

Accommodation and Bone Marrow Engraftment

The accommodation in bone marrow transplantation occurs even in the case of an ABO incompatible recipient and the donor. The blood group of the recipients are changed to the donor's blood group after donor-derived bone marrow infusion. Still, the original tissue type of organs is conserved by the recipients. Accommodation occurs irrespective of ABO incompatibility in successful bone marrow engraftment [49]. In solid organ transplantation, the recipient's ABO blood type regulates the graft's ABO blood type. In contrast, in bone marrow transplantation, the new ABO blood group type from the donor suppresses and regulates the ABO blood type in the recipient organs. Gene therapy, autologous bone marrow cells, and autologous lymphocytes administered before transplantation may induce active tolerance induction. These cells can be manipulated to express incompatible transplantation carbohydrate antigens by introducing the corresponding glycosyltransferase gene [50, 51].

Desensitization After ABOI Transplantation

The desensitization protocol for ABOI transplantation varies widely across the centers in the countries and across the globe. However, most strategies share the common principles as shown in Box 1:

1. Reducing the B lymphocytes by using the anti-CD20 antibody, Rituximab, or rarely, splenectomy.
2. Reducing the anti-A and anti-B antibodies at the time of transplantation using one or a combination of the techniques, plasmapheresis (PP), double-filtration PP/membrane filtration, or selective or unselective immunoadsorption (IA).

3. Immunomodulation of the immune system of the organ recipients by using intravenous immunoglobulins (IVIG).
4. Preventing the harmful effects of complement activation after anti-A and anti-B antibody binding to the allograft endothelium.

B-Cell Depletion by Splenectomy or Rituximab

Splenectomy, an integral part of ABOi KT to deplete B cells in the past, is rarely performed nowadays, mainly due to the comorbidities associated with surgery and the increased risk of infection post-splenectomy. The anti-CD20 antibody rituximab has gradually substituted for it [52]. The possibility of ABOi KT without anti-B cell therapies has also been explored. Initial success with a full dose of Rituximab 375 mg/m^2 10–14 days before the transplantation and a higher risk of infection with this dose prompted the use of a low-dose Rituximab and Rituximab-free protocol ABOI [53]. A study comprising 37 patients who underwent ABOI-KT only after antibody removal by PP without Rituximab, along with standard immunosuppressive therapy, demonstrated 100% patient and graft survival after a median follow-up period of 26 months, with only two episodes of antibody-mediated rejection that were successfully reversed [54]. A rituximab low-dose desensitization protocol was also tried with a 100 mg single-dose rituximab. In ABOI-KT patients with low anti-A/B antibody titers, the desensitization protocol with Rituximab in a single dose prevented acute antibody-mediated rejection without increasing the risk of other adverse events [55]. Moreover, a higher rate of death-censored graft loss was reported in ABOI-KT recipients when Rituximab was not used [56]. We feel that Rituximab should be given, as memory response after graft transplantation may vary, and rebound of anti-A/B may be robustly high [57].

Removal of Antibody by Extracorporeal Therapy

The method of antibody removal practice varies depending on the center's experience and convenience. Antibody removal strategies include the complete removal of plasma proteins such as plasmapheresis (PP), double filtration PP, which depletes portions of the plasma proteins, including the immunoglobulins, and unselective or selective immunoadsorption (IA). PP for desensitization is primarily practised in the USA, membrane separation methods are used in Japan, and unselective and selective IAs are primarily practised in Europe. Selective removal of anti-A and anti-B antibodies is possible using Glycosorb columns. The Glycosorb column contains synthetic terminal trisaccharide A or B blood group antigens linked to a Sepharose matrix, which binds antibodies and gets removed from the body. In addition, it also removes, to some extent, total IgG and IgG against polysaccharide antigens such as anti-Pneumococcus immunoglobulin (IgG) [58]. Moreover, unselective IA is more effective, with a 59% reduction compared to selective IA, which removes anti-A/B antibody columns and anti-A/B IgG, resulting in a 28% median decrease

from baseline [48]. In contrast, unselective IA was less effective in the removal of anti-A/B antibodies. Unselective immunoadsorption (IA) is less effective than selective IA in removing specific anti-A/B antibodies, particularly IgM and IgG3 subclasses. Unselective IA removed 74% of IgM and 72% of IgG3, while selective IA removed 30% of IgM and 42% of IgG3. Possibly a combination of membrane separation with unselective IA may be the most effective tool for removing IgG and IgM antibodies, as well as effector molecules, such as the complement C1q component [59, 60]. However, the clinical outcomes such as survival, kidney function, rejection episodes, or complications do not differ with the use of different IA strategies [61] (Fig. 13.2).

Intravenous Immunoglobulins (IVIG)

IVIG is a key component of desensitization protocols, which aim to reduce antibody levels to the extent that allows for transplantation to happen safely. IVIG contains a variety of antibodies, including those that can neutralize or block the action of anti-A and anti-B antibodies [52]. It helps lower the levels of anti-A and anti-B antibodies in the recipient's blood, which can damage the transplanted organ, thereby mitigating the risk of allograft rejection—a significant barrier in ABO-incompatible transplantation. Many commercially available IVIG products contain detectable

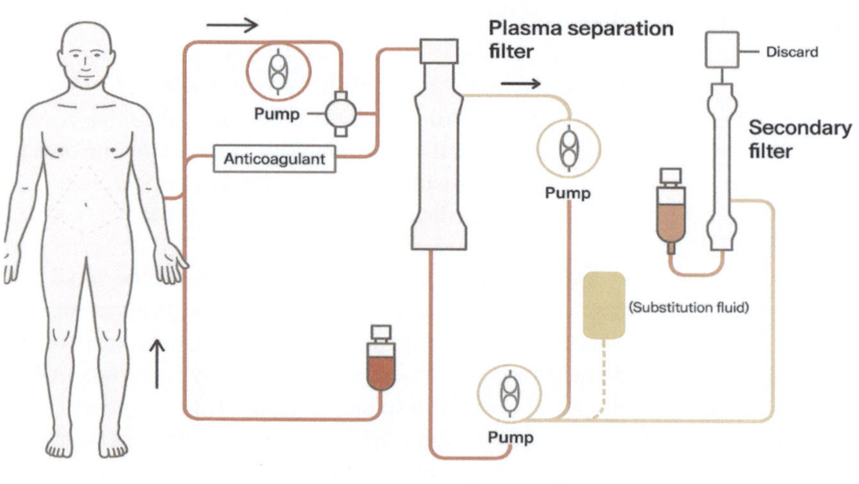

Fig. 13.2 Diagrammatic depiction of double filtration plasmapheresis. Double Filtration Plasmapheresis (DFPP) is a therapeutic apheresis technique that uses two filters to selectively remove large molecules from plasma while returning smaller, beneficial molecules to the patient. The first filter, the plasma separator filter, separates blood into plasma and cellular components. The separated plasma then passes through a second filter, the plasma fractionator, which allows smaller molecules like albumin and specific immunoglobulins to pass through and return to the patient, while larger molecules, such as anti-ABO antibodies, are discarded

levels of anti-A and anti-B antibodies, which can potentially lead to hemolysis [62]. It is given before ABOI-KT and after PP to prevent bouncing back of the anti-A/B antibody in the early transplantation period. It may also neutralize the antibodies, which rebound after PP. IVIG infusion replenishes some immunoglobulin and decreases infectious complications to some extent. However, commercial IVIG preparations may contain IgG antibodies directed against ABO antigens and may also increase anti-A/B antibody titers after IVIG infusion [61, 63].

Inhibition of Complement Activation

The complement inhibitors used in ABOI transplantation are inconclusive. A concept is emerging that, after binding antibodies to the endothelium, complement is activated. Inhibiting complement activation in such a situation may be beneficial. In a study of simultaneous ABOI-KT and pancreas transplantation [64], severe rebound of anti-A/B antibody led to a severe antibody-mediated rejection. Eculizumab successfully reversed rejection by inhibiting the formation of the membrane attack complex. Eculizumab has been used for the prevention of antibody-mediated rejection in ABO-incompatible kidney transplant recipients before transplantation [65]. However, evidence for such use is still evolving [52].

Desensitization Protocols and Pre-conditioning Before Transplantation

Figure 13.3 summarizes the desensitization protocols for ABOI-KT used at several transplant centers across the world, including ours. Most centers have modified the original successful protocol to suit their availability and local resources. Desensitization protocols of different institutions from Japan (1); Germany (2); Sweden (3); Johns Hopkins, USA (4); Mayo Clinic, USA (5); Korea (6), and our institute's SGPGIMS protocol (7) are illustrated in Fig. 13.3 [66–71]. The protocols are summarized based on the individual centers' publication with little modifications; moreover, there is a possibility of change in protocols since publication. The number of PP and the titers at which transplantations are performed also varied widely. We modified the Johns Hopkins protocol and successfully performed ABOI KT on approximately 20% of the total transplant population. In summary, most centers, including ours, use Rituximab and start immunosuppression (Tacrolimus and Mycophenolate Mofetil) day 10–14 before scheduled transplantation, plasmapheresis/double filtration/Selective IA, followed by low-dose IVIG in case of PP. We replace half the amount with plasma from a person with an AB blood group, devoid of anti-A/B antibodies to avoid coagulopathy, and half with albumin. Induction is preferably Basiliximab if Rituximab has been given for desensitization. Transplantation is preferably done at an IgG titer of 1:8.

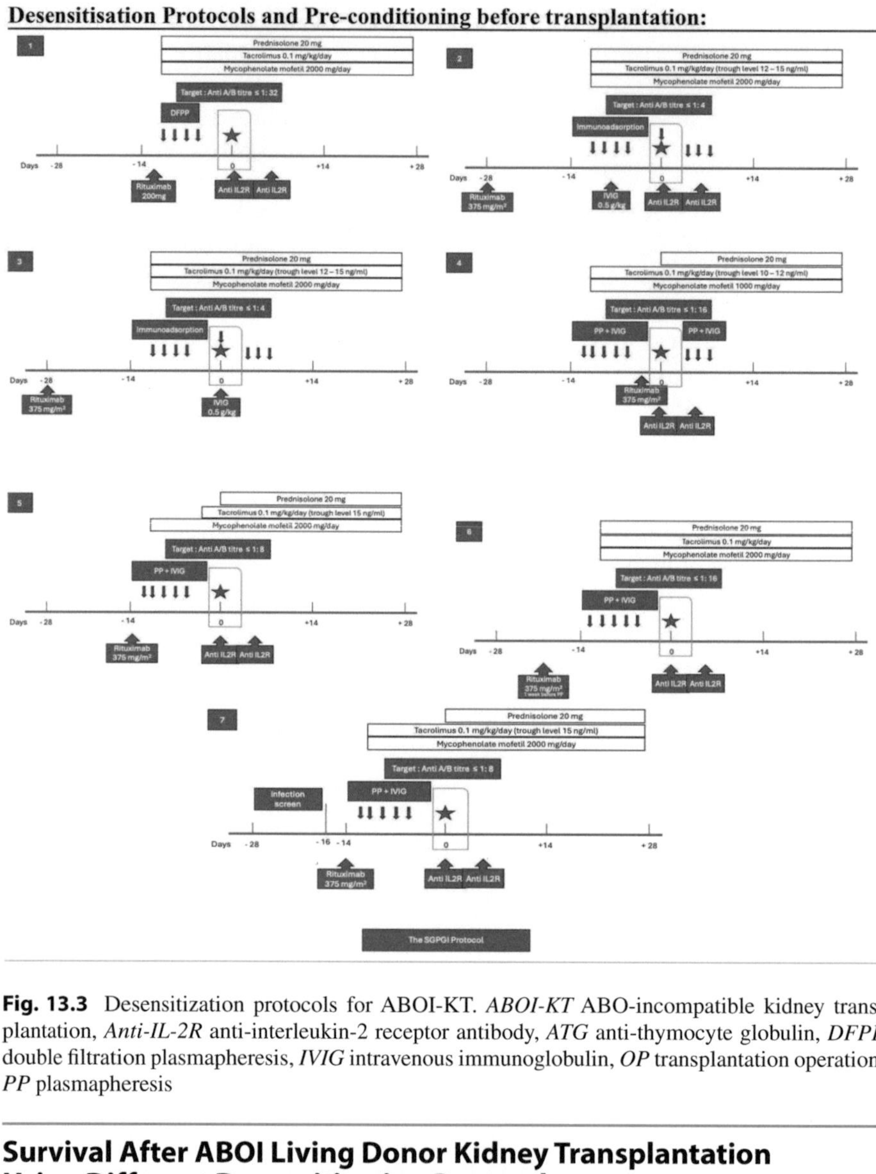

Fig. 13.3 Desensitization protocols for ABOI-KT. *ABOI-KT* ABO-incompatible kidney transplantation, *Anti-IL-2R* anti-interleukin-2 receptor antibody, *ATG* anti-thymocyte globulin, *DFPP* double filtration plasmapheresis, *IVIG* intravenous immunoglobulin, *OP* transplantation operation, *PP* plasmapheresis

Survival After ABOI Living Donor Kidney Transplantation Using Different Desensitization Protocols

Now, data are available from the largest cohort with the most extensive follow-up of 20 years from Japan, utilizing the current practice of double-filtration PP and the use of Rituximab, an anti-CD20 antibody, which has replaced splenectomy. An excellent patient and graft survival rate of 98% and 96% for the first year, and 91% and 83% after 9 years, respectively, was reported in a study of 1427 patients from Japan [72]. In the USA, desensitization was mainly performed using PP and

low-dose IVIG, and a study of 738 patients revealed first-year graft loss of 5.9% in ABOI-KT and only 2.9% in ABOC-KT transplantations, primarily due to rejection within the first 2 weeks after surgery [73].

In early 2003, a Swedish protocol for desensitization in ABOI-KT using selective IA with Glycosorb columns and the anti-CD20 antibody rituximab in place of splenectomy was reported [74]. The success story of the new protocol led to the resurgence of ABOI-KT in Europe, and a larger experience with this protocol was further published [75], revealing an overall patient and graft survival of 93% and 91%, respectively, after 4.5 years of mean follow-up, without experiencing early acute antibody-mediated rejection associated with anti-A/B antibodies. Freiburg group from Germany [76] used the Stockholm protocol and achieved an excellent 10-year patient and death-censored graft survival of 99% and 94%, respectively in 100 ABOI recipients, which did not differ significantly from the 80% and 88% survival rates, respectively, in 248 ABOC transplantations with similar antibody and T cell mediated rejection during the same timeframe. A study from the UK also showed similar death-censored graft survival of 84% after 3 years of ABOI-KT, with a comparison to ABOC-KT. However, a lower patient survival was noticed due to death because of higher infectious complications, mostly *Pneumocystis jirovecii* pneumonia [77].

Since 2006, the Heidelberg group has used a similar protocol to the Swedish protocol for desensitization of ABOI KT [78]. The significant difference was the use of unselective instead of selective IA, which successfully desensitized HLA-incompatible patients for KT. One of the significant differences was the avoidance of IVIG and variable immunoadsorption depending on anti-A/B antibody titer. A minimum of one extra plasmapheresis treatment was performed the day prior to surgery to remove IgM-class anti-A or anti-B antibodies efficiently [79]. A successful desensitization was achieved after a median of six IA treatments in 12 patients [61], with comparable Anti-A/B titer reduction between unselective and selective immunoadsorption. Another study compared desensitization using unselective immunoadsorption in 34 ABOI-KT to 68 matched standard-risk living donor kidney recipients. [79] and showed a graft survival in ABOI-KT recipients was not significantly lower compared to standard risk recipients on 22 months of follow-up. Two deaths occurred, and one of the deaths in the ABOI-KT recipients occurred due to *P. jirovecii* pneumonia at 169 postoperative day which occurred because of intensified immunosuppression, particularly Rituximab used during desensitization. A higher incidence of BK virus replication, considered as $>10^4$ copies/mL in plasma, was observed in 20% in ABOI compared to 6% in standard risk. A greater prevalence of multidrug-resistant bacterial colonization was also observed in about 15% of ABOI compared to that of 1% in ABOC transplantation. In a study of 1420 ABOI recipients comparing ABOC recipients from 101 different centers between 2005 and 2012 [56], overall graft, death-censored graft survival, and patient survival were similar. Early infection-associated death resulted in lower patient survival in ABOI-KT recipients. Specifically, one additional death per 100 patients occurred in the first year after ABOI-KT from an infectious complication [56]. A trend toward better 3-year death-censored graft survival in patients receiving anti-CD20 therapy

after longer follow-up suggested a need for Rituximab in cases of ABOI transplantation. A multicenter retrospective cohort study from India, comparing 1759 living donor ABOI KT and 33,157 ABOC-KT recipients, showed a similar outcome, with little difference in infection and rejection between groups [80]. Another study from India revealed a feasibility of ABOI KT during COVID pandemic without outcome difference [81]. We have shown that the ABOI-KT is feasible in high-risk HLA sensitized patients with the same desensitization protocols [82], and it is possible with PP-based protocol and non-selective IA; however, anti-HLA DSA is associated with poor survival [83].

Infection and Malignancy

Due to a greater immunosuppression in the early phase of ABOI, infectious complications are thought to be high; however, the results are conflicting. Reportedly, there is a higher incidence of viral infections, such as Cytomegalovirus, Herpes simplex, Zoster virus, and BK virus, as well as P. jirovecii pneumonia, wound infections, and urinary tract infections in ABOI-KT [84]. Collaborative Transplant study (CTS) and the Heidelberg cohort reported an additional patient death in 100 ABOI recipients during the first year after surgery due to infections [56, 79]. Other studies have also noted a greater incidence of BK virus replication and BK virus-associated nephropathy in ABOI transplantation [79, 85]. Sharif et al. reported a three times higher incidence of BKV nephropathy in ABOI patients compared to that of the HLAI antibodies, despite a similar degree of immunosuppression [86]. It is speculated that this may have occurred due to the differing binding capacity of viral pathogen receptors to sialic acid on the tubular cells of the kidney. In terms of malignancy outcomes, there was no difference in cancer risk when comparing cancer risk in ABOI-KT recipients and ABOC-KT [87, 88].

Bleeding Risks Involved with ABOI

Coagulopathy due to plasmapheresis as part of desensitization protocol and removal of coagulation factors may lead to bleeding-associated complications. Peri and post-transplant bleeding risk was associated with the number of pre-transplant apheresis treatments [89]. The USRDS data revealed a two times higher risk of early haemorrhage in 119 ABOI-KT recipients compared to ABOC-KT [90]. A transplant center in a cohort of pediatric KT recipients reported two major bleeding episodes out of three patients, attributing them to the nonspecific binding of coagulation factors during IA [90].

Other Surgical Complications with ABOI

A higher surgical complications rate after ABOI-KT due to early intensified immunosuppression and removal of coagulation factors by apheresis has been reported. ABOI-KT patients experienced a higher incidence of lymphoceles compared to ABOC-KT controls (33% vs. 15%), reported by the Freiburg group [76]. This finding required surgical revisions in 20% and 8% of patients, respectively. There was a need for a higher surgical revision rate in ABOI-KT patients compared to ABOC-KT controls (38% vs. 24%).

Future Directions

Endo-beta-galactosidase ex-vivo infusion can effectively remove 82% of A and 95% of B antigens from human A and B red blood cells, thereby suppressing anti-A and anti-B antibody binding and complement activation. It may be one of the future strategies to generalize the feasibility of ABOI without other barriers [91]. A large, paired kidney exchange may be an alternative approach to bypass the ABO antibody barrier completely. However, despite utilizing altruistic donors, there is an accumulation of blood group O recipients on the waiting list in most transplant programs [92]. Desensitization for ABOI-KT remains the only viable option for transplantation to these patients within a reasonable timeframe.

Conclusion

ABOI KT is a feasible approach to increase the living kidney donors pool, and approximately 30% of living donors will be eligible to donate their kidneys for a successful ABOIKT. Initial reports of higher risk of early rejection, infection, and infection-associated death reduced significantly with newer strategies using low-dose Rituximab and immunoadsorption for ABOI. However, a meta-analysis revealed that despite enormous progress in desensitization protocols, higher mortality and loss of kidney grafts occur with ABOI KT compared with ABOC KT within the first 3 years. Most importantly, the long-term outcomes after 5 years yielded equivalent survival rates and organ function [93]. Still, ABOC transplantation should be preferred whenever possible, and alternatively, if feasible, swapping may be offered in select cases. Moreover, ABOI KT is feasible and may be offered to suitable candidates for the benefit of KT, with similar long-term outcomes.

Acknowledgements We acknowledge Dr. Jeyakumar Meyyappan, Assistant Professor in Department of Nephrology, for sketching figures for the book chapter.

References

1. Dunea G, Nakamoto S, Straffon RA, et al. Renal homotransplantation in 24 patients. Br Med J. 1965;1(5426):7–13.
2. Starzl TE, Marchioro TL, Holmes JH, Hermann G, Brittain RS, Stonington OH, Talmage DW, Waddell WR. Renal homografts in patients with major donor-recipient blood group incompatibilities. Surgery. 1964;55(2):195–200.
3. Chung BH, Lee JY, Kang SH, Sun IO, Choi SR, Park HS, Kim J-I, Moon IS, Choi BS, Park CW, Kim Y-S, Yang CW. Comparison of clinical outcome between high and low baseline anti-ABO antibody titers in ABO-incompatible kidney transplantation. Ren Fail. 2011;33(2):150–8.
4. Thielke J, Kaplan B, Benedetti E. The role of ABO-incompatible living donors in kidney transplantation: state of the art. Semin Nephrol. 2007;27(4):408–13. https://doi.org/10.1016/j.semnephrol.2007.03.11].
5. Genberg H, Kumlien G, Wennberg L, Tydén G. Isoagglutinin adsorption in ABO-incompatible transplantation. Transfus Apher Sci. 2010;43(2):231–5. https://doi.org/10.1016/j.transci.2010.07.01612].
6. Sonnenday CJ, Warren DS, Cooper M, Samaniego M, Haas M, King KE, Shirey RS, Simpkins CE, Montgomery RA. Plasmapheresis, CMV hyperimmune globulin, and anti-CD20 allow ABO-incompatible renal transplantation without splenectomy. Am J Transplant. 2004;4(8):1315–22. https://doi.org/10.1111/j.1600-6143.2004.00507.x.2.
7. Ishida H, Tanabe K, Toma H, Akiba T. Therapeutic apheresis therapy for ABO-incompatible renal transplantations. Ther Apher. 2003;7(6):520–8. https://doi.org/10.1046/j.1526-0968.2003.00099.x.13.
8. Genberg H, Kumlien G, Wennberg L, Tydén G. Isoagglutinin adsorption in ABO-incompatible transplantation. Transfus Apher Sci. 2010;43(2):231–5. https://doi.org/10.1016/j.transci.2010.07.016.
9. Jha PK, Bansal SB, Rana A, Nandwani A, Kher A, Sethi S, Jain M, Bansal D, Yadav DK, Gadde A, Mahapatra AK, Sodhi P, Ahlawat R, Kher V. ABO-incompatible kidney transplantation in India: a single-center experience of first hundred cases. Indian J Nephrol. 2022;32(1):42–6.
10. Bhalla AK, Anil KBT, Chauhan M, Das P, Gandhi B, Hegde U, Jeloka T, Mali M, Jha PK, Kher V, Kher A, Mukkavilli KK, Ramachandran R, Jha V. ABO-incompatible kidney transplantation: Indian Working Group Recommendations. Indian J Transplant. 2019;13(4):252–8. https://doi.org/10.4103/ijot.ijot_39_19.
11. Begum NAS, Kashem TS, Nobi F, Arefin SU, Rashid HU. Experiences of performing ABO-incompatible kidney transplantation in Bangladesh. Korean J Transplant. 2022;36(2):111–8. https://doi.org/10.4285/kjt.22.0014.
12. Rudmann SV. Textbook of blood banking and transfusion medicine. WB Saunders; 1995. p. 73–5.
13. Elnour MA, Ali NY, Mohammed HA, Hummeda SA, Alshazally WA, Omer NA. Frequency of the A2-subgroup among blood group A and blood group AB among students of faculty of medicine and health sciences at Alimam Almahadi University, White Nile, Sudan. Hematol Transfus Int J. 2015;1(4):00022.
14. Giriyan SS, Agrawal A, Bajpai R, Nirala NK. A1 and A2 sub-types of blood group 'A': a reflection of their prevalence in North Karnataka region. J Clin Diagn Res. 2017;11(5):EC40–2. https://doi.org/10.7860/JCDR/2017/26772.9893.
15. Breimer ME, Molne J, Norden G, Rydberg L, Thiel G, Svalander CT. Blood group A and B antigen expression in human kidneys correlated to A1/A2/B, Lewis, and secretor status. Transplantation. 2006;82(4):479–85. https://doi.org/10.1097/01.tp.0000231697.15817.51.
16. Alkhunaizi AM, de Mattos AM, Barry JM, Bennett WM, Norman DJ. Renal transplantation across the ABO barrier using A2 kidneys. Transplantation. 1999;67(10):1319–24.
17. Breimer ME, Samuelsson BE. The specific distribution of glycolipid-based blood group A antigens in human kidney related to A1/A2, Lewis, and secretor status of single individuals. A

possible molecular explanation for the successful transplantation of A2 kidneys into O recipients. Transplantation. 1986;42(1):88–91.
18. Krishnan NS, Fleetwood P, Higgins RM, Hathaway M, Zehnder D, Mitchell D, et al. Application of flow cytometry to monitor antibody levels in ABO incompatible kidney transplantation. Transplantation. 2008;86(3):474–7. https://doi.org/10.1097/TP.0b013e31817c4c4c.
19. Contreras M, Armitage SE, Hewitt PE. Response to immunization with A and B human glycoproteins for the procurement of blood grouping reagents. Vox Sang. 1984;47(3):224–35. https://doi.org/10.1111/j.1423-0410.1984.tb01590.x.
20. Stussi G, Huggel K, Lutz HU, Schanz U, Rieben R, Seebach JD. Isotype-specific detection of ABO blood group antibodies using a novel flow cytometric method. Br J Haematol. 2005;130(6):954–63. https://doi.org/10.1111/j.1365-2141.2005.05705.x.
21. Toki D, Ishida H, Horita S, Yamaguchi Y, Tanabe K. Blood group O recipients associated with early graft deterioration in living ABO-incompatible kidney transplantation. Transplantation. 2009;88(10):1186–93. https://doi.org/10.1097/TP.0b013e3181ba07ec.
22. Scornik JC, Salomon DR, Lim PB, Howard RJ, Pfaff WW. Posttransplant antidonor antibodies and graft rejection. Evaluation by two-color flow cytometry. Transplantation. 1989;47(2):287–90. https://doi.org/10.1097/00007890-198902000-00018.
23. Bryan CF, Nelson PW, Helling TS, Shield CF, Beck M, Bryan CF. Current experience with renal transplantation across the ABO barrier. Am J Surg. 1992;164(5):541–4. https://doi.org/10.1016/s0002-9610(05)81197-3.
24. Mironov AA, Savin MA, Zaitseva AV, Dimov ID, Sesorova IS. Mechanisms of formation of antibodies against blood group antigens that do not exist in the body. Int J Mol Sci. 2023;24(20):15044. https://doi.org/10.3390/ijms242015044.
25. Yamamoto F, Clausen H, White T, Marken J, Hakomori S. Molecular genetic basis of the histo-blood group ABO system. Nature. 1990;345:229–33.
26. Hur M, Moon HW, Kwo SW. ABO-incompatible kidney transplantation. In: Understanding the complexities of kidney transplantation. INTECH; 2011. https://doi.org/10.5772/20514.
27. Kenta I, Takaaki K. Molecular mechanisms of antibody-mediated rejection and accommodation in organ Transplantation. Nephron. 2020;144(Suppl 1):2–6. https://doi.org/10.1159/000510747.
28. Stites E, Le Quintrec M, Thurman JM. The complement system and antibody-mediated transplant rejection. J Immunol. 2015;195(12):5525–31.
29. Shim H, Hwang JH, Kang SJ, Seo HS, Park EY, Park KU, Kong SY. Comparison of ABO isoagglutinin titres by three different methods: tube haemagglutination, micro-column agglutination and automated immunohematology analyzer based on erythrocyte-magnetized technology. Vox Sang. 2020;115(3):233–40. https://doi.org/10.1111/vox.12878.
30. Datta SS, Basu S, Reddy M, Gupta K, Sinha S. Comparative evaluation of the conventional tube test and column agglutination technology for ABO antibody titration in healthy individuals: a report from India. Immunohematology. 2021;37(1):25–32. https://doi.org/10.21307/immunohematology-2021-006.
31. Nayak S, Makroo RN, Prakash B, Chandra T, Agrawal S, Chowdhry M, Mohapatra A. Comparative evaluation of five different methods of anti-ABO antibody titration: an aid for ABO-incompatible organ transplants. Ther Apher Dial. 2019;23(1):86–91. https://doi.org/10.1111/1744-9987.12745.
32. Schneider D, Vicarioto M, Coluzzi S, Matteocci A, Revelli N, Foglieni B, Artusi P, Londero D, Quaglietta A, Barrotta G, Visceglie D, Portararo G, Gilsdorf J. ABO antibody titres: a multisite comparative study of equivalency and reproducibility for automated solid-phase and haemagglutination titration, and manual dilution with gel column agglutination technology. Blood Transfus. 2022;20(4):329–37. https://doi.org/10.2450/2022.0197-21.
33. de Mattos G, Barbosa M, Cascalho M, Platt JL. Accommodation in ABO-incompatible organ transplants. Xenotransplantation. 2018;25(3):e12418. https://doi.org/10.1111/xen.12418.
34. Lynch RJ, Platt JL. Accommodation in organ transplantation. Curr Opin Organ Transplant. 2008;13(2):165–70.

35. Park WD, Grande JP, Ninova D, Nath KA, Platt JL, Gloor JM, Stegall MD. Accommodation in ABO-incompatible kidney allografts, a novel mechanism of self-protection against antibody-mediated injury. Am J Transplant. 2003;3(8):952–60. https://doi.org/10.1034/j.1600-6143.2003.00179.x.
36. Forbes RC, Feurer ID, Shaffer D. A2 incompatible kidney transplantation does not adversely affect graft or patient survival. Clin Transpl. 2016;30:589–97.
37. Iwasaki K, Miwa Y, Ogawa H, Yazaki S, Iwamoto M, Furusawa T, et al. Comparative study on signal transduction in endothelial cells after anti-a/b and human leukocyte antigen antibody reaction: implication of accommodation. Transplantation. 2012;93(4):390–7.
38. Vidarsson G, Dekkers G, Rispens T. IgG subclasses and allotypes: from structure to effector functions. Front Immunol. 2014;20(5):520. https://doi.org/10.3389/fimmu.2014.00520.
39. Wasowska BA. Mechanisms involved in antibody- and complement-mediated allograft rejection. Immunol Res. 2010;47(1-3):25–44. https://doi.org/10.1007/s12026-009-8136-3.
40. Murata K, Baldwin WM 3rd. Mechanisms of complement activation, C4d deposition, and their contribution to the pathogenesis of antibody-mediated rejection. Transplant Rev (Orlando). 2009;23(3):139–50. https://doi.org/10.1016/j.trre.2009.02.005.
41. Datta SR, et al. Akt phosphorylation of BAD couples survival signals to the cell-intrinsic death machinery. Cell. 1997;91:231–41.
42. Jin YP, Fishbein MC, Said JW, Jindra PT, Rajalingam R, Rozengurt E, Reed EF. Anti-HLA class I antibody-mediated activation of the PI3K/Akt signaling pathway and induction of Bcl-2 and Bcl-xL expression in endothelial cells. Hum Immunol. 2004;65(4):291–302.
43. Lim BJ. C4d puzzle in ABO-incompatible kidney transplantation. Kidney Res Clin Pract. 2024;43(1):6–7. https://doi.org/10.23876/j.krcp.23.139.
44. Jindra PT, Zhang X, Mulder A, et al. Anti-HLA antibodies can induce endothelial cell survival or proliferation depending on their concentration. Transplantation. 2006;82:S33–5.
45. Hew EY, Kessaris N, Stojanovic J, Jones H, Christian M, Edwards A, Milford DV, Ognjanovic M, Shenoy M, Baker RJ, Marks SD. Successful ABO and HLA incompatible kidney transplantation in children in the UK. Pediatr Nephrol. 2023;38(2):529–35.
46. Jadaun SS, Agarwal S, Gupta S, Saigal S. Strategies for ABO incompatible liver transplantation. J Clin Exp Hepatol. 2023;13(4):698–706. https://doi.org/10.1016/j.jceh.2022.12.008.
47. Galili U. Antibody production and tolerance to the α-gal epitope as models for understanding and preventing the immune response to incompatible ABO carbohydrate antigens and for α-gal therapies. Front Mol Biosci. 2023;28(10):1209974.
48. Bach FH, Ferran C, Hechenleitner P, et al. Accommodation of vascularised xenografts: expression of "protective genes" by donor endothelial cells in a host Th2 cytokine environment. Nat Med. 1997;3:196–204. https://doi.org/10.1038/nm0297-196.
49. Takahashi K. A new concept of accommodation in ABO-incompatible kidney transplantation. Clin Transpl. 2005;19(Suppl 14):76–85.
50. Stussi G, Halter J, Schanz U, Seebach JD. ABO-histo blood group incompatibility in hematopoietic stem cell and solid organ transplantation. Transfus Apher Sci. 2006;35(1):59–69.
51. Galili U. Xenotransplantation and ABO incompatible transplantation: the similarities they share. Transfus Apher Sci. 2006;35(1):45–58.
52. Koo TY, Yang J. Current progress in ABO-incompatible kidney transplantation. Kidney Res Clin Pract. 2015;34(3):170–9.
53. Lee HR, Kim K, Lee SW, Song JH, Lee JH, Hwang SD. Effect of rituximab dose on induction therapy in ABO-incompatible living kidney transplantation: a network meta-analysis. Medicine (Baltimore). 2021;100(10):e24853.
54. Flint SM, Walker RG, Hogan C, Haeusler MN, Robertson A, Francis DM, Millar R, Finlay M, Landgren A, Cohney SJ. Successful ABO-incompatible kidney transplantation with antibody removal and standard immunosuppression. Am J Transplant. 2011;11(5):1016–24.
55. Okada M, Narumi S, Hasegawa Y, Futamura K, Hiramitsu T, Ichimori T, Goto N, Kobayashi T, Uchida K, Takeda A, Watarai Y. Optimal dose of rituximab in ABO-incompatible kidney transplantation in patients with low anti-A/B antibody titers: a single-center retrospective cohort study. Clin Transpl. 2023;37(2):e14915.

56. Opelz G, Morath C, Susal C, Tran TH, Zeier M, Dohler B. Three-year outcomes following 1420 ABO-incompatible living-donor kidney transplants performed after ABO antibody reduction: results from 101 centers. Transplantation. 2015;99(2):400–4.
57. Lee J, Lee JG, Kim S, Song SH, Kim BS, Kim HO, Kim MS, Kim SI, Kim YS, Huh KH. The effect of rituximab dose on infectious complications in ABO-incompatible kidney transplantation. Nephrol Dialysis Transplant. 2016;31(6):1013–21.
58. Valli PV, Puga Yung G, Fehr T, Schulz-Huotari C, Kaup N, Gungor T, et al. Changes of circulating antibody levels induced by ABO antibody adsorption for ABO-incompatible kidney transplantation. Am J Transplant. 2009;9(5):1072–80.
59. Wahrmann M, Schiemann M, Marinova L, Kormoczi GF, Derfler K, Fehr T, et al. Anti-A/B antibody depletion by semiselective versus ABO blood group-specific immunoadsorption. Nephrol Dial Transplant. 2012;27(5):2122–9.
60. Eskandary F, Wahrmann M, Biesenbach P, Sandurkov C, Konig F, Schwaiger E, et al. ABO antibody and complement depletion by immunoadsorption combined with membrane filtration—a randomized, controlled, cross-over trial. Nephrol Dial Transplant. 2014;29(3):706–14.
61. Morath C, Becker LE, Leo A, Beimler J, Klein K, Seckinger J, et al. ABO-incompatible kidney transplantation enabled by non-antigen-specific immunoadsorption. Transplantation. 2012;93(8):827–34.
62. Quinti I, Pulvirenti F, Milito C, Granata G, Giovannetti G, La Marra F, Pesce AM, et al. Hemolysis in patients with antibody deficiencies on immunoglobulin replacement treatment. Transfusion. 2015;55(5):1067–74.
63. Staley EM, Carruba SS, Manning M, Pham HP, Williams LA III, Marques MB, et al. Anti-blood group antibodies in intravenous immunoglobulin may complicate interpretation of antibody titers in ABO-incompatible transplantation. Am J Transplant. 2016;16(8):2483–6.
64. Biglarnia AR, Nilsson B, Nilsson T, von Zur-Muhlen B, Wagner M, Berne C, et al. Prompt reversal of a severe complement activation by eculizumab in a patient undergoing intentional ABO-incompatible pancreas and kidney transplantation. Transpl Int. 2011;24(8):e61–6.
65. Lee JH, Lee H, Kim K, Lee SW, Song JH, Hwang SD. Acute anti-A/B antibody-mediated rejection after ABO-incompatible kidney transplantation treated with bortezomib and plasmapheresis: a case report. Transplant Proc. 2022;54(2):540–3. https://doi.org/10.1016/j.transproceed.2021.10.027.
66. Toki D, Ishida H, Setoguchi K, Shimizu T, Omoto K, Shirakawa H, Iida S, Horita S, Furusawa M, Ishizuka T, Yamaguchi Y, Tanabe K. Acute antibody-mediated rejection in living ABO-incompatible kidney transplantation: long-term impact and risk factors. Am J Transplant. 2009;9:567–77.
67. Gloor JM, Lager DJ, Fidler ME, Grande JP, Moore SB, Winters JL, Kremers WK, Stegall MD. A comparison of splenectomy versus intensive posttransplant antidonor blood group antibody monitoring without splenectomy in ABO-incompatible kidney transplantation. Transplantation. 2005;80:1572–7.
68. Genberg H, Kumlien G, Wennberg L, Berg U, Tyden G. ABO- incompatible kidney transplantation using antigen-specific immunoadsorption and rituximab: a 3-year follow-up. Transplantation. 2008;85:1745–54.
69. Geyer M, Fischer KG, Drognitz O, Walz G, Pisarski P, Wilpert J. ABO-incompatible kidney transplantation with antigen-specific immunoadsorption and Rituximab—insights and uncertainties. Contrib Nephrol. 2009;162:47–60.
70. Takahashi K, Saito K, Takahara S, Okuyama A, Tanabe K, Toma H, Uchida K, Hasegawa A, Yoshimura N, Kamiryo Y. Excellent long-term outcome of ABO-incompatible living donor kidney transplantation in Japan. Am J Transplant. 2004;4:1089–96.
71. Montgomery RA, Locke JE, King KE, Segev DL, Warren DS, Kraus ES, Cooper M, Simpkins CE, Singer AL, Stewart ZA, Melancon JK, Ratner L, Zachary AA, Haas M. ABO incompatible renal transplantation: a paradigm ready for broad implementation. Transplantation. 2009;87:1246–55.
72. Takahashi K, Saito K. ABO-incompatible kidney transplantation. Transplant Rev. 2013;27(1):1–8. https://doi.org/10.1016/j.trre.2012.07.003).

73. Montgomery JR, Berger JC, Warren DS, James NT, Montgomery RA, Segev DL. Outcomes of ABO-incompatible kidney transplantation in the United States. Transplantation. 2012;93(6):603–9. https://doi.org/10.1097/TP.0b013e318245b2af.
74. Tydén G, Kumlien G, Fehrman I. Successful ABO-incompatible kidney transplantations without splenectomy using antigen-specific immunoadsorption and rituximab. Transplantation. 2003;76(4):730–1.
75. Genberg H, Kumlien G, Wennberg L, Tyden G. The efficacy of antigen-specific immunoadsorption and rebound of anti-A/B antibodies in ABO-incompatible kidney transplantation. Nephrol Dial Transplant. 2011;26(7):2394–400.
76. Zschiedrich S, Janigen B, Dimova D, Neumann A, Seidl M, Hils S, et al. One hundred ABO-incompatible kidney transplantations between 2004 and 2014: a single-centre experience. Nephrol Dial Transplant. 2016;31(4):663–71.
77. Barnett AN, Manook M, Nagendran M, Kenchayikoppad S, Vaughan R, Dorling A, et al. Tailored desensitization strategies in ABO blood group antibody incompatible renal transplantation. Transpl Int. 2014;27(2):187–96.
78. Becker LE, Susal C, Morath C. Kidney transplantation across HLA and ABO antibody barriers. Curr Opin Organ Transplant. 2013;18(4):445–54.
79. Becker LE, Siebert D, Susal C, Opelz G, Leo A, Waldherr R, et al. Outcomes following ABO-incompatible kidney transplantation performed after desensitization by nonantigen-specific immunoadsorption. Transplantation. 2015;99(11):2364–71.
80. Kute VB, Pathak V, Ray DS, Bhalla AK, Godara SM, Narayanan S, Hegde U, Das P, Jha PK, Kher V, Dalal S, Bahadur MM, Gang S, Sinha VK, Patel HV, Deshpande R, Mali M, Sharma A, Das SS, Thukral S, Shingare A, Bt AK, Hafeeq B, Aziz F, Aboobacker IN, Gopinathan JC, Dave RM, Bansal D, Anandh U, Singh S, Kriplani J, Bavikar S, Siddini V, Balan S, Singla M, Chauhan M, Tripathi V, Patwari D, Abraham AM, Chauhan S, Meshram HS. A multicenter retrospective cohort study on management protocols and clinical outcomes after ABO-incompatible kidney transplantation in India. Transplantation. 2024;108(2):545–55. https://doi.org/10.1097/TP.0000000000004789.
81. Kute V, Ray DS, Dalal S, Hegde U, Godara S, Pathak V, Bahadur MM, Khullar D, Guleria S, Vishwanath S, Singhare A, Yadav D, Bansal SB, Chauhan S, Meshram HS. A Multicenter cohort study from India of ABO-incompatible kidney transplantation in post-COVID-19 patients. Transplant Proc. 2022;54(10):2652–7.
82. Prasad N, Chellapan A, Srivastava A, Bhadauria D, Kaul A, Patel M, Gupta A. ABO-incompatible repeat kidney transplantation: coping with the 'twin immunological barrier'. Indian J Nephrol. 2022;32(1):82–6. https://doi.org/10.4103/ijn.IJN_64_20.
83. Mehrotra S, Sharma RK, Vishwakarma K, Prasad N, Gupta A, Bhadauria DS, Kaul A. Donor specific anti HLA sensitization is associated with inferior short term outcome in ABO- incompatible renal transplantation. Transplant Rep. 2020;5(3):100059. https://doi.org/10.1016/j.tpr.2020.100059.
84. Habicht A, Broker V, Blume C, Lorenzen J, Schiffer M, Richter N, et al. Increase of infectious complications in ABO-incompatible kidney transplant recipients—a single centre experience. Nephrol Dial Transplant. 2011;26(12):4124–31.
85. Lentine KL, Axelrod D, Klein C, Simpkins C, Xiao H, Schnitzler MA, et al. Early clinical complications after ABO-incompatible live-donor kidney transplantation: a national study of medicare-insured recipients. Transplantation. 2014;98(1):54–65.
86. Sharif A, Alachkar N, Bagnasco S, Geetha D, Gupta G, Womer K, et al. Incidence and outcomes of BK virus allograft nephropathy among ABO- and HLA-incompatible kidney transplant recipients. Clin J Am Soc Nephrol. 2012;7(8):1320–7.
87. Bentall A, Neil D, Sharif A, Ball S. ABO-incompatible kidney transplantation is a novel risk factor for BK nephropathy. Transplantation. 2015;99(2):e8–9. https://doi.org/10.1097/TP.0000000000000483.
88. Hall EC, Engels EA, Montgomery RA, Segev DL. Cancer risk after ABO-incompatible living-donor kidney transplantation. Transplantation. 2013;96(5):476–9.

89. de Weerd AE, van Agteren M, Leebeek FW, Ijzermans JN, Weimar W, Betjes MG. ABO-incompatible kidney transplant recipients have a higher bleeding risk after antigen-specific immunoadsorption. Transpl Int. 2015;28(1):25–33.
90. Schaefer B, Tonshoff B, Schmidt J, Golriz M, Mehrabi A, Gombos P, et al. Bleeding complications in pediatric ABO-incompatible kidney transplantation. Pediatr Nephrol. 2013;28(2):327–32.
91. Kobayashi T, Liu D, Ogawa H, Miwa Y, Nagasaka T, Maruyama S, et al. Removal of blood group A/B antigen in organs by ex vivo and in vivo administration of endo-beta-galactosidase (ABase) for ABO-incompatible transplantation. Transpl Immunol. 2009;20(3):132–8.
92. Roodnat JI, van de Wetering J, Claas FH, Ijzermans J, Weimar W. Persistently low transplantation rate of ABO blood type O and highly sensitized patients despite alternative transplantation programs. Transpl Int. 2012;25(9):987–93.
93. Scurt FG, Ewert L, Mertens PR, Haller H, Schmidt BMW, Chatzikyrkou C. Clinical outcomes after ABO-incompatible renal transplantation: a systematic review and meta-analysis. Lancet. 2019;393(10185):2059–72. https://doi.org/10.1016/S0140-6736(18)32091-9.

Approach to Lung Transplant Immunology

14

Soumitra Sinha Roy and Georgi Abraham

Case Study: A Complex Journey of Lung Transplantation and Antibody-Mediated Rejection

Mr. P, a 67-year-old former chain smoker, presented a challenging clinical picture that encapsulates many of the complexities of modern lung transplantation. With a history of severe COPD requiring high-flow oxygen, insulin-dependent type 2 diabetes, and complete atherosclerotic blockage of his left vertebral artery, he had previously undergone endobronchial valve placement 6 years prior to his eventual bilateral lung transplantation 2 years ago at our center. His pre-transplant condition was particularly dire, with a BODE index of 10 reflecting the severity of his disease.

The initial post-transplant course was remarkably successful, with Mr. P achieving excellent pulmonary function and returning to an active lifestyle that included mountaineering. However, this success story took a dramatic turn when he contracted influenza while attending a crowded cricket match in a stadium 2 months ago. The viral infection triggered a severe antibody-mediated rejection (AMR), characterized by a precipitous decline in lung function to 50% of predicted FEV1, accompanied by obstructive changes on pulmonary function testing.

Diagnostic evaluation revealed concerning findings: lung biopsy demonstrated A3B3 rejection (Figs. 14.1 and 14.2), C4d staining was positive, and sensitive single bead antigen testing detected multiple donor-specific antibodies (DSAs) with high mean fluorescence intensity (MFI ≈ 2500) against HLA-A and HLA-DQ antigens. Initial treatment with methylprednisolone bolus and tapering steroids proved inadequate, prompting escalation to rituximab over four weekly cycles and followed by intravenous immunoglobulin (IVIG).

S. S. Roy (✉)
Lung Transplant Unit, MGM Healthcare, Chennai, Tamil Nadu, India

G. Abraham
The Madras Medical Mission Hospital, Chennai, Tamil Nadu, India

G. Abraham et al. (eds.), *Immunological Challenges in Organ Transplantation*, https://doi.org/10.1007/978-981-95-1533-2_14

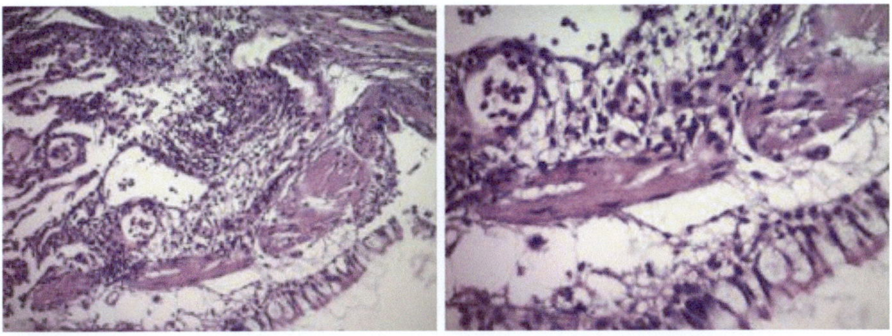

Fig. 14.1 Bronchial epithelium intact but sub epithelial lymphocytes extending into deeper muscle layer—Diagnosis—ACR—A3B3. (Photomicrograph of lung biopsy of Mr P)

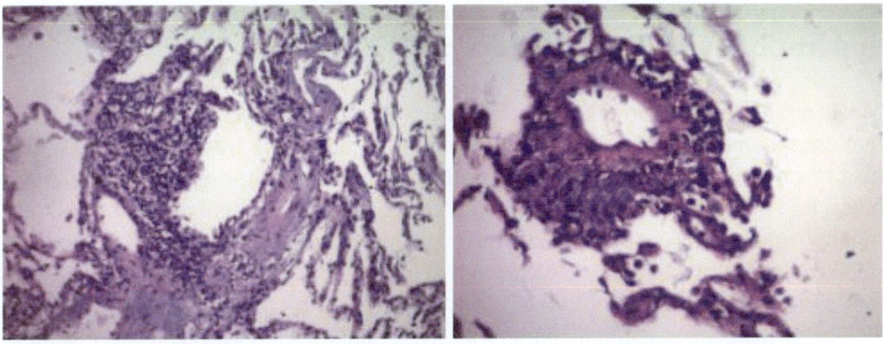

Fig. 14.2 Perivascular lymphocytes 4-5 layers forming a cuff around the capillaries. (Photomicrograph of lung biopsy of Mr P)

When these measures failed to produce significant improvement, and he remains breathless with mild exertion, FEV1 remaining at around 50% with obstructive changes and persistent high DSAs: the clinical team initiated an aggressive course of plasmapheresis consisting of five daily cycles. This intervention proved pivotal, yielding both subjective symptomatic relief and objective improvement in pulmonary function to approximately 75% of predicted FEV1. Laboratory monitoring showed resolution of all DSAs except for one persistent HLA-A antibody but with substantially reduced MFI. A final "mop-up" cycle of rituximab was administered to consolidate these gains, and at most recent follow-up, Mr. P continues to do well.

Introduction: The Evolving Landscape of Lung Transplantation

Since the first successful lung transplantation in 1963, the procedure has become an established treatment for various end-stage pulmonary diseases, including idiopathic pulmonary fibrosis, connective tissue disease-associated interstitial lung

disease, chronic obstructive pulmonary disease, pulmonary hypertension, cystic fibrosis, and both primary ciliary dyskinesia and non-cystic fibrosis bronchiectasis. In India and other regions with high tuberculosis prevalence, post-tuberculosis lung destruction represents an additional important indication.

Despite significant technical and pharmacological advances, long-term outcomes remain suboptimal compared to other solid organ transplants, with a median post-transplant survival of just 6.7 years. This disappointing statistic primarily reflects the challenges of acute and chronic allograft rejection, against which conventional immunosuppressive regimens often prove inadequate [1, 2]. Recent insights into the lung's unique immunological properties have begun to explain these limitations while simultaneously suggesting novel therapeutic approaches [3].

This chapter will systematically examine current immunosuppressive strategies in lung transplantation, focusing on their distinctive aspects compared to other organ transplants. We will then explore recent advances in our understanding of rejection mechanisms before surveying emerging therapeutic modalities that hold promise for improving long-term outcomes. These include innovative preservation techniques, cellular therapies, targeted anti-inflammatory agents, memory T-cell modulation strategies, and tolerance induction protocols (Fig. 14.3).

Induction Immunosuppression: Balancing Risk and Benefit

Approximately 40–50% of transplant centers worldwide employ induction immunosuppression, an intensive perioperative strategy designed to mitigate early rejection risk while and thus allowing delayed introduction of nephrotoxic calcineurin inhibitors. Current International Society for Heart and Lung Transplantation (ISHLT) registry data indicate that interleukin-2 receptor antagonists (IL2RAs), particularly basiliximab, represent the most commonly used induction agents [5].

Basiliximab's mechanism of action involves selective blockade of the CD25 α-subunit of the IL-2 receptor on activated T lymphocytes, thereby inhibiting their proliferation. Typically administered as 20 mg intravenous infusions on postoperative days 0 and 4, it finds particular use when delayed calcineurin inhibitor initiation is desired and it is also used in Burkholderia cepacia-positive recipients and thus it is not used routinely for onduction therapy.

Anti-thymocyte globulin (ATG), available in both rabbit (rATG) and equine (eATG) formulations, ranks as the second most prevalent induction choice and used by about 20% of the centers [5]. These polyclonal antibody preparations induce profound T-cell depletion and are often reserved for sensitized patients with positive virtual crossmatches, typically administered at 3–5 mg/kg over 3–5 days following plasmapheresis.

The humanized anti-CD52 monoclonal antibody alemtuzumab has emerged as a more recent addition to the induction armamentarium. By targeting CD52-expressing lymphocytes (including T cells, B cells, natural killer cells, and monocytes), it achieves comprehensive and prolonged lymphocyte depletion, with B cell recovery occurring within 3–6 months and T cell recovery >12 months [6, 7]. This

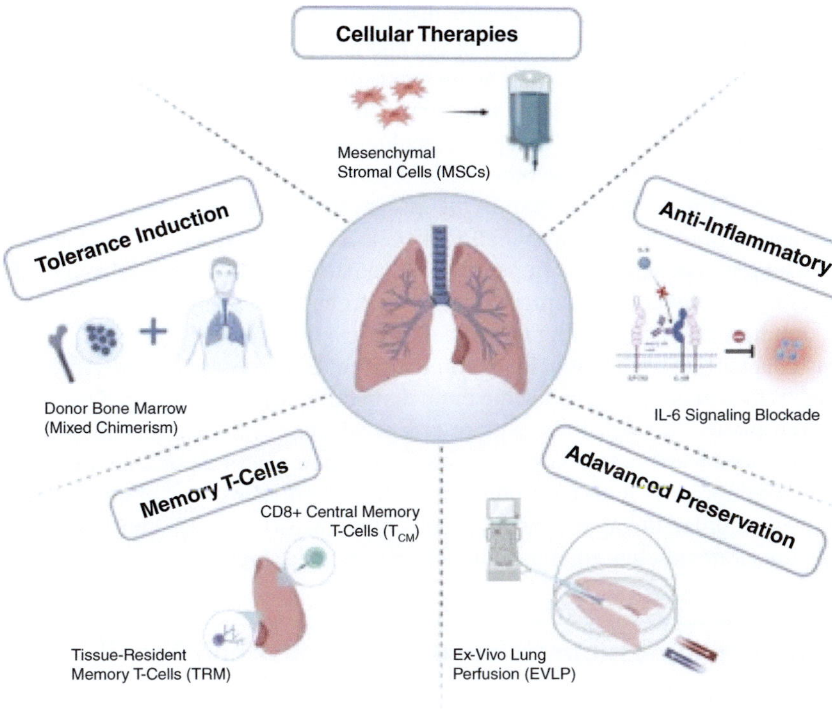

Fig. 14.3 Novel approaches to achieve long-term lung transplant survival. (Represented with permission: Miller et al. [4])

extended immunosuppressive effect allows for significant reduction in maintenance therapy, with many centers withdrawing mycophenolate mofetil or azathioprine for the first post-transplant year while maintaining lower-dose calcineurin inhibitors.

Clinical outcomes data reveal nuanced differences between induction strategies. A 2008 ISHLT registry analysis of 3970 recipients demonstrated 4-year graft survival rates of 64% with IL2RAs, 60% with ATG, and 57% without induction ($p = 0.0067$) [8]. More recent single-center studies suggest alemtuzumab-based regimens may achieve comparable or superior outcomes [9, 10], though large randomized controlled trials are needed to confirm these observations and establish optimal protocols.

The decision to employ induction therapy requires careful consideration of competing risks. While reducing acute rejection, these potent regimens increase vulnerability to infections and malignancies with infections being particularly hazardous in Indian circumstances. This risk-benefit calculus underscores the need for well-designed clinical trials comparing outcomes across induction strategies, particularly with respect to bronchiolitis obliterans syndrome development, acute rejection, graft and patient survival, and infectious complications.

Maintenance Immunosuppression in Lung Transplantation

The foundation of long-term graft survival lies in carefully calibrated maintenance immunosuppression, a delicate balancing act between preventing rejection and avoiding medication toxicity. Since the early days of lung transplantation, when cyclosporine and azathioprine formed the therapeutic backbone, we have witnessed significant evolution in our pharmacological approach. The contemporary standard of care now centers on a three-pronged strategy combining calcineurin inhibitors, antiproliferative agents, and corticosteroids—each addressing distinct pathways in the immune response. Current standard-of-care regimens predominantly utilize tacrolimus, mycophenolate mofetil (MMF), and prednisolone due to their superior efficacy over historical alternatives. Selection of regimens is largely protocolized and based on studies from other types of organ transplantation as well as currently available literature in lung transplant, and center-specific outcomes and experience.

Calcineurin Inhibitors

- **Tacrolimus:** At the heart of modern regimens stands tacrolimus, a calcineurin inhibitor whose superior efficacy in preventing chronic lung allograft dysfunction [11] has made it the agent of choice in most transplant centers worldwide. Its remarkable potency, estimated at 10–100 times greater than cyclosporine, comes with unique challenges including unpredictable absorption and numerous drug interactions that demand vigilant therapeutic monitoring. These considerations take on particular significance in regions like India, where the frequent need for concurrent antifungal and anti-tuberculosis therapy creates complex pharmacological landscapes that clinicians must navigate. Target trough levels (7–20 ng/mL) are center-specific and adjusted based on post-transplant duration, rejection history, and concurrent medications. Sublingual administration (50% oral dose) is an alternative for patients with oral intolerance.
- **Cyclosporine:** Modified formulations (e.g., Neoral®) offer improved bioavailability. Therapeutic drug monitoring employs trough (C0: 150–350 ng/mL) or 2-h post-dose (C2: 800–1200 ng/mL) levels, with C2 correlating better with AUC in Lung Transplantation in comparison to renal transplant [12] and potentially reducing nephrotoxicity [13, 14]. Intravenous dosing is typically one-third of oral equivalents.

Antiproliferative Agents

- **Mycofenolate:** The second pillar of maintenance therapy, mycophenolate mofetil, has largely supplanted azathioprine due to its more favourable efficacy profile. It is available as Mycophenolate mofetil (for IV or NG route usage) and mycophenolate sodium (for oral route usage). By selectively targeting T and B lymphocyte proliferation through inhibition of inosine monophosphate dehydro-

genase, it achieves potent immunosuppression while minimizing effects on other cell lineages. Dosing requires careful adjustment in special populations, particularly those with Burkholderia cepacia infection where higher doses may pose unacceptable risks. Dosing ranges from 1–1.5 g/day (divided BID), reduced to 750 mg/day in *Burkholderia cepacia*-positive recipients. Mycofenolate is preferred over azathioprine in sensitized (PRA+) or heart-lung transplant patients.

- **Azathioprine:** Now limited only to non-sensitized (DSA-negative) recipients, typically dosed at 1.5–2 mg/kg/day (or 1 mg/kg/day in *B. cepacia* cases), with WBC monitoring (>3.0 × 10^3/μL). It can be used IV, only if NPO (1:1 conversion).

Corticosteroids (CS) complete the therapeutic triad, their enduring role a testament to their broad anti-inflammatory effects that range from NF-κB pathway inhibition to cytokine modulation. While the quest to minimize steroid exposure has led some centers to attempt withdrawal in stable patients after about 3 years, this approach remains controversial and requires meticulous monitoring. It remains a corner stone of both induction and maintenance immunosuppression but they are also used to treat acute cellular rejection (ACR). The most commonly used CS in solid organ transplant are methylprednisolone and prednisone and Methylprednisolone (500–1000 mg IV intraoperatively) is usually tapered to oral Prednisolone 5–10 mg/day by 3 months,

Emerging strategies have introduced mTOR inhibitors like everolimus into the maintenance arsenal, particularly for their antifibrotic properties that may slow the progression of bronchiolitis obliterans syndrome [15]. However, their use demands careful timing due to early postoperative risks in the first 3 months for impaired wound healing leading to anastomosis dehiscence and pneumonitis that can mimic chronic rejection. It is possible to reduce or replace completely the CNI drugs later on towards renal preservation by introduction of these drugs in the regimen. Everolimus has comparable efficacy as with Tacrolimus. However, immunosuppressive regimens containing low dose CNI have superior survival outcomes, highlighting the importance of retaining a CNI even in low dose along with Everolimus wherever possible.

Despite these advances, significant challenges persist. The specters of acute rejection and chronic lung allograft dysfunction continue to haunt long-term outcomes, reminding us that current regimens remain imperfect. This reality underscores the importance of ongoing research into personalized approaches that may 1 day allow us to tailor immunosuppression to individual patient profiles, optimizing efficacy while minimizing toxicity.

(Figure 14.4 illustrates the molecular targets and mechanisms of these immunosuppressive agents, providing visual reinforcement of these concepts.)

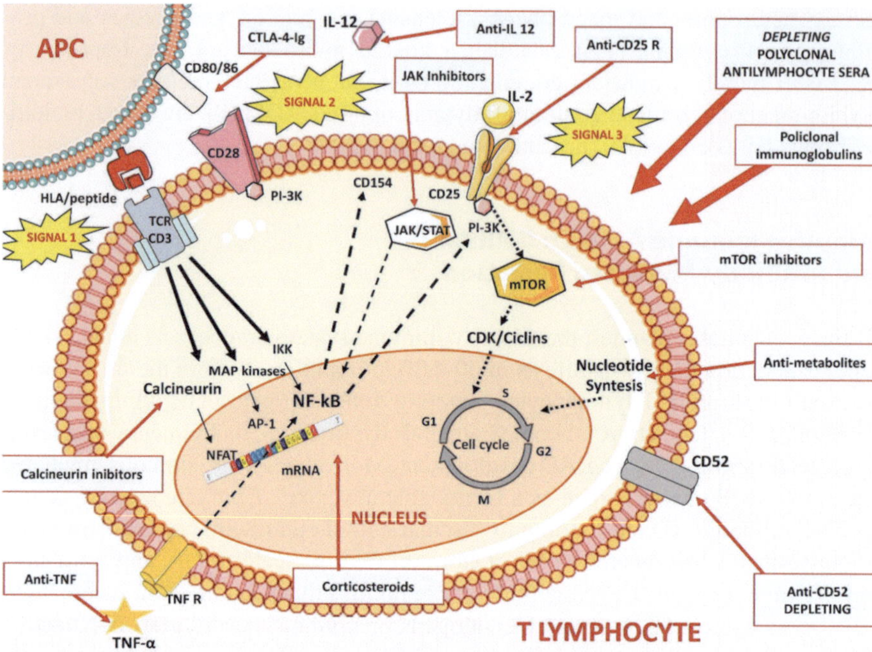

Fig. 14.4 Mechanism of action of immunosuppressive agents. Key: TCR—T cell receptor, NFAT—Nuclear factor of activated T cell, mTOR—mamillian target of Rapamycin, IKK—Inhibitor of kappa B kinase, NF-kB—Nuclear factor kappa B, CTLA-4-Ig—Abatacept, JAK inhibitors-Tofacitinib, Baricitinib, Anti CD 25 R—Daclizumab, Baciliximab, mTor inhibitors—Everolimus, Sirolimus, Anti metabolites—Azathioprine, MMF, Anti CD52—Alemtuzumab, Anti TNF—Etanercept, Infliximab, Adalimumab, Calcineurin inhibitors—Cyclosporin, Tacrolimus. (Represented with permission: Meneghini et al. [16])

Mechanisms of Lung Allograft Failure: A Multifaceted Challenge

The pathobiology of lung allograft failure unfolds through distinct yet interconnected phases, each contributing to the sobering reality that lung transplants have the poorest long-term survival among solid organ transplants. This complex process begins immediately post-transplantation and evolves through progressive immunological stages that ultimately threaten graft viability.

Primary Graft Dysfunction: The Initial Insult

The inaugural threat emerges as primary graft dysfunction (PGD), affecting up to 25% of lung transplants [17]. This severe form of acute lung injury stems from ischemia-reperfusion injury, where the delicate pulmonary endothelium suffers oxidative damage during the transplant process. The resulting inflammatory cascade

recruits neutrophils and macrophages, releasing reactive oxygen species and pro-inflammatory cytokines that establish a hostile microenvironment. Importantly, PGD isn't merely a transient complication—it independently predicts subsequent development of chronic lung allograft dysfunction (CLAD) [18], creating a biological bridge between early injury and late graft failure.

Adaptive Immune Assault: Cellular and Antibody-Mediated Rejection

As the acute phase subsides, the adaptive immune system launches its attack. Acute cellular rejection (ACR) manifests in 30–50% of recipients within the first year [9, 10], with infiltrating T lymphocytes targeting donor antigens. The clinical significance of ACR extends beyond immediate graft dysfunction; its frequency and severity correlate strongly with CLAD development. More insidiously, antibody-mediated rejection (AMR) has emerged as a formidable adversary. Characterized by donor-specific antibodies (DSA) binding to vascular endothelium with complement deposition (C4d+), AMR creates a perfect storm of microvascular injury and acute lung parenchymal damage. The case of Mr. P exemplifies this phenomenon, where high-MFI DSAs against HLA-A and DQ antigens precipitated severe graft dysfunction that proved refractory to initial therapies. AMR also predispose to CLAD later on.

Chronic Lung Allograft Dysfunction: The Final Common Pathway

The relentless progression to CLAD represents the principal barrier to long-term success, affecting 50% of recipients by 5 years [19]. This fibrotic transformation manifests through distinct phenotypes:

- *Bronchiolitis obliterans syndrome (BOS)*: Characterized by concentric fibrosis obliterating small airways
- *Restrictive allograft syndrome (RAS)*: Marked by alveolar and pleural fibrosis
- *Mixed phenotype*: Combining features of both patterns

The pathogenesis of CLAD reflects a complex interplay between alloimmune responses (both cellular and antibody-mediated) and non-immune factors including respiratory bacterial and viral infections and gastroesophageal reflux. Notably, some cases develop without any identifiable triggers, suggesting alternative pathways to chronic rejection.

The Lung's Immunological Uniqueness

Several distinctive features explain why lungs are particularly vulnerable to rejection:

1. **Environmental Exposure**: As the only transplant organ continuously exposed to the external environment with each breath, the lung faces constant antigenic challenges that stimulate toll-like receptors and perpetuate inflammation towards activating innate and adaptive immune responses, thereby promoting rejection.
2. **Vascular Complexity**: The lung's vast endothelial surface area provides abundant targets for immune attack while increasing susceptibility to ischemia-reperfusion injury and complement-mediated injury, leading to an influx of neutrophils/macrophages into the graft with early T-cell activation and consequent rejection [20].
3. **Lymphoid Richness**: Bronchus-associated lymphoid tissue (BALT) and resident immune cells can prime local immune responses without requiring lymphocyte trafficking to secondary lymphoid organs.
4. **Memory T-Cell Dynamics**: Tissue-resident memory T-cells (TRMs) and CD8+ memory populations play dual roles—both driving rejection and potentially promoting tolerance [21, 22].

Therapeutic Implications and Future Directions

Current immunosuppressive strategies, largely adapted from other organ transplants, fail to address the above-described lung's unique immunobiology [3]. Early graft injury from ischemia-reperfusion and infection has shown to activate innate immune pathways within the lung allograft, which in turn trigger alloantigen-specific T-cell proliferation. During ischemia-reperfusion injury, resident donor monocytes in the lung secrete chemotactic and pro-inflammatory cytokines which facilitate neutrophil entry into the graft, enhancing CD4+ T-cell responses to donor antigens and resulting in PGD. Moreover, lung monocytes can interact with Th17 cells and contribute to the development of CLAD [23].

The predominance of T-cell focused regimens overlooks critical contributions from innate immunity, humoral responses, and local inflammatory processes. This therapeutic gap underscores the urgent need for:

- Novel agents targeting lung-specific rejection pathways
- Strategies to preserve regulatory immune populations
- Biomarkers to predict rejection before functional decline
- Tolerance induction protocols adapted for pulmonary immunity

The case of Mr. P's response to plasmapheresis highlights the importance of addressing antibody-mediated mechanisms, while also illustrating how current therapies often represent salvage rather than preventive approaches. As our understanding of these complex immunological interactions deepens, so too must our therapeutic paradigms evolve to improve long-term outcomes in lung transplantation.

Current Treatment Paradigms for Lung Allograft Rejection

The management of post-transplant rejection remains one of the most challenging aspects of lung transplantation, with distinct therapeutic approaches required for different rejection types. While protocols vary between institutions, current strategies reflect an evolving understanding of rejection pathophysiology.

Acute Cellular Rejection (ACR) Management

For initial episodes of minimal-to-mild ACR, high-dose corticosteroid pulses (methylprednisolone 10–15 mg/kg IV daily for 3–5 days) remain first-line therapy. However, more severe cases (Grade ≥ A2) or steroid-resistant rejection often require lymphocyte-depleting agents. Anti-thymocyte globulin (ATG) administered at 1.5 mg/kg daily for 3–5 days provides broad T-cell suppression, while alemtuzumab (30 mg single dose) offers more profound and prolonged lymphocyte depletion through its anti-CD52 activity. The choice between these agents often depends on center experience and the patient's infection risk profile.

Antibody-Mediated Rejection (AMR) Strategies

The multifaceted approach to AMR reflects its complex pathogenesis:

1. *Plasmapheresis* (5–6 cycles) serves as the cornerstone for rapid DSA removal
2. *IV immunoglobulin* (1–2 g/kg over 3–6 days) modulates complement activation and Fc receptor signalling
3. *B-cell targeting* with rituximab (either 375 mg/m^2 weekly×4 or 1 g biweekly × 2) disrupts antibody production
4. *Plasma cell depletion* using bortezomib (1–1.3 mg/m^2 every 72 h × 4 doses) addresses established antibody factories

Mr. P's case exemplifies this approach, where plasmapheresis achieved what steroids and rituximab/IVIG alone could not—demonstrating the importance of timely, aggressive intervention in AMR.

Chronic Lung Allograft Dysfunction (CLAD) Therapies

The irreversible nature of CLAD has spurred various strategies:

- *Immunomodulation*: Azithromycin's anti-inflammatory effects benefit approximately 30–40% of BOS patients
- *Extracorporeal photopheresis* shows promise in stabilizing lung function

- *Antifibrotics*: Pirfenidone's TGF-β inhibition may slow progression in RAS-predominant cases
- *mTOR inhibitors*: Everolimus/sirolimus offer potential dual benefit through immunosuppression and antifibrotic activity
- Substitution of mycophenolate for Azathioprine or of tacrolimus for cyclosporine

Currently no agent available to date that reverses the process of CLAD and restores lung function, other than re-transplantation, when possible, which itself is extremely challenging both surgically and immunologically.

Emerging Considerations

1. *Precision monitoring*: DSA characterization, donor-derived cell-free DNA and using a sentinel skin flap transplant may guide as early warning for pre-emptive therapy
2. *Maintenance optimization*: Tacrolimus troughs >10 ng/mL and MMF doses >1 g BID may reduce rejection risk
3. *Adjunctive therapies*: Statins and montelukast may attenuate inflammatory pathways in CLAD

Unmet Needs

The lack of CLAD-reversing therapies underscores the importance of:

- Early DSA detection and intervention
- Personalized immunosuppression based on immune-risk profiling
- Novel agents targeting fibrotic pathways (e.g., nintedanib)
- Enhanced infection prophylaxis (encompassing viral, bacterial, and fungal) during rejection treatment

This therapeutic landscape continues to evolve as we better understand the intricate balance between alloimmunity, infection risk, and drug toxicity in lung transplant recipients.

Future Directions in Lung Transplant

Preservation: Advancing Beyond Cold Storage

The critical challenge of ischemia-reperfusion injury (IRI) continues to impact up to 25% of lung transplant recipients, serving as a key mediator of primary graft dysfunction (PGD) and a harbinger of chronic lung allograft dysfunction (CLAD). This persistent clinical problem has driven remarkable innovation in organ

preservation techniques, with ex vivo lung perfusion (EVLP) emerging as a transformative technology that addresses both the quality and quantity limitations of donor lungs.

The Limitations of Conventional Preservation

Traditional cold static preservation, while logistically simple, creates a metabolic crisis for donor lungs:

- Hypothermic (4 °C) storage slows but doesn't halt cellular metabolism
- Anaerobic conditions lead to ATP depletion and ionic imbalance, cell swelling, and impaired enzymatic activity [23].
- Reperfusion triggers a burst of oxidative stress and inflammatory cascades

These pathophysiological changes manifest clinically as PGD, which remains strongly associated with poorer long-term outcomes.

EVLP: A Paradigm Shift in Organ Preservation

EVLP involves machine preservation of donor lungs in a perfused, ventilated, normothermic condition, thereby reducing tissue injury resulting from anaerobic metabolism and hypothermia [24]. EVLP technology represents a fundamental departure from cold storage by maintaining near-physiologic conditions:

- Normothermic perfusion preserves aerobic metabolism
- Ventilation maintains alveolar recruitment
- Continuous monitoring enables functional assessment

Since first used clinically in 1991, EVLP has undergone modifications primarily relating to technical parameters of the platform (static vs. portable) and composition of the perfusate (cellular vs. acellular). Other proposed modifications involve donor lung positioning (supine vs. prone) [25], mode of ventilation (positive vs. negative-pressure) [26], and perfusate temperature (normothermic vs. subnormothermic) [27]. Currently, three different EVLP platforms are approved for clinical use and the major difference between these platforms is that they are either mobile, enabling donor lungs to be placed on EVLP immediately after procurement, or fixed, requiring transport of donor lungs to a specialized perfusion center.

Clinical Applications and Benefits

1. **Marginal Lung Assessment**: EVLP provides a 4–6-h window to evaluate questionable grafts using parameters like:

- Pulmonary vascular resistance
- Oxygenation capacity (PaO_2:FiO_2 ratio)
- Compliance and edema formation

2. **Therapeutic Platform**: Perhaps the most significant advantage of EVLP is the possibility of its usage as therapeutic platform for administration of localized, targeted lung-specific therapies [28, 29].

   - Anti-inflammatory agents (e.g., α1-antitrypsin)
   - Cytokine modulators
   - Cellular therapies (e.g., mesenchymal stem cells)
   - Antivirals and antibiotics

3. **Extended Preservation**: Modified protocols now enable:

   - Safe preservation for 12+ h. EVLP using the Toronto Protocol has been shown to safely maintain lungs for >12 h [30], with extended preservation to 24 h using perfusate modifications that reduce metabolite accumulation and electrolyte imbalances known to occur during prolonged EVLP [31]
     - Intercontinental organ transport
     - Flexible surgical scheduling

Current Evidence and Applications

While randomized trials like INSPIRE showed EVLP reduced severe PGD (PGD3) by 50% without impacting survival [32], its true value may lie in enabling use of extended criteria donors. Till date, EVLP has not evidenced superior survival in clinical studies of standard criteria donor lungs [32, 33] but, remarkably, EVLP-rehabilitated marginal lungs demonstrate outcomes comparable to standard criteria organs, effectively expanding the donor pool by 20–30%. Also, EVLP has not impacted short-term clinical outcomes between groups (duration of intubation, length of intensive care unit (ICU) and hospital stay) [33].

Innovations on the Horizon

1. **Perfusate Optimization**: Hydrogen-enriched solutions in place of usual Perfadex (XVIVO Perfusion AB, Gothenburg, Sweden) solution which is low-potassium dextran solution show promise in reducing oxidative damage
2. **Xenogeneic Cross-Circulation**: Early experiments demonstrate successful recovery of severely injured lungs using swine support
3. **Cellular Reprogramming**: EVLP delivery of gene therapies to modify graft immunogenicity

Global Implementation Challenges

Though EVLP technology is available in India, but it faces significant barriers to its widespread adoption:

- High capital equipment costs (~$250,000 per system)
- Substantial operational expenses (~$15,000 per case)
- Need for specialized training programs
- Limited reimbursement frameworks

Conclusion

EVLP has matured from experimental technology to clinical reality, offering solutions to two fundamental challenges in lung transplantation: improving graft quality and expanding donor availability. While current evidence supports its use primarily for marginal lungs, ongoing refinements in perfusion protocols and therapeutic adjuncts may soon establish EVLP as standard care for all donor lungs. The next decade will likely see this technology evolve from an assessment tool to an active treatment platform capable of modifying graft immunogenicity and function before implantation (Fig. 14.5).

Key Features of MSCs in Lung Transplantation

Mesenchymal stromal cells (MSCs) have emerged as a promising therapeutic option in lung transplantation due to their regenerative, immunomodulatory, and anti-inflammatory properties. Here is a summary of their key features and applications:

Mesenchymal stromal cells (MSCs) have emerged as a highly promising therapeutic approach in lung transplantation due to their unique biological properties and multifaceted mechanisms of action. Originally identified in bone marrow, these cells have since been found as resident populations in nearly every organ system,

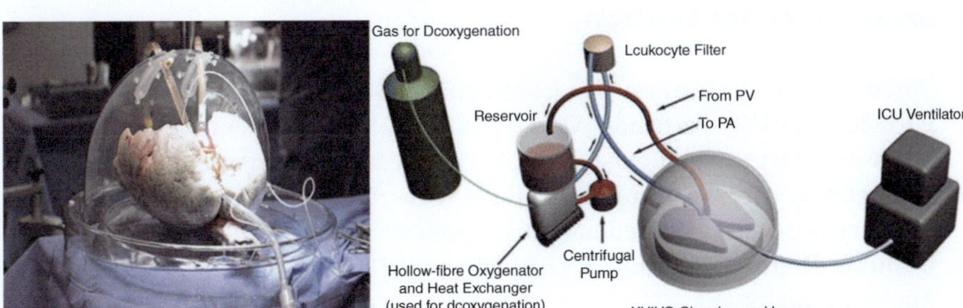

Fig. 14.5 Ex vivo lung perfusion machine and schematics. (Represented with permission: Ex vivo lung perfusion machine and lung: Cypel and Keshavjee [34])

including the lungs, where they play crucial roles in tissue maintenance and repair. In their native microenvironment, MSCs create specialized niches through the secretion of trophic factors and regulatory molecules—in bone marrow they support hematopoiesis, while in lungs they appear to maintain bronchoalveolar stem cell populations and facilitate tissue regeneration.

Several distinctive characteristics make MSCs particularly attractive for lung transplant applications. First, their remarkable longevity and self-renewal capacity is evidenced by studies of gender-mismatched transplants showing donor-derived MSCs persisting years after transplantation. Second, their immunological properties are extraordinary—they express minimal levels of MHC and co-stimulatory molecules, rendering them poorly immunogenic. This has enabled successful transplantation across not just HLA barriers but even species lines, paving the way for off-the-shelf allogeneic therapies using standardized cell lines, often derived from neonatal sources. Third, and perhaps most importantly, MSCs possess potent immunomodulatory capabilities that vary depending on their tissue origin. In vitro studies demonstrate their ability to suppress effector T-cell activation while promoting regulatory T-cell populations, with these effects being amplified in inflammatory environments through mechanisms like TNF-TNFR2 signaling. Their secretome includes a diverse array of cytokines, chemokines, and growth factors with anti-apoptotic, anti-oxidative, and tissue-protective effects.

The delivery and targeting of MSCs presents unique advantages in lung transplantation. When administered intravenously, their relatively large size causes them to become trapped in pulmonary vasculature, creating a natural first-pass effect. More recently, ex vivo lung perfusion (EVLP) systems have emerged as an innovative platform for targeted MSC delivery, allowing pretreatment of donor lungs while avoiding potential complications of systemic administration. This approach may be particularly valuable for preventing ischemia-reperfusion injury before it occurs.

Preclinical studies using MSC therapies in lung injury models have shown remarkably consistent benefits across various injury paradigms including hyperoxia, ventilator-induced damage, endotoxin challenge, and bleomycin toxicity. These investigations demonstrate improved survival, reduced histological injury, and decreased edema formation. The therapeutic effects appear to be mediated both by direct cellular actions and through paracrine mechanisms involving extracellular vesicles and secreted factors.

Clinical translation is now underway, with several ongoing trials registered on clinicaltrials.gov investigating MSC therapy (Table 14.1) for complications like primary graft dysfunction (PGD) and bronchiolitis obliterans syndrome (BOS). Early phase studies have established the safety profile of MSC administration in lung transplant recipients [35, 36], with doses up to 100 million cells/kg being well tolerated.

Table 14.1 Mesenchymal stem cell therapy

Lead institution	Type	Phase	Patients	MSC source	Intervention	Primary outcome	NCT	Status
Rigshospitalet, Denmark	Double blind	1/2	All lung transplant	Allogeneic adipose	100 million vs. 200 million vs. placebo	PGD	NCT04714801	Recruiting
Mayo clinic, MN	Non-randomized	1	BOS+	Allogeneic bone marrow	0.5 million vs 1 million ± 1 million booster	Safety, changes in PFTs	NCT02181712	Completed 2021
University of Queensland, Australia	Randomized	2	CLAD+	Allogeneic bone marrow	8 million/kg vs. placebo	Progression-free survival from CLAD	NCT02709343	Recruiting

PGD primary graft dysfunction, *BOS* bronchiolitis obliterans syndrome, *PFT* pulmonary function test, *CLAD* chronic lung allograft dysfunction

Other Novel Directions—IL6 Pathway and Tolerance

Beyond MSC therapy itself, related innovations are emerging in the field. IL-6 pathway inhibitors like IL-6 inhibitors (clazakizumab, siltuximab) or the IL-6 receptor inhibitors (tocilizumab, sarilumab) show promise for modulating transplant rejection, given IL-6's central role in both inflammatory responses and adaptive immunity which is responsible for both acute and chronic rejections [37, 38]. Another exciting development involves bronchus-associated lymphoid tissue (BALT), where the formation of tertiary lymphoid structures enriched in regulatory T cells appears correlated with the development of transplant tolerance. These Treg populations not only mediate local immunoregulation but may also exert systemic effects through lymphatic drainage.

The ultimate goal of transplantation immunology—the induction of true immune tolerance—remains particularly challenging for lung allografts, which are considered among the most "tolerance-resistant" organs [39]. While protocols using donor bone marrow transplantation to create mixed chimerism have achieved tolerance in kidney transplantation [40], similar approaches for lung recipients remain experimental. Current research continues to explore combination strategies that may 1 day make this goal attainable, potentially through the synergistic use of MSCs, Treg therapies, and targeted immunomodulation.

Summary and Conclusion

Despite significant advancements in lung transplantation over recent decades, long-term outcomes still lag behind those of kidney and liver transplants. The lung's unique immunologic properties—such as constant environmental exposure and heightened alloimmune responses—make it particularly susceptible to rejection and chronic dysfunction, limiting graft survival even with conventional immunosuppression.

Emerging strategies aim to overcome these challenges through:

- **Targeted drug/cell delivery** using ex vivo lung perfusion (EVLP), enhancing therapeutic precision while minimizing systemic side effects.
- **Novel anti-inflammatory approaches**, including IL-6 pathway blockade (e.g., tocilizumab), to mitigate acute and chronic rejection.
- **Cell-based therapies**, particularly mesenchymal stromal cells (MSCs), which offer immunomodulation, tissue repair, and potential tolerance induction.
- **Regulatory T-cell (Treg) manipulation**, leveraging lung-specific memory T-cell populations to promote graft-protective immunity.
- **Tolerance induction protocols**, building on lessons from kidney transplantation (e.g., mixed chimerism) but adapted to the lung's tolerance-resistant nature.

These innovations, combined with ongoing research into biomarkers and personalized immunosuppression, provide hope for significantly improving lung

transplant outcomes. While challenges remain—such as Increasing EVLP usage inspite of its huge cost towards emulating human microbiome environment than cold storage towards reducing PGD, standardizing MSC therapies and achieving reliable tolerance—the integration of advanced therapeutics into clinical practice promises a transformative impact on graft longevity and patient survival. The future of lung transplantation lies in these multidisciplinary approaches, moving beyond broad immunosuppression toward precision medicine and immune modulation.

References

1. Bos S, Vos R, Van Raemdonck DE, Verleden GM. Survival in adult lung transplantation: where are we in 2020? Curr Opin Organ Transplant. 2020;25(3):268–73. https://doi.org/10.1097/MOT.0000000000000753.
2. Chung PA, Dilling DF. Immunosuppressive strategies in lung transplantation. Ann Transl Med. 2020;8(6):409. https://doi.org/10.21037/atm.2019.12.117.
3. Shepherd HM, Gauthier JM, Kreisel D. Tolerance, immunosuppression, and immune modulation: impacts on lung allograft survival. Curr Opin Organ Transplant. 2021;26(3):328–32. https://doi.org/10.1097/MOT.0000000000000871.
4. Miller CL, Jane MO, Allan JS, Madsen JC. Novel approaches for long-term lung transplant survival. https://www.frontiersin.org/journals/immunology/articles/10.3389/fimmu.2022.931251. https://doi.org/10.3389/fimmu.2022.931251.
5. Christie JD, Edwards LB, Kucheryavaya AY, et al. The registry of the International Society for Heart and Lung Transplantation: 29th adult lung and heart-lung transplant report-2012. International Society of Heart and Lung Transplantation. J Heart Lung Transplant. 2012;31:1073–86.
6. Heidt S, Hester J, Shankar S, et al. B cell repopulation after alemtuzumab induction-transient increase in transitional B cells and long-term dominance of naïve B cells. Am J Transplant. 2012;12:1784–92.
7. Noris M, Casiraghi F, Todeschini M, et al. Regulatory T cells and T cell depletion: role of immunosuppressive drugs. J Am Soc Nephrol. 2007;18:1007–18.
8. Hachem RR, Edwards LB, Yusen RD, et al. The impact of induction on survival after lung transplantation: an analysis of the international society for heart and lung transplantation registry. Clin Transpl. 2008;22:603–8.
9. Shyu S, Dew MA, Pilewski JM, et al. Five-year outcomes with alemtuzumab induction after lung transplantation. J Heart Lung Transplant. 2011;30:743–54.
10. Benazzo A, Auner S, Boehm PM, Morscher A, Schwarz S, Vidmar B, Dzubur F, Schweiger T, Hoda AM, Moser B, Matilla JR, Muraközy G, Lang G, Taghavi S, Klepetko W, Hoetzenecker K, Jaksch P. Outcomes with alemtuzumab induction therapy in lung transplantation: a comprehensive large-scale single-center analysis. Transpl Int. 2021;34:2633–43. https://doi.org/10.1111/tri.14153.
11. Treede H, Glanville AR, Klepetko W, et al. Tacrolimus and cyclosporine have differential effects on the risk of development of bronchiolitis obliterans syndrome: results of a prospective, randomized international trial in lung transplantation. J Heart Lung Transplant. 2012;31:797–804. https://doi.org/10.1016/j.healun.2012.03.008.
12. Grevel J, Welsh MS, Kahan BD. Cyclosporine monitoring in renal transplantation: area under the curve monitoring is superior to trough-level monitoring. Ther Drug Monit. 1989;11:246–8.
13. Jaksch P, Kocher A, Neuhauser P, et al. Monitoring C2 level predicts exposure in maintenance lung transplant patients receiving the microemulsion formulation of cyclosporine (Neoral). J Heart Lung Transplant. 2005;24:1076–80. https://doi.org/10.1016/j.healun.2003.05.0020.

14. Morton JM, Aboyoun CL, Malouf MA, et al. Enhanced clinical utility of de novo cyclosporine C2 monitoring after lung transplantation. J Heart Lung Transplant. 2004;23:1035–9. https://doi.org/10.1016/j.healun.2003.08.008.
15. Azzola A, Havryk A, Chhajed P, et al. Everolimus and mycophenolate mofetil are potent inhibitors of fibroblast proliferation after lung transplantation. Transplantation. 2004;77:275–80. https://doi.org/10.1097/01.TP.0000101822.50960.AB.
16. Meneghini M, Bestard O, Grinyo JM. Immunosuppressive drugs modes of action. Best Pract Res Clin Gastroenterol. 2021. pp. 54–5. https://doi.org/10.1016/j.bpg.2021.101757.
17. Lee JC, Christie JD. Primary graft dysfunction. Proc Am Thorac Soc. 2009;6(1):39–46. https://doi.org/10.1513/pats.200808-082GO.
18. Kreisel D, Krupnick AS, Puri V, Guthrie TJ, Trulock EP, Meyers BF, et al. Short- and long-term outcomes of 1000 adult lung transplant recipients at a single center. J Thorac Cardiovasc Surg. 2011;141(1):215–22. https://doi.org/10.1016/j.jtcvs.2010.09.009.
19. Khush KK, Potena L, Cherikh WS, Chambers DC, Harhay MO, Hayes D Jr, et al. The international thoracic organ transplant registry of the international society for heart and lung transplantation: 37th adult heart transplantation report-2020; focus on deceased donor characteristics. J Heart Lung Transplant. 2020;39(10):1003–15. https://doi.org/10.1016/j.healun.2020.07.010.
20. Witt CA, Puri V, Gelman AE, Krupnick AS, Kreisel D. Lung transplant immunosuppression—time for a new approach? Expert Rev Clin Immunol. 2014;10(11):1419–21. https://doi.org/10.1586/1744666X.2014.959499.
21. Krupnick AS, Lin X, Li W, Higashikubo R, Zinselmeyer BH, Hartzler H, et al. Central memory Cd8+ T lymphocytes mediate lung allograft acceptance. J Clin Invest. 2014;124(3):1130–43. https://doi.org/10.1172/JCI71359.
22. Takahashi T, Hsiao HM, Tanaka S, Li W, Higashikubo R, Scozzi D, et al. Pd-1 expression on Cd8(+) T cells regulates their differentiation within lung allografts and is critical for tolerance induction. Am J Transplant. 2018;18(1):216–25. https://doi.org/10.1111/ajt.14437.
23. den Hengst WA, Gielis JF, Lin JY, Van Schil PE, De Windt LJ, Moens AL. Lung ischemia-reperfusion injury: a molecular and clinical view on a complex pathophysiological process. Am J Physiol Heart Circ Physiol. 2010;299(5):H1283–99. https://doi.org/10.1152/ajpheart.00251.2010.
24. Cypel M, Rubacha M, Yeung J, Hirayama S, Torbicki K, Madonik M, et al. Normothermic ex vivo perfusion prevents lung injury compared to extended cold preservation for transplantation. Am J Transplant. 2009;9(10):2262–9. https://doi.org/10.1111/j.1600-6143.2009.02775.x.
25. Niikawa H, Okamoto T, Ayyat KS, Itoda Y, Farver CF, McCurry KR. The protective effect of prone lung position on ischemia-reperfusion injury and lung function in an ex vivo porcine lung model. J Thorac Cardiovasc Surg. 2019;157(1):425–33. https://doi.org/10.1016/j.jtcvs.2018.08.101.
26. Buchko MT, Boroumand N, Cheng JC, Hirji A, Halloran K, Freed DH, et al. Clinical transplantation using negative pressure ventilation ex situ lung perfusion with extended criteria donor lungs. Nat Commun. 2020;11(1):5765. https://doi.org/10.1038/s41467-020-19581-4.
27. Gloria JN, Yerxa J, Kesseli SJ, Davis RP, Samoylova ML, Barbas AS, et al. Subnormothermic ex vivo lung perfusion attenuates graft inflammation in a rat transplant model. J Thorac Cardiovasc Surg. 2021;S0022-5223(21):00165–3. https://doi.org/10.1016/j.jtcvs.2021.01.066.
28. Iske J, Hinze CA, Salman J, Haverich A, Tullius SG, Ius F. The potential of ex vivo lung perfusion on improving organ quality and ameliorating ischemia reperfusion injury. Am J Transplant. 2021;21(12):3831–9. https://doi.org/10.1111/ajt.16784.
29. Prasad NK, Pasrija C, Talaie T, Krupnick AS, Zhao Y, Lau CL. Ex vivo lung perfusion: current achievements and future directions. Transplantation. 2021;105(5):979–85. https://doi.org/10.1097/TP.0000000000003483.
30. Yeung JC, Krueger T, Yasufuku K, de Perrot M, Pierre AF, Waddell TK, et al. Outcomes after transplantation of lungs preserved for more than 12 h: a retrospective study. Lancet Respir Med. 2017;5(2):119–24. https://doi.org/10.1016/S2213-2600(16)30323-X.

31. Takahashi M, Andrew Cheung HY, Watanabe T, Zamel R, Cypel M, Liu M, et al. Strategies to prolong homeostasis of ex vivo perfused lungs. J Thorac Cardiovasc Surg. 2021;161(6):1963–73. https://doi.org/10.1016/j.jtcvs.2020.07.104.
32. Warnecke G, Van Raemdonck D, Smith MA, Massard G, Kukreja J, Rea F, et al. Normothermic ex-vivo preservation with the portable organ care system lung device for bilateral lung transplantation (Inspire): a randomised, open-label, non-inferiority, phase 3 study. Lancet Respir Med. 2018;6(5):357–67. https://doi.org/10.1016/S2213-2600(18)30136-X.
33. Slama A, Schillab L, Barta M, Benedek A, Mitterbauer A, Hoetzenecker K, et al. Standard donor lung procurement with normothermic ex vivo lung perfusion: a prospective randomized clinical trial. J Heart Lung Transplant. 2017;36(7):744–53. https://doi.org/10.1016/j.healun.2017.02.011.
34. Cypel M, Keshavjee S. The clinical potential of ex vivo lung perfusion. Expert Rev Respir Med. 2012;6(1):27–35. https://doi.org/10.1586/ers.11.93.
35. Keller CA, Gonwa TA, Hodge DO, Hei DJ, Centanni JM, Zubair AC. Feasibility, safety, and tolerance of mesenchymal stem cell therapy for obstructive chronic lung allograft dysfunction. Stem Cells Transl Med. 2018;7(2):161–7. https://doi.org/10.1002/sctm.17-0198.
36. Chambers DC, Enever D, Lawrence S, Sturm MJ, Herrmann R, Yerkovich S, et al. Mesenchymal stromal cell therapy for chronic lung allograft dysfunction: results of a first-in-Man study. Stem Cells Transl Med. 2017;6(4):1152–7. https://doi.org/10.1002/sctm.16-0372.
37. Rizzo M, SivaSai KS, Smith MA, Trulock EP, Lynch JP, Patterson GA, et al. Increased expression of inflammatory cytokines and adhesion molecules by alveolar macrophages of human lung allograft recipients with acute rejection: decline with resolution of rejection. J Heart Lung Transplant. 2000;19(9):858–65. https://doi.org/10.1016/s1053-2498(00)00165-0.
38. Yoshida Y, Iwaki Y, Pham S, Dauber JH, Yousem SA, Zeevi A, et al. Benefits of posttransplantation monitoring of interleukin 6 in lung transplantation. Ann Thorac Surg. 1993;55(1):89–93. https://doi.org/10.1016/0003-4975(93)90479-2.
39. Madariaga ML, Kreisel D, Madsen JC. Organ-specific differences in achieving tolerance. Curr Opin Organ Transplant. 2015;20(4):392–9. https://doi.org/10.1097/MOT.0000000000000206.
40. Kawai T, Cosimi AB, Spitzer TR, Tolkoff-Rubin N, Suthanthiran M, Saidman SL, et al. HLA-mismatched renal transplantation without maintenance immunosuppression. N Engl J Med. 2008;358(4):353–61. https://doi.org/10.1056/NEJMoa071074.

Immunology of Heart Transplantation 15

Sanjana Adig, Kareem Ballut, Sharon Sairah Joy, and Sudakshina Gosh

Case Report

A 34-year-old male with genetic dilated cardiomyopathy presented in acute decompensated heart failure to the emergency room. He was diagnosed initially when he was 27 and had been on guideline directed medical therapy since then. However, over time, the dose for some of his medications were reduced due to intolerance. His current medications included 6.25 mg q12 Carvedilol, 10 mg q24 Empagliflozin, 49–51 mg q12 Sacubitril-Valsartan, and 12.5 mg q24 Spironolactone. Over the last 10 months, he has progressively declined despite being compliant with medications, salt and fluid restrictions, recommendations, and follow-up. His Echo revealed an LVEF of around 20% with an LVEDD of 7.1 cm. Cardiac MRI revealed LVEF of 18% with patchy mid-septal LGE. In the outpatient setting, patient had a cardiopulmonary exercise test that showed VO_2 max of 11.2 mL/Kg/min.

After his admission to the hospital, patient had a RHC that showed RA 18 mmHg, PCW 27 mmHg, PVR 2.9 WU, and TD CI 1.7. He was started on IV inotropes and IV diuretics. In the next 4 days patient continued to deteriorate despite vasoactive agents and required VA-ECMO for stabilization. He was listed for orthotopic heart transplantation, a donor heart match was received on day 11. He underwent a

S. Adig (✉)
Department of Cardiology, Indiana University, Indianapolis, IN, USA
e-mail: sadig@iu.edu

K. Ballut
Department of Cardiology, University of Michigan, Ann Arbor, MI, USA
e-mail: kballut@med.umichigan.edu

S. S. Joy
Madras Medical Mission Hospital, Chennai, Tamil Nadu, India

S. Gosh
Muhimbili University of Health and Allied Sciences, Upanga West, Tanzania

G. Abraham et al. (eds.), *Immunological Challenges in Organ Transplantation*,
https://doi.org/10.1007/978-981-95-1533-2_15

successful heart transplant using a bicaval anastomosis. His discharge medications included Tacrolimus, prednisone, and mycophenolate mofetil. He was followed in the clinic with weekly echocardiograms and endomyocardial biopsies. On the fourth week after transplant, patient presented to the ER with low grade fever and shortness of breath. Physical exam revealed respiratory distress, S3 gallop, and elevated JVP. Echocardiogram revealed LVEF dropped from 60% to 40% with global hypokinesis. NT-pro BNP was 1250 pg/mL. Endomyocardial biopsy was performed that revealed ISHLT Grade 2R acute cellular rejection. Histology demonstrated multifocal infiltrates of activated lymphocytes associated with myocyte damage (Fig. 15.1). There was no evidence of antibody-mediated rejection on immunofluorescence or C4d staining. Donor-specific antibodies were negative.

Treatment

Patient was continued on Mycophenolate mofetil and started on 1 g IV Methyl prednisone for 3 days followed by a prednisone taper. Tacrolimus was continued with trough levels optimized to higher end of therapeutic range (10–12 ng/mL). The patient showed clinical improvement on the third day with a decline in NT-pro BNP levels and improvement in symptoms. Echocardiogram was repeated and showed improvement in LVEF from 40% to 50%. Endomyocardial biopsy was repeated on the 11th day and revealed resolution of rejection (Grade 0R) with occasional scattered lymphocytes and no active myocyte injury. Patient was discharged to close follow-up in the transplant clinic.

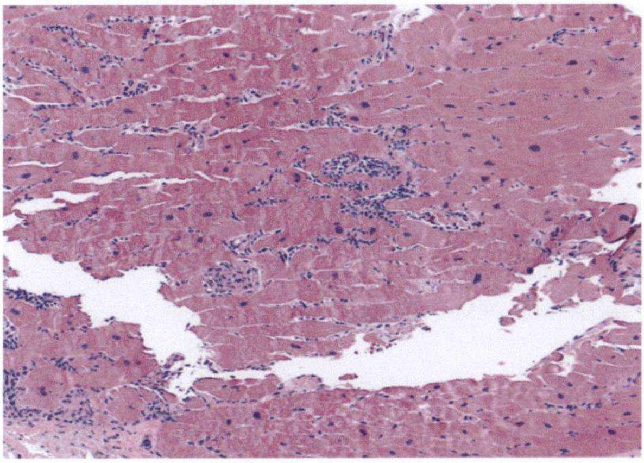

Fig. 15.1 Acute cellular rejection

Acute Cellular Rejection

Acute cellular rejection (ACR) is a significant early complication of heart transplantation. It is an immune-mediated process that occurs when the recipient's immune system mounts a T-cell driven response against the donor allograft [1, 2]. Two pathways are involved; the direct pathway in which the recipient T-cells directly recognize intact donor MHC molecules on donor antigen presenting cells and the indirect pathway, where recipient antigen-presenting cells process and present donor-derived peptides to T-cells. CD 4+ helper T-cells are activated and produce inflammatory cytokines that recruit and activate additional immune cells. CD8+ cytotoxic T-cells directly attack donor myocytes leading to myocardial damage. This leads to infiltration of the allograft by lymphocytes and macrophage, endothelial activation, cytokine production, and subsequent myocyte necrosis. This immune activity leads to significant tissue injury and eventual graft failure.

Mild ACR episodes are asymptomatic and detected only via surveillance endomyocardial biopsies. Moderate and severe grades may present with hemodynamic compromise. Treatment is based on clinical severity, ISHLT grading, and associated findings. Immunosuppression with corticosteroids is first line therapy. Immediate treatment is required to prevent graft dysfunction and long-term sequalae.

Biomarkers

Monitoring for rejection after heart transplantation remains a critical aspect of postoperative care. While endomyocardial biopsy is the current gold standard, it is invasive, carries procedural risk, and has limitations including sampling error and inter-observer variability [3]. To address these limitations, several biomarkers have been used as alternatives for rejection surveillance.

AlloMap Gene Expression Profiling: AlloMap is a non-invasive blood test approved for use in stable heart transplant patients (>55 days post-transplant). It measures the expression of 11 genes related to immune activation and correlates with risk of rejection. A low AlloMap score suggests low risk of rejection, potentially allowing biopsy avoidance in stable patients.

Donor-Derived Cell-Free DNA (dd-cfDNA): This emerging biomarker measures the proportion of donor DNA fragments in the recipient's plasma [4]. Elevated dd-cfDNA levels correlate with both cellular and antibody-mediated rejection. It is especially useful for early detection and may be more sensitive than traditional biomarkers.

Gene Expression and Molecular Microscope (MMDx): This advanced tool uses microarray analysis to evaluate gene expression profiles from biopsy specimens, offering an objective assessment of immune activity. MMDx provides insights into the type and intensity of rejection and can identify subclinical graft injury.

Inflammatory Cytokines and Chemokines: Markers such as interleukin-6 (IL-6), interferon-gamma (IFN-γ), and CXCL9 have been studied for their roles in

immune-mediated injury. Elevated levels often correlate with rejection, but lack of specificity limits their clinical use.

Immunophenotyping and T-cell Monitoring: Flow cytometry-based analysis of circulating T-cell subsets, including regulatory T cells (Tregs), and activated HLA-DR+ T cells may provide real-time insights into immune status. However, these assays require standardization.

MicroRNA (miRNA) Signatures: MicroRNAs are small, non-coding RNAs involved in post-transcriptional gene regulation. Specific miRNA profiles have been linked to rejection states and may offer future utility as minimally invasive biomarkers.

Anti-HLA Antibodies and Donor-Specific Antibodies (DSA): The presence of circulating DSA is a key indicator of antibody-mediated rejection (AMR). Luminex-based assays quantify both Class I and Class II HLA antibodies. Monitoring DSA trends is critical in patients with AMR risk.

Metabolomic and Proteomic Signatures: Mass spectrometry-based profiling of metabolites and proteins in blood or urine samples is a developing field. Early studies show promise in identifying rejection-specific metabolic patterns.

B-type Natriuretic Peptide (BNP) and NT-proBNP: These peptides are released from ventricular myocardium in response to wall stress and volume overload. Elevated levels may indicate graft dysfunction and correlate with rejection severity. However, BNP can be non-specific and elevated in other conditions such as infection, renal dysfunction, or volume overload.

Troponins (cTnI and cTnT): Cardiac troponins are markers of myocyte injury. Elevated levels have been associated with acute cellular rejection, particularly higher grades. Troponin elevations may precede echocardiographic changes but are also influenced by procedural factors and comorbidities [5, 6].

Biomarkers offer the potential for earlier, less invasive detection of rejection, and graft dysfunction [7, 8]. While none have yet replaced biopsy as the gold standard, their integration into clinical practice enhances surveillance, reduces procedural burden, and personalizes immunosuppression [8]. Ongoing research and validation in multicenter cohorts are essential to refine these tools and establish robust clinical thresholds for intervention.

Introduction

Solid organ transplantation has come a long way with the advent of immunosuppressive medications and with the ability to combat opportunistic infections with newer antibiotics.

A heart transplant is performed where a diseased or failing heart is replaced with a healthy heart from a brain dead donor who is either donation after cardiac death (DCD) or donation after brain death (DBD). End stage heart failure where medical therapy has failed to improve patient's condition, a cardiac transplant is the treatment of choice [2].

Heart transplants are usually considered for individuals with severe, irreversible heart conditions that don't respond to other treatments. The conditions include like congestive heart failure, coronary artery disease, or certain genetic heart disorders.

Data from 2023 indicates that India performed approximately 221 heart transplants with a steady increase in the number [9].

After a heart transplant, patients require lifelong medical care to prevent complications like rejection of the new heart and to manage other potential issues.

Like all surgeries, heart transplants have risks, including rejection of the new heart, infection, and other complications.

The first successful human-to-human heart transplant was performed on December 3, 1967, at Groote Schuur Hospital in Cape Town, South Africa from a deceased 25-year-old woman into a 53-year-old man.

Pre-transplant compatibility testing is crucial to assess the risk of rejection, and various strategies are used to manage antibodies and induce tolerance [10, 11]. The immune system including HLA and non-HLA antibodies, and ABO incompatibility can lead to acute (cellular or antibody-mediated) or chronic (cardiac allograft vasculopathy) [12] (Fig. 15.2a, b).

A good HLA match between donor and recipient reduces the risk of rejection. Endocardial myocardial biopsy (EMB) is the gold standard for monitoring rejection [2, 13]. Strategies to reduce the risk of antibody-mediated rejection in sensitized recipients, including plasmapheresis and rituximab. Cardiac allograft vasculopathy is a significant long-term complication that affects graft survival. Overall, the most significant advance in the history of heart transplantation was the approval of cyclosporine (CyA) for heart transplantation 34 years ago.

Presence of anti HLA antibodies or allosensitization in patients awaiting solid organ transplantation is on the rise due to various sensitization events such as transfusions of blood products, childbirth, procedures, or surgeries where foreign objects are implanted in the body, ventricular assist devices, grafts, stents, and meshes to name a few. This poses a major barrier to transplantation due to significant challenge in finding a suitable donor with prolongation of the wait time before a suitable donor can be identified with HLA compatibility and thereby increased waitlist

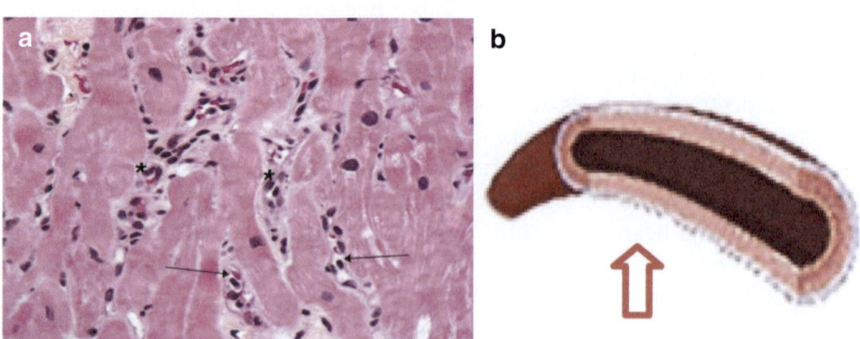

Fig. 15.2 Different types of rejection. (**a**) Antibody mediated, (**b**) chronic allograft vasculopathy

mortality for heart transplantation. It increases the risk of post-transplant rejection, reduced graft survival, increased incidences of allograft vasculopathy, and higher incidences of opportunistic infections due to the use of much intense immunosuppression such as rabbit anti thymoglobulin after and desensitization protocols prior to transplantation.

Our immune system is a complex interplay of cellular, humoral, and cytokine responses, all of which are designed to protect the host against foreign pathogens and to protect the self. In solid organ transplantation, activation of adaptive and innate immune responses leads to rejection of the transplanted organ. This is mediated through antibodies to donor antigens in the Major Histocompatibility complex, also known as the Human Leukocyte Antigen or HLA [10, 14]. Non HLA antigens also constitute some forms of rejection.

HLA antibody testing can be done by Luminex based (membrane based peripheral leukocytes or cell lines) and by crossmatch via CDC (complement dependent cytotoxicity) or flow cytometry based methods. The CDC lacks sensitivity for all HLAs, cannot reliably identify HLA class II antibodies, and may not be able to distinguish IgG from IgM. The Luminex based method is a solid phase assay and is usually done pretransplant as part of the transplant workup to detect Class I and Class II HLA antibodies. It also quantifies the antibody levels as median fluorescence intensity (or) MFI and has a relatively rapid turnaround time. Panel reactive antibodies is a measure of allosensitization risk, and the calculated PRA is a calculated score that estimates the likelihood of a recipient with preformed antibodies finding a compatible donor based on the prevalence of those antigens in the local donor pool. cPRA estimates the percentage of donor heart in a given population to which a transplant recipient has significant antibodies. HLA donor frequencies are determined from thousands of renal transplant donors.

The results of flow crossmatch are reported as median channel shifts (MCS) where in the amount of recipient antibody bound to the beads is expressed as mean/median channel fluorescence and compared with the mean/median channel fluorescence of the negative control (serum with no known HLA antibodies) run at the same time. In addition to allogeneic crossmatch (donor cells vs. recipient serum), autologous crossmatch (recipient serum vs. recipient cells) is also performed due to an approximately 5–8% autoreactivity. This value can range from no change to a 100-fold change and identifies the risk of the potential transplant.

For negative crossmatch in T and B cell flow crossmatch, each lab has to do its own standardization protocols and come up with values to categorize as negative, gray zone and positive. The decision to go ahead with the transplant in such situations is based on the clinical status of the patient, the degree of median channel shift, and the quality of the donor organ that is available, and various other donor organ factors such as the heart mass ratio compatibility, etc. A discussion between the clinicians, surgeons, and the immunologist is very crucial in such circumstances and needs prior discussion and up-to-date awareness of the patient's condition (such as being on mechanical support in heart transplant recipients, degree of inotrope use, presence of any sensitizing event such as infection, blood transfusion) their wait time in the transplant list etc.

Once a patient listed for an organ transplant has significant (>10% PRA) preformed antibodies to potential donor HLA antigens, (based on the frequency of the HLA in the donor pool as well as the intensity of the antibody), desensitization protocols are planned to reduce the HLA antibodies, and to increase the chances of finding a suitable and compatible donor.

Desensitization has to be tailored to individual patients based on their risk profile. There is no guarantee that two patients with HLA antibodies will respond in the same way to a desensitization protocol [12]. There is a real need for controlled clinical trial in this area.

One such is therapeutic plasma exchange (TPE) with intravenous immunoglobulin(IVIG). TPE separates plasma from extracorporeal cellular components. It is typically performed to exchange 1–2 times the plasma volume with albumin or fresh frozen plasma for replacement. TPE does not remove allo antibodies from the extravascular compartment or stop the antibody production. This method is not effective with high titer of antibodies, and the antibodies can rebound after cessation of the therapy. This form of desensitization is employed if there is a suspicion of a recent sensitizing event with prior 0% PRA and if the antibody production is felt to be related to a transient event. During TPE, antihypertensives that work on the renin angiotensin pathway are held on the day of the procedure to avoid hypotension.

Intravenous immunoglobulin is pooled IgG from thousands of donors and is supposed to act via inhibition of the complement pathway. It results in negative feedback on the plasma cells by blocking the FcReceptor, and subsequently downregulation of the B cell receptors which produce the antibodies. IVIG is more effective against HLA class I antibodies than class II, and takes up to 3–4 weeks to see effect. Similar to TPE, cessation of IVIG therapy can also cause a rebound in antibody production.

Rituximab, anti CD 20 chimeric antibody, has been used in resistant HLA antibodies in the recipients of solid organ transplantation. This kills the mature memory B cells, however does not kill the plasma cells or the circulating antibodies. It is usually administered with TPE and/or IVIG. This typically stays in the system for up to 6 months and can be a cause of positive B cell crossmatches.

Velcade (or) Bortezomib and carfilzomib are proteasome inhibitors and are used with TPE to increase plasma cell sensitivity [15]. These agents are also more effective against class I HLA antibodies than class II. They seem to have a longer lasting effect for up to 2 years and are more resistant to rebound antibody production. It is given as an intravenous bolus or as a subcutaneous injection. However, side effects include development of or worsening of peripheral neuropathy, increased risk of bleeding with thrombocytopenia, necessitating cessation of antiplatelet therapy for a few days prior, and heart failure exacerbation. In general, IV IG is to be avoided prior to Bortezomib infusions. Concomitant CYP3A4 inhibitors usage should be checked in patients receiving bortezomib due to drug interaction increasing the exposure to bortezomib leading to increased side effects. Antibody levels are checked after 3 weeks to see if there is a reduction in the levels.

Eculizumab is one of the newer agents targeting terminal complement inhibition for the prevention of antibody mediated rejection in immunologically high risk allograft recipients [16]. In a single center study of 36 patients at Cedar Sinai Heart Institute, for patients more than 18 years of age and with at least one value of PRA more than 70% in the lifetime, and with MFI (median fluorescence intensity) of more than 5000, a total of nine infusions of eculizumab were given for patients in the first 2 months' post-transplant along with cytolytic induction in the immediate post-transplant period. It consisted of 1200 mg at release of cross clamp following transplantation in the operating room followed by 900 mg on post operative day 1 and then, 900 mg weekly thereafter for 3 weeks. The dose was escalated to 1200 mg weeks 4, 6, and 8. Along with this regimen, rabbit ATG and the routine therapies with tacrolimus, mycophenolate, and steroids were given. Out of 36 patients enrolled, 20 patients received this therapy [17, 18].

Ten to twelve percent of transplanted heart recipients develop cardiac allograft vasculopathy (CAV) each year. This is due to chronic rejection and is a leading cause of death in patients over 5 years post-transplant [19].

References

1. Kittleson MM, Garg S. Solid gold, or liquid gold?: towards a new diagnostic standard for heart transplant rejection. Circulation. 2021;143:1198–201.
2. Kolata G. Organ shortage clouds new transplant era. Science. 1983;221:32–3. https://doi.org/10.1126/science.6857260.
3. Kobashigawa J, Hall S, Shah P, Fine B, Halloran P, Jackson AM, Khush KK, Margulies KB, Sani MM, Patel JK, Patel N, Peyster E. The evolving use of biomarkers in heart transplantation: Consensus of an expert panel. Am J Transplant. 2023;23(6):727–35. https://doi.org/10.1016/j.ajt.2023.02.025.
4. Daly KP. Circulating donor-derived cell-free DNA: a true biomarker for cardiac allograft rejection? Ann Transl Med. 2015;3(4):47.
5. Garrido IP, García-Lara J, Pinar E, Pastor-Pérez F, Sánchez-Mas J, Valdés-Chavarri M, et al. Optical coherence tomography and highly sensitivity troponin T for evaluating cardiac allograft vasculopathy. Am J Cardiol. 2012;110:655–61.
6. Patel PC, Hill DA, Ayers CR, Lavingia B, Kaiser P, Dyer AK, et al. High sensitivity cardiac troponin I assay to screen for acute rejection in patients with heart transplant. Circ Heart Fail. 2014;7:463–9.
7. Kobashigawa J, Colvin M, Potena L, et al. The management of antibodies in heart transplantation: an ISHLT consensus document. J Heart Lung Transplant. 2018;37(5):537–47.
8. Snyder TM, Khush KK, Valantine HA, Quake SR. Universal noninvasive detection of solid organ transplant rejection. Proc Natl Acad Sci USA. 2011;108(15):6229–34.
9. Khush KK, Cherikh WS, Chambers DC, Harhay MO, Hayes D, Hsich E, et al. The international thoracic organ transplant registry of the International Society for Heart and Lung Transplantation: thirty-sixth adult heart transplantation report—2019; focus theme: donor and recipient size match. J Heart Lung Transplant. 2019;38:1056–66.
10. Kobashigawa JA. The search for a gold standard to detect rejection in heart transplant patients: are we there yet? Circulation. 2017;135(10):936–8.
11. Loupy A, Duong Van Huyen JP, Hidalgo L, et al. Gene expression profiling for the identification and classification of antibody-mediated heart rejection. Circulation. 2017;135(10):917–35.

12. Colvin MM, Cook JL, Chang P, et al. Antibody-mediated rejection in cardiac transplantation: emerging knowledge in diagnosis and management: a scientific statement from the American Heart Association. Circulation. 2015;131:1608–39.
13. Peyster EG, Madabhushi A, Margulies KB. Advanced morphologic analysis for diagnosing allograft rejection: the case of cardiac transplant rejection. Transplantation. 2018;102(8):1230–9.
14. Setia G, Nishihara K, Singer-Englar T, Zhang X, Patel J, Kobashigawa J. Crossing low/moderate-level donor-specific antibodies during heart transplantation. Clin Transpl. 2021;35(3):e14196.
15. Brinkley DM, Mangione M, Fossey SC, Harrison KM, Zalawadiya S, Rali AS, Siddiqi HK, Lindenfeld J, Schlendorf KH. Efficacy of bortezomib desensitization among heart transplant candidates. Clin Transpl. 2023;37(4):e14907. https://doi.org/10.1111/ctr.14907.
16. Kittleson MM, Patel N, Chang DH, Kransdorf EP, Kobashigawa JA, Patel JK. Eculizumab for antibody-mediated rejection in heart transplantation: a case-control study. Clin Transpl. 2021;35(12):e14454. https://doi.org/10.1111/ctr.14454.
17. Kobashigawa J, Miller L, Renlund D, et al. A randomized active-controlled trial of mycophenolate mofetil in heart transplant recipients. Transplantation. 1998;66:507–15. https://doi.org/10.1097/00007890-199808270-00016.
18. Reichart B, Meiser B, Vigano M, et al. European Multicenter tacrolimus (FK506) heart pilot study: one-year results—European Tacrolimus Multicenter Heart Study Group. J Heart Lung Transplant. 1998;17:775–81.
19. Ludhwani D, Abraham J, Sharma S, et al. Heart Transplantation Rejection. [Updated 2024 Aug 11]. In: StatPearls. Treasure Island, FL: StatPearls Publishing; 2025. https://www.ncbi.nlm.nih.gov/books/NBK537057/.

Approach to Liver Transplant: The Immunomodulatory Role of the Liver and Its Mesenchymal Stem Cells

16

Timucin Taner

Case Report

A 56-year-old woman presented for a kidney transplant evaluation with her daughter, who wanted to be a living donor. The patient developed end-stage renal disease due to diabetic nephropathy and had been on dialysis for a year. She was highly sensitized against her daughter, with significantly high titers of donor-specific HLA antibodies (DSA) and a positive flow cytometric cross-match (Channel shifts of 550 for T cell and 568 for B cell), that would have precluded the kidney transplant. During the work-up, the patient was found to have cirrhosis with portal hypertension. The daughter requested to be evaluated as a living liver donor, and after a satisfactory evaluation, she donated her right liver lobe to her mother. Three days after the transplant, the cross-match became negative without any additional DSA-targeting therapy (Fig. 16.1). The cross-match stayed negative at 1 year, when the same donor donated a kidney to the recipient. Both grafts have excellent function to date, with over 3 years of follow-up.

Introduction

The liver is the second most commonly transplanted solid organ after the kidney. Its unique inherent immunomodulatory properties set it apart from other solid organs [1]. In addition, most liver transplants are done without induction immunosuppression. This chapter explores the distinctive immunomodulatory functions of the liver, its cellular components, and the role of mesenchymal stem cells (MSCs) in shaping the immune response post-transplantation.

T. Taner (✉)
Mayo Clinic, Rochester, MN, USA
e-mail: taner.timucin@mayo.edu

G. Abraham et al. (eds.), *Immunological Challenges in Organ Transplantation*, https://doi.org/10.1007/978-981-95-1533-2_16

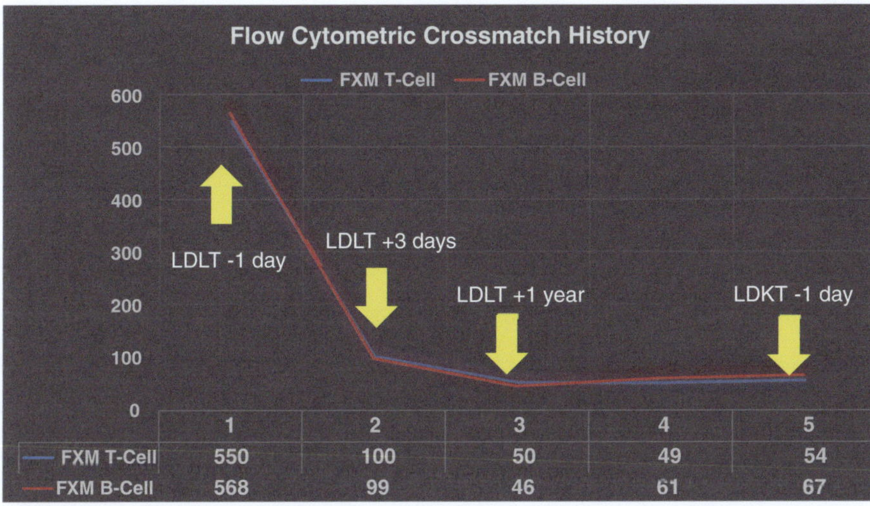

Fig. 16.1 Flow cytometric crossmatch course of the case presented. *LDLT* living-donor liver transplantation, *LDKT* living-donor kidney transplantation, *FXM* flow cytometric cross-match

Immunologic Challenges in Solid Organ Transplantation and Mechanisms of Immune Response in Transplantation

Transplantation of any solid organ poses a significant immunologic challenge due to the host's strong "non-self" immune response. The host immune system mounts both innate and adaptive responses against the allograft. Early after a solid organ transplant, the ischemia-reperfusion injury exacerbates the initial immune response that is dominated by the innate immune system. At this stage macrophages, neutrophils and natural killer (NK) cells form the bulk of the graft infiltrating cells. The release of pro-inflammatory cytokines leads to further graft inflammation through chemotaxis. Complement system activation can also contribute to early graft damage. Subsequently, adaptive immune responses, mediated by T cells and HLA antibodies, ensue. The T-cell-mediated responses are primarily driven by CD4-positive and CD8-positive T-cells. B-cell immediate responses lead to the production of DSA. These responses, if not controlled with immunosuppression, can lead to acute and chronic rejection. Untreated rejection or frequent episodes of rejection can lead to long-term graft dysfunction. Rejection episodes, in turn, are treated with increasing immunosuppression.

Liver Transplantation and Immune Tolerance

Liver transplantation occupies a unique position in the field of organ transplantation due to its ability to modulate immune responses. Unlike kidney and heart transplants, liver recipients require lower doses of immunosuppressive medications and

have a lower incidence of chronic rejection. In about 20% of liver transplant recipients who are carefully selected, immunosuppression could be discontinued completely [2]. Antibody-mediated rejection (AMR) after liver transplantation is a rare finding, even in highly sensitized patients. Several factors contribute to the liver's tolerogenic properties:

- Anatomical location: The liver is strategically located between the gastrointestinal tract and the systemic circulation, constantly exposed to gut-derived endotoxins and dietary antigens. As such, it is the front-line immune organ that evolved to detect and clear pathogens entering the body through the gut and the portal system. This low-level, continuous exposure fosters an immune microenvironment that favors tolerance.
- Cellular composition: The liver contains a diverse array of immune-active cells that contribute to its immunomodulatory effects. As detailed below, most of these immune active cells act differently than their counterparts anywhere else in the body, favoring immune suppression.
- Microenvironmental factors: The liver's unique cytokine milieu and cellular interactions promote an immune-tolerant state both under steady-state conditions and after transplantation. The sinusoids, which form the majority of the microvasculature in the liver, are lined by sinusoidal endothelial cells and Kupffer cells that scavenge antigens and regulate immune responses.

Cellular Components of the Liver with Immunologic Functions

A variety of liver-resident cells contribute to its immunologic function:

- Kupffer Cells: Liver-resident macrophages account for 80% of all the macrophages in the body. Similar to their counterparts, they take up antigens through phagocytosis. They express HLA class I and II, and co-stimulatory molecules. Thus, they can act as antigen-presenting cells (APCs). Unlike the latter, however, they secrete anti-inflammatory cytokines, such as IL-10, PEG2, and TGF-β. Therefore, at a steady state, Kupffer cells induce cytotoxic T-cell apoptosis and stimulate regulatory T-cells (Tregs).
- Liver Sinusoidal Endothelial Cells (LSEC): These are specialized endothelial cells that line the walls of the sinusoids. They are fenestrated with minimal basement membrane. They express scavenger receptors and have a very high capacity for endocytosis. They have low levels of class I and class II HLA as well as co-stimulatory molecules on their surface. At the same time, they do not produce IL-12. With that combination, they act as APCs with regulatory capacity, inducing T-cell tolerance and promoting Tregs.
- Dendritic Cells (DC): The liver contains both myeloid and plasmacytoid DCs as professional APCs. The liver DCs are more immature compared to DCs from the spleen, expressing low levels of HLA-II and costimulatory molecules. Compared to their counterparts elsewhere in the body, liver DCs secrete more anti-

inflammatory cytokines, such as IL-10 and TGF-β, which are ineffective in promoting effector T cells and more effective in promoting Treg generation.

- Liver resident T Cells: In the liver, CD8-positive T-cells are more abundant than CD4-positive T-cells. Activation of the CD8-positive T-cells within the liver often leads to their deletion or anergy. Similarly, the liver microenvironment promotes the skewing of CD4-positive T-cells into T Helper 2 (Th2) and Treg lineages rather than Th1 or Th17 lineages. These cells contribute to immune surveillance and tolerance mechanisms.
- Mesenchymal Stem Cells (MSC): Play a pivotal role in the liver's ability to regulate immune responses. A detailed account of these cells and their functions can be found later in this chapter.

Studying the Liver's Impact on Systemic Alloimmunity

The tolerogenic microenvironment of the liver at a steady state is retained after transplantation, allowing for an interesting phenomenon of altered immune responses to donor-specific antigens. This phenomenon was initially recognized in animal models. Evidence for this phenomenon in humans has been provided initially by clinical observations and corroborated by more detailed molecular and genetic analysis, as described below.

- Clinical Observations: The increased number of combined organ transplants including the liver over the past two decades has allowed researchers to observe the outcomes of the other transplanted organs and compare these to the outcomes after solitary organ transplantation. In simultaneous liver and kidney transplantation (SLK), whereby both organs come from the same deceased donor, the incidence of T-cell mediated rejection in the kidney based on protocol biopsies is much lower than the incidence of such rejection after solitary kidney transplantation in people that are otherwise matched for immunological risk [3]. Similarly, the incidence of AMR, transplant glomerulopathy, and subclinical inflammation in the kidneys of the SLK recipients is significantly lower than that after solitary kidney transplantation. As a result, the kidney allograft function in SLK patients tends to stay stable in the long term. The immunoprotective effect of the liver is not unique to the kidney. In simultaneous liver and heart transplant patients, both the T-cell-mediated and antibody-mediated rejection of the heart is much less frequent than seen after solitary part transplantation [4]. When the liver and the other organ confirm different donors, as often is the case in non-simultaneous transplantation, this phenomenon is not seen, suggesting that donor specificity plays a major role in liver-driven immuno protection.
- Molecular and Genetic Analyses: The transcriptomic analysis of kidney allografts from SLK patients demonstrates a distinctive profile of downregulation of inflammation-related genes and genesets and upregulation of homeostatic jeans and genesets [5]. In contrast, kidney allografts of patients who underwent solitary kidney transplantation have upregulated inflammation-related genes and

genesets, especially in recipients with circulating donor-specific antibodies. Remarkably, the latter is seen in kidney allografts with no histological evidence of inflammation, highlighting that the genetic changes can be detected before the inflammation sets in, and a simultaneously transplanted liver allograft appears to modulate these genetic changes.

- Systemic alloimmune responses: Analysis of the peripheral blood mononuclear cells (PBMCs) from SLK and solitary kidney transplant recipients for phenotypic and functional changes demonstrate that the circulating T cells of the SLK patients have and reached subpopulations with exhaustion markers. When stimulated in vitro with donor antigens, T-cells of SLK patients appear hyporesponsive compared to T cells of solitary kidney patients [6]. When the T cells of the SLK patients are stimulated with third-party antigens, however, they proliferate and undergo clonal expansion as expected. These observations support the theory that antigen-specific training of the T cells occurs after liver transplantation and the simultaneously transplanted "other" organs from the same donor benefit from this active immunomodulation.

Mesenchymal Stem Cells (MSC) and Their Immunomodulatory Role

In search for components of the liver that could be adoptively transferred to induce the immunomodulation stated above without the need for liver transplantation, investigators have been studying the immunosuppressive roles of different liver resident cells in the past decade. Among the liver's immunomodulatory components, mesenchymal stem cells (MSC) play a particularly important role. MSCs are nonhematopoietic multipotent adult stem cells that can differentiate into various cell types including bone, cartilage, fat, and muscle cells. They are found in the perivascular stroma of many tissues in the body and can promote the repair and regeneration of damaged tissues. They also have remarkable immunomodulatory properties. In fact, MSC isolated from bone marrow or adipose tissue have been used in multiple clinical trials for the treatment of autoimmune disorders.

Liver-derived MSCs exhibit more enhanced immunosuppression compared to MSCs from other sources such as bone marrow and adipose tissue. Liver MSCs express characteristic markers such as CD90, CD73, and CD105 while lacking hematopoietic markers like CD45 and HLA-II [7]. In a head-to-head comparison, liver MSCs cultured in vitro express several surface markers with immunoinhibitory functions, including CD200 and CD 274 (a.k.a PD-L1), at a higher concentration than MSCs isolated from other tissues. Similarly, the transcriptomic analysis of liver MSCs demonstrates that they are different from other MSCs, with significant enrichment of genes and gene sets associated with immunoregulation. Liver MSCs inhibit alloreactive T-cell proliferation and reduce the frequency of interferon-gamma (IFN-gamma) producing T-cells better than other MSCs. The anti-proliferative impact of liver MSCs is not dependent on the cell-to-cell contact and could be reversed incompletely by blocking programmed death ligand 1 and requires

a higher concentration of the competitive inhibitor indoleamine 2,3-dioxygenase for complete reversal.

Liver MSC secretome is uniquely different from that of other MSCs, with enriched protein sets involved in IFN-gamma responses and signaling [8]. When co-cultured with primary human NK cells, liver MSCs, but not other MSCs, decrease surface expression of activating receptors NKp44 and NKG2D. Liver MSCs also decrease IFN-gamma secretion by interleukin-2 (IL-2)—stimulated NK cells more effectively than other MSCs. The cytolytic function of NK cells is reduced significantly when co-cultured with liver MSCs, whereas other MSCs do not have a major impact. Mechanistic studies show that the liver MSC-mediated reduction in NK cell cytotoxicity is not through changes in secretion of the cytotoxic proteins Perforin, Granzyme A or B, but through increased production of HLA-C1 found in liver MSC secretome that inhibits NK cells by stimulating their inhibitory receptor KIRDL2/3.

When infused into mice with unilateral renal artery stenosis (RAS), an established model of kidney inflammation, liver MSCs reduce the levels of plasma renin and proteinuria. Serum creatinine improves in mice treated with liver MSC [9]. While all types of MSCs migrate to and are retained within the stenotic kidneys after systemic infusion, liver MSC retention is significantly higher than other MSCs. Although all MSC-treated mice display reduced overall inflammation and macrophage counts in stenotic kidneys, liver MSCs show superior potency in-vivo at localizing to the site of inflammation and inducing M2 (reparative) macrophage polarization to reduce inflammatory changes.

Clinical Implications and Future Directions

Liver's Role in Allograft Protection

The ability of the liver to withstand circulating DSA while also protecting simultaneously transplanted organs from both antibody-mediated and T-cell-mediated injury has significant clinical implications. Although combined abdominal or thoracic organ transplantation with the liver in the setting of a normal liver function for the purpose of reducing the risk of rejection has not yet become an established treatment option, this was discussed as a potential option at two recent consensus conferences [10]. In the meanwhile, innovative techniques such as heart-after-liver transplantation in patients needing both organs have shown efficacy in sensitized patients by leveraging liver transplantation's immunologic benefits, resulting in reduced DSA and improved cardiac outcomes.

Potential for MSC-Based Cellular Therapy

Given their potent immunomodulatory effects, liver MSCs are promising candidates for cellular therapy in transplantation medicine. Current research is exploring the use of MSCs, including clinical trials that involve infusing MSCs directly into

kidney allografts with subclinical rejection (clinical trial IND#27685, NCT05456243). The results of this and similar trials will likely pave the way for new treatment options in transplantation.

Conclusion

The liver is a unique immunologic organ with the capability to modulate alloimmune responses post-transplantation. Ongoing research into the mechanisms of liver-mediated immunomodulation and the potential of MSC-based therapies holds great promise for advancing transplant medicine. Understanding these intricate immune interactions will be crucial for developing novel strategies to improve transplant outcomes and patient quality of life.

References

1. Thomson AW, Vionnet J, Sanchez-Fueyo A. Understanding, predicting and achieving liver transplant tolerance: from bench to bedside. Nat Rev Gastroenterol Hepatol. 2020;17(12):719–39.
2. Thomson AW, Geller DA, Gandhi C, Murase N, Demetris AJ, Beer-Stolz D. Hepatic antigen-presenting cells and regulation of liver transplant outcome. Immunol Res. 2011;50(2–3):221–7.
3. Taner T, Heimbach JK, Rosen CB, Nyberg SL, Park WD, Stegall MD. Decreased chronic cellular and antibody-mediated injury in the kidney following simultaneous liver-kidney transplantation. Kidney Int. 2016;89(4):909–17.
4. Wong TW, Gandhi MJ, Daly RC, et al. Liver allograft provides immunoprotection for the cardiac allograft in combined heart-liver transplantation. Am J Transplant. 2016;16(12):3522–31.
5. Taner T, Park WD, Stegall MD. Unique molecular changes in kidney allografts after simultaneous liver-kidney compared with solitary kidney transplantation. Kidney Int. 2017;91(5):1193–202.
6. Taner T, Gustafson MP, Hansen MJ, et al. Donor-specific hypo-responsiveness occurs in simultaneous liver-kidney transplant recipients after the first year. Kidney Int. 2018;93(6):1465–74.
7. Taner T, Abrol N, Park WD, et al. Phenotypic, transcriptional, and functional analysis of liver mesenchymal stromal cells and their immunomodulatory properties. Liver Transpl. 2020;26(4):549–63.
8. Yigitbilek F, Ozdogan E, Abrol N, Park WD, Hansen MJ, Dasari S, Stegall MD, Taner T. Liver mesenchymal stromal cells are superior inhibitors of NK cell functions through differences in their secretome compared to other mesenchymal stem cells. Front Immunol. 2022;13:952262.
9. Liang Y, Ozdogan E, Hansen MJ, Tang H, Saadiq I, Jordan KL, Krier JD, Gandhi DB, Grande JP, Lerman LO, Taner T. Human liver-derived mesenchymal stromal cells ameliorate murine ischemia-induced inflammation through macrophage polarization. Front Immunol. 2024;15:1448092.
10. Taner T, Biggins SW, Cummins N, Daly RC, Dietz AB, Emamaullee J, Gandhi MJ, Heimbach JK, Patel JK, Pereira NL, Rosenbaum A, Sanchez-Fueyo A, Shingina A, Stegall MD, Villavicencio Theoduloz MA, Wald JW, Kushwaha SS, Consensus Conference Participants. Summary of a consensus conference on the management of highly sensitized multiorgan transplant candidates. Mayo Clin Proc. 2025;100(4):700–11.

Challenges in Bone Marrow Transplant: Tale of Two Women

17

Gopinathan Mathiyazhagan and Jovita M. Daniel

Introduction

The number of allogeneic HSCTs performed globally and across the country has increased sharply over the last decade, from 560 in 2013 to 1665 in 2022. The increase in transplant centers, national and international stem registries sourcing matched unrelated donors (MUD) has certainly helped this progress [1]. Nevertheless, survival outcomes have also improved with better supportive care, refinement of pretransplant conditioning regimens, apheresis techniques, increased availability of leucodepleted and irradiated blood products, and higher antibiotics. Effective stem cell mobilization regimens have eased collection techniques, transitioning from painful iliac crest punctures performed historically over long hours to peripheral venesection completed in a short time. Over the years, advancements in cellular therapy have paved the way for graft manipulation techniques, including post-transplant adoptive cell therapy for refractory leukemias and viral infections.

A deeper understanding of disease pathophysiology has helped demarcate transplant principles for benign and malignant disorders. While in benign transplants, the goal is to restore stable long-term donor hematopoiesis, in malignant cases, the graft-versus-leukemia (GVL) effect from the donor immune system is essential for sustaining remission with minimal GVHD. The two cases presented sequentially highlight the importance of performing HSCT in complete remission and on time, with the need for vigilant follow-up.

G. Mathiyazhagan (✉)
Department of Hematology and Bone Marrow Transplant, MGM Health Care and Cancer Institute, Chennai, Tamil Nadu, India

J. M. Daniel
Department of Medical Oncology, MGM Health Care and Cancer Institute, Chennai, Tamil Nadu, India
e-mail: jovita.md@mgmcancerinstitute.in

G. Abraham et al. (eds.), *Immunological Challenges in Organ Transplantation*, https://doi.org/10.1007/978-981-95-1533-2_17

Case Report 1

A 34-year-old breastfeeding mother presented with low-grade fever and neck pain for a week. She self-referred to our center due to high white blood cell counts, where workup confirmed high-risk AML due to the presence of an FLT3-ITD mutation and NSD-NUP98 fusion. She was offered induction chemotherapy followed by HSCT with an appropriate donor for long-term disease control. She received two cycles of fludarabine, cytarabine, and idarubicin-based chemotherapy, resulting in remission with minimal transfusion support.

The prospective family donors included her sister, parents, and son, as no matched donors were available in the national MUD registry. Among the haploidentical donor (HID) pool in her family, her sister was chosen over her parents, who were in their early 60s, and her son, who was only 2 years old. The HLA match was 8/10 except at HLA-DQ and DR, and the donor was negative for donor-specific antibodies (DSA). Both the recipient and donor were CMV seropositive and had a major blood group (BG) mismatch (Recipient: O+, Donor: A+). We proceeded with myeloablative conditioning (MAC) consisting of fludarabine, busulfan, and low-dose cyclophosphamide (Flu-Bu-LD Cy) with post-transplantation cyclophosphamide (PTCy), tacrolimus, and mycophenolate mofetil (MMF) as GVHD prophylaxis.

The immunohematologist performed a stringent apheresis procedure to minimize RBC contamination due to the major BG mismatch, resulting in a final graft composition of CD34 9.84×10^6 cells/kg and CD3 1.9×10^8 cells/kg of recipient weight. Despite precautions, she developed an allergic reaction with minimal hemolysis during stem cell infusion, which was managed conservatively. She had her neutrophil and platelet engraftment on day 14, 19 respectively. She developed nausea, low-grade fever, and palpable hepatosplenomegaly with rising ferritin on day 22, attributed to cytokine release syndrome (CRS)/endotheliopathy of HID transplant, which was managed with steroids.

On day 28, she developed grade 3 hemorrhagic BK viral cystitis, which was managed with bladder irrigation, urinary analgesics, and IV cidofovir. The challenge was balancing prolonged cytokine release with liver dysfunction while managing viral reactivations, including CMV, on day 32. This was cautiously handled by modulating steroid doses, withholding MMF, and switching tacrolimus to sirolimus with adequate antiviral coverage. Her lymphocyte counts recovered on day 40 with waning viral copies, but liver dysfunction and hyperferritenemia recurred whenever steroids were tapered. Given the malignant transplant context, where early immunosuppression tapering is required for the GVL effect, she was switched to ruxolitinib, which normalized her liver function within 6 months. She was reimmunized against all childhood vaccine-preventable diseases 1-year post-transplant, as the reconstituted immune system required vaccine memory. She was followed up periodically with bone marrow examinations confirming remission 18 months post-transplant with 100% donor chimerism and seroconversion to the donor blood group.

Case Report 2

A 28-year-old unmarried woman, living with her single mother and younger brother, presented with a history of mucosal bleeding and thrombocytopenia for 1.5 years. She was treated in the periphery with variable courses of steroids and over-the-counter (OTC) indigenous medications. Unfortunately, when evaluated for persistent symptoms, her marrow was found to be dysplastic with myeloid blasts at 32%. She was classified as high risk AML due to background dysplasia and a RUNX1 mutation.

In young patients, intensive induction chemotherapy is typically offered. However, given her dental infection and marrow dysplasia, azacytidine-venetoclax (7, 14 days) was chosen, even before molecular reports arrived. Despite receiving low-intensity induction, her blood count recovery took 42 days and was complicated by perianal infection, fungal pneumonia, and high transfusion requirements. A second course of single-agent azacytidine (7 days) was administered, but she developed bloodstream infections and perianal infection flares, which required granulocyte transfusions, indicating poor tolerance.

A donor search revealed a 10/10 MUD alongside HID family donors. Given the history of AML with dysplasia following prolonged thrombocytopenia and mucosal bleeding, family donors were also screened for RUNX1 mutation penetrance. We counselled the family about the difficulty in giving complete induction pretransplant, to our surprise marrow was cleared of blasts but remained hypoplastic. She had a partially healed fungal pneumonia but with no choices went ahead with HSCT from 10/10 MUD donor.

She received Flu-Bu3-rabbit ATG (rATG)-based conditioning with tacrolimus and methotrexate for GVHD prophylaxis. She received a cell dose of 6.9×10^6 cells/kg, CD3 2.4×10^8 cells/kg of recipient weight, with neutrophil and platelet engraftment on days 15 and 18, respectively. After mild engraftment syndrome, she achieved normal blood counts by day 21. However, on day 28, she developed high CMV titres (5600 copies) requiring intravenous ganciclovir.

On day 28, her chimerism was only 87% donor, prompting dose reduction of immunosuppression to allow graft expansion. A donor lymphocyte infusion (DLI) was considered, but the MUD donor was unavailable, and her chimerism dropped to 75%, leading to transfusion-dependent cytopenias. She faced complications successively such as intracranial bleed, nasal mucormycosis, and perianal folliculitis which were meticulously managed. If not salvaged by an urgent second transplant, she would succumb to infections and bleeding. The family was left only with the brother HID donor and was counselled about the grave prognosis of taking second haploidentical transplant with an active perianal infection. She was given Flu-Treosulphan-r ATG- 2Gy TBI conditioning and T cell deplete transplant from her brother. However, she developed worsening septic shock and succumbed on day −1 of transplant.

Discussion

The above cases illustrate that the striking difference in outcomes, despite high-risk disease biology in both, is contributed by pretransplant factors, including the time of presentation, disease status, high transfusion burden, active infections, and performance status.

The historical safety data with HSCT takes precedence in the order of fully HLA-matched donor (MSD), followed by 10/10 MUD and HID, but the possibility of finding a MSD within the family is only 25%. These donors are invariably siblings and rarely parents in ethnicities where consanguinity is very high; otherwise, they are haploidentical donors (HID). National and international MUD registries provide a donor pool but come with added costs, an increased risk of GVHD compared to MSD, waiting time for donor confirmation, and uncertainty regarding post-transplant procedures like DLI, as in case 2. The first circle of family members are potential HIDs who are always motivated, and they are increasingly being adopted across many centers over time as outcomes have become comparable between MUD and HID transplants with evolving transplant techniques [2]. However, while transplanting for a benign hematological disorder, a matched donor should be chosen in all possible ways, as GVL is undesired and GVHD is unacceptable. The 2-year overall survival rate of haplotransplantation in the Indian context has increased from 50% to 90% over the years and depends on various factors like the age of the donor and recipient, disease status, CMV serostatus, accompanying infections, and comorbidities [1].

The first step to a successful haploidentical transplant is choosing an appropriate donor, considering factors such as age, gender, CMV serostatus, parity (if female), and blood group. In malignant transplants, we also calculate the killer-cell immunoglobulin-like receptor (KIR) mismatch score and choose the most immunoreactive donor to take advantage of the GVL effect [3]. Preferably, a young male donor is ideal; if female, a nulliparous donor is preferred over a parous one for better outcomes. In case 1, parents were avoided due to possible age-related mutations that could compromise stem cell quality, and the child was ruled out due to donor-to-recipient weight disparity. In case 2, a readily available 10/10 MUD was chosen over a 5/10 HID brother, considering the conditioning intensity and suspected germline RUNX1 mutation [4].

The conditioning chemotherapy given before stem cell infusion, aimed at eradicating diseased bone marrow and providing a conducive environment for recipient stem cells to engraft, is usually myeloablative, except in certain scenarios such as old age or active infections. Although both cases received adequate MAC, we could only deliver 4 mg/kg of r-ATG instead of the standard 4.5–6 mg/kg in case 2, fearing a flare-up of a partially healed fungal infection. Mucositis and veno-occlusive disease (VOD) are common at such high doses of alkylators like busulfan, necessitating pharmacokinetic monitoring, but its availability is restricted to only a few centers across the country. In second transplants after graft failure, as seen in case 2, a high risk of VOD is anticipated due to significant endothelial damage incurred

during the first transplant, where newer agents like treosulfan serve as a relatively safer option [5].

With current advances in apheresis techniques, blood group mismatch is not a major concern. However, case 1 experienced anaphylaxis, possibly due to blood cell contamination. Historically, stem cell grafts were obtained from iliac crest bone marrow punctures, but with advancements in mobilization techniques, stem cells are now harvested from peripheral blood via major vascular access. The composition of the graft should be optimal, with a stem cell dose (CD34) of 6 to 8×10^6 cells/kg and a CD3 dose of 2 to 4×10^6 cells/kg in HID transplants to facilitate engraftment. While lower CD3 counts result in rejection and higher counts increase GVHD risk, achieving a balance in dosing relative to CD34 counts is imperative [6].

Although case 1 engrafted on time, she experienced delayed cytokine release syndrome following HID transplantation, which persisted for a longer period, along with liver dysfunction and hyperferritenemia. A liver biopsy to rule out GVHD is ideal, but transplant physicians try to avoid such an invasive procedure in the early post-transplant period unless absolutely indicated. This period is crucial, as viral infections can manifest simultaneously with GVHD, but their treatment poles are wide apart—one requiring a decrease in immunosuppression and the other requiring an escalation. A strategic immunosuppression plan until immune recovery, particularly lymphocyte subset recovery to combat viral infections, is essential.

The outcome in case 2 was disheartening, as this young girl, despite recovering counts in the third week, did not attain full chimerism. Multiple transfusions pre-transplant leading to alloimmunisation, an unfavourable marrow microenvironment with dysplasia and OTC native medications, fungal infection necessitating a lower ATG dose, and early CMV viremia with high copy numbers were contributing factors. Though tapering immunosuppression is the norm in malignant transplants with a heavy transfusion burden and inadequate pretransplant chemotherapy exposure, applying the same approach in this case remains debatable. A DLI could have potentially salvaged the situation, but unfortunately, despite multiple attempts to contact the donor through the registry, he was untraceable. This raised concerns among transplant physicians about litigation risks and a growing preference for HID donors.

She faced multiple complications due to progressive graft loss, but meticulous supportive care enabled her to undergo a second transplant [7]. We must note that the recipient had rejected a graft previously, but with an active perianal infection she had to undergo a second transplant from her HID brother as the donor. The conditioning regimen should be more lymphoablative than myeloablative, with careful dose adjustments due to the high risk of developing VOD in a second transplant. However, early engraftment is also necessary due to the active perianal infection, which is difficult to achieve with the PTCy platform. Hence, a T-cell-depleted graft with a Flu-Treosulfan-rATG-2Gy TBI conditioning regimen was chosen. The incidence of graft failure and post-transplant viral infections is high with this approach due to delayed immune recovery, but memory cell add-backs and virus-specific cytotoxic T lymphocytes (CTLs) can serve as compensatory strategies.

Conclusion

The learning curve of HSCT is steadily evolving with newer drugs, cell manipulation techniques, and individualized post-transplant care. In the coming years, with disease-specific management groups and increased knowledge-sharing platforms, we hope that the challenges in bone marrow transplantation will become obsolete.

References

1. Mishra K, Naithani R. Haploidentical hematopoietic stem cell transplantation in India-bridging the gap. Indian J Hematol Blood Transfus. 2024;40:371–4.
2. Ciurea SO, Zhang M-J, Bacigalupo AA, Bashey A, Appelbaum FR, Aljitawi OS, et al. Haploidentical transplant with posttransplant cyclophosphamide vs matched unrelated donor transplant for acute myeloid leukemia. Blood. 2015;126(8):1033–40.
3. Chang YJ, Luznik L, Fuchs EJ, et al. How do we choose the best donor for T-cell-replete, HLA-haploidentical transplantation? J Hematol Oncol. 2016;9:35. https://doi.org/10.1186/s13045-016-0265-2.
4. Klco JM, Mullighan CG. Advances in germline predisposition to acute leukaemias and myeloid neoplasms. Nat Rev Cancer. 2021;21(2):122–37.
5. Beelen DW, Stelljes M, Reményi P, Wagner-Drouet EM, Dreger P, Bethge W, et al. Treosulfan compared with reduced-intensity busulfan improves allogeneic hematopoietic cell transplantation outcomes of older acute myeloid leukemia and myelodysplastic syndrome patients: final analysis of a prospective randomized trial. Am J Hematol. 2022;97(8):1023–34.
6. Hexner E. The role of T cells in hematopoietic stem cell engraftment. ScientificWorldJournal. 2006;6:246–53.
7. Giammarco S, Raiola AM, Di Grazia C, Bregante S, Gualandi F, Varaldo R, Chiusolo P, Sora F, Sica S, Laurenti L, Metafuni E, Innocenti I, Autore F, Murgia B, Bacigalupo A, Angelucci E. Second haploidentical stem cell transplantation for primary graft failure. Bone Marrow Transplant. 2021;56(6):1291–6.

Immunological Evaluation and Management of Pediatric Kidney Transplant Recipients

18

Priya Pais, Jitendra Meena, R. V. Deepthi, Kalaivani Ganeshan, and Georgi Abraham

Clinical Case

A 9-year-old girl with neurogenic bladder and bilateral grade 5 reflux with scarred kidneys presented in CKD stage 5. She was operated for myelomeningocele at birth but was subsequently lost to follow-up. The child was initiated on maintenance hemodialysis as well as clean intermittent catheterization (CIC) for bladder drainage. Once the child was medically stabilized and her bladder was found to be safe for transplant, a living donor transplant was planned. Her mother who was the same blood group as her daughter, completed her evaluation as a living donor.

Patient's Blood group: O positive, Mother—O positive. Weight-28 kg
Immunological Testing:

HLA typing by Luminex- HLA-1 mismatch in class I. No mismatch in Class II
Negative for preformed donor specific antibodies in both HLA class I and 2
Flow Cytometry T and B cell Crossmatch: Negative

P. Pais (✉)
Department of Pediatric Nephrology, St. John's Medical College, St. John's National Academy of Health Sciences, Bengaluru, Karnataka, India

J. Meena
Division of Pediatric Nephrology, Department of Pediatrics, All India Institute of Medical Sciences, New Delhi, India

R. V. Deepthi
Department of Paediatric Nephrology, CMC Hospital, Vellore, Tamil Nadu, India

K. Ganeshan
Department of Pediatric Nephrology, Mehta Multispecialty Hospitals India Pvt. Ltd., Chennai, Tamil Nadu, India

G. Abraham
The Madras Medical Mission Hospital, Chennai, Tamil Nadu, India

G. Abraham et al. (eds.), *Immunological Challenges in Organ Transplantation*,
https://doi.org/10.1007/978-981-95-1533-2_18

CDC crossmatching—< 5%: Negative
CMV D+/R+

Induction Therapy:

Induction with IV Basiliximab 10 mg on Day of transplant and post-op Day 4

IV Methylprednisolone 10 mg/kg on day of transplant, prior to the release of clamps

Maintenance Therapy

- Tacrolimus 1.5 mg (0.1 mg/kg) twice a day on day 1. The tacrolimus trough levels on Day 3 was low at 3.8 ng/mL. Therefore the dose of tacrolimus was increased to 2 mg twice a day to target trough levels of 10 ng/mL.
- Mycophenolate mofetil 800 mg–1000 mg/m^2 until Tacrolimus trough levels were at goal, then 600 mg/m^2.
- Oral steroids at 1 mg/kg × 4 weeks, thereafter planned for a regular taper every 3 days until 0.1 mg/kg/day at 3 months (if graft function stable).

Infection Prophylaxis

- For pneumocystis jirovecii pneumonia with trimethoprim sulfamethoxazole.
- For CMV with valganciclovir was initiated on Day 5 post-transplant.
- No UTI prophylaxis was initiated as the patient was doing well with no evidence of a UTI.

Introduction

Kidney transplantation (KT) remains the best management strategy for kidney failure in children, significantly improving survival and quality of life. While the general consensus for pediatric transplantation is that children should weigh at least 10 kg, some transplant centers have proceeded with transplants in infants as small as 6 kg. During history taking all relevant details of sensitizing events like blood product transfusion, including platelets, clinical events that can impact panel reactive antibody (PRA) like routine childhood, optional and catch up vaccination, withdrawal of immunosuppression, transplant nephrectomy, significant infection, systemic sepsis, pregnancy or miscarriage and communicated to the human leukocyte antigen (HLA) laboratory. However, successful transplantation hinges on the delicate balance between immune tolerance and alloimmune activation. Human leukocyte antigen (HLA) compatibility typing and immunological monitoring are cornerstones of optimizing graft outcomes. This chapter explores the advanced principles and clinical application of HLA testing, immune risk stratification, and post-transplant immunological monitoring in the pediatric population. Children differ from adults in terms of good thymic activity, low expression of CD40L on T cells, decreased mature macrophages, dendritic and HLA-alloreactive B cells.

HLA Testing: Principles and Clinical Significance from a Pediatric Perspective

The HLA system, encoded by genes on chromosome 6p21, plays a pivotal role in antigen presentation and immune response modulation. The classical HLA loci important for transplantation include:

- **Class I**: HLA-A, -B, -C—expressed on nearly all nucleated cells, platelets and occasionally on red blood cells.
- **Class II**: HLA-DR, -DQ, -DP—primarily expressed by a subgroup of specific antigen-presenting cells namely B-cells, activated T-cell, dendritic cells macrophages, and thymic cells. Under the influence of interferon gamma, other cells can also express class II HLA.

Gene complex encoding HLA class I and II is among the most polymorphic loci in human genomes which allow variations of peptide binding abilities and specificities contributing to individual's immunological repertoire. Since HLA antigens function to initiate the adaptive immune response by presenting foreign antigen, they play major role in immunological response following solid organ transplantation. This is chiefly based on the fact that proteins on donor organs perceived as foreign, "non-self" antigen. Pediatric recipients, due to developing immune systems and often naive immunological exposure, present a unique landscape for HLA matching considerations compared to adults [1].

HLA Typing Methodologies

HLA matching between donor and recipient is crucial to improve graft outcomes. Mismatching between donor and recipient for HLA-A, HLA-B, and HLA-DR is especially associated with high risk of HLA sensitization. Although with advancement in technology and testing methods now wider typing is feasible for all HLA loci (*HLA-A*, *HLA-B*, *HLA-C*, *HLA-DRB1*, *HLA-DRB3/4/5*, *HLA-DQA*, *HLA-DQB*, and *HLA-DPB*). This is more crucial for children in order to minimize the sensitization as they likely to require second transplant [1].

Traditionally HLA typing was performed by serological methods which involve reference sera with antibodies to all HLA antigens and mixing donor or recipients lymphocytes. After incubation period, interpretation was based on the cell lysis. Comparison of the serologic specificities of the different sera that reacted allows one to assign the HLA type. High-resolution typing using next-generation sequencing (NGS) and sequence-based typing (SBT) has largely supplanted older serologic and low-resolution DNA-based techniques. NGS allows unambiguous allele-level typing, crucial for precise donor-recipient matching and epitope-level compatibility assessment. As NGS and SBT testing takes more time, real-time polymerase chain reaction (RT-PCR) based typing is preferred for deceased donors. Virtual crossmatching integrates HLA antibody specificity data with donor typing, reducing cold

ischemia time and enhancing allocation efficiency, especially important in pediatric programs where ischemia impacts long-term outcomes more substantially.

Pediatric recipients benefit particularly from better HLA matching, with data indicating prolonged graft half-life in HLA-identical or well-matched living donor transplants. **Epitope matching**, focusing on shared eplets rather than full antigens, is gaining ground in pediatric allocation schemes to better reflect immunological risk. HLA-C and HLA-DQ loci, historically underemphasized, are now recognized for their contribution to alloimmune sensitization and chronic rejection risk [2].

Alloimmune Sensitization and Donor-Specific Antibodies (DSAs)

Sensitization in Pediatric Populations

Sensitization refers to the presence of preformed anti-HLA antibodies, usually secondary to prior transplants, blood transfusions, or maternal alloantigens (in utero sensitization). Pediatric patients, especially those with congenital anomalies of the kidney and urinary tract (CAKUT), often have fewer sensitization events pre-transplant. However, older children or those with multiple prior exposures (e.g., dialysis, transfusions) are at heightened risk.

Crossmatching Strategies

Complement-Dependent Cytotoxicity (CDC) Crossmatch

Although less sensitive than flow-based methods, CDC remains relevant in certain contexts for detecting complement-activating antibodies. A positive CDC crossmatch contraindicates transplantation due to risk of hyperacute rejection.

Flow Cytometric Crossmatch (FCXM)

FCXM detects IgG antibodies binding to donor T and B cells and is more sensitive than CDC. B-cell positivity without T-cell reactivity often reflects Class II DSAs and requires nuanced interpretation.

Virtual Crossmatch

A cornerstone in deceased donor allocation, the virtual crossmatch incorporates recipient HLA antibody profiles with donor HLA typing, enabling rapid organ acceptance decisions and lowering ischemic times. **Luminex-based single antigen**

bead (SAB) assays remain the gold standard for detecting anti-HLA antibodies with high specificity and sensitivity. Key metrics include:

- **Mean fluorescence intensity (MFI)**: Semiquantitative measure of antibody strength, though non-linear and assay-dependent.
- **Complement-binding assays** (e.g., C1q or C3d binding) to assess functional potential of DSAs.
- **IgG subclass analysis**: Certain subclasses (e.g., IgG3) are more pathogenic and linked with antibody-mediated rejection (AMR).

Preformed DSAs are associated with increased risk of hyperacute rejection and inferior graft survival. Hence, their identification informs both donor selection and immunosuppression tailoring.

Immunological Monitoring Post-Transplant

Given the risks of both rejection and over-immunosuppression, dynamic immune monitoring is essential in pediatric allograft recipients. Monitoring strategies following kidney transplantations include:

- **DSA surveillance**: Monitoring for de novo DSA (dnDSA), particularly within the first 1–2 years, predicts AMR and chronic graft injury.
- **Protocol biopsies**: Though more established in adults, some pediatric centers employ them to detect subclinical rejection or chronic allograft injury, including transplant glomerulopathy.
- **Non-invasive biomarkers**: Emerging tools includes Donor-derived cell-free DNA (dd-cfDNA), gene expression profiling and urinary chemokines. A key concept in using dd-cfDNA arises from that acute rejection will increase cell apoptosis and release of cf-DNA derived from donor cells. Recent data in adults support the routine clinical use of dd-cfDNA for detection of antibody mediated and T-cell mediated rejection [3]. It can also be used to assess the clinical significance of denovo DSA in children. Certain urinary chemokines (CXCL9, CXCL10) have been found to be useful in predicting severity and resolution of acute rejection. **Gene expression profiling (e.g., AlloMap)** although investigational in pediatrics, but likely to offers a non-invasive window into immune activation in coming years [4].

Immunosuppression for Pediatric Kidney Transplant Recipients

Induction Therapy

Similar to adults, pediatric kidney transplant recipients' immunosuppression protocols include both induction and maintenance therapies. The risks of acute rejection

are higher in the absence of induction therapy. Currently, the most commonly used induction agents are lymphocyte depleting agents (anti-thymocyte globulin or recombinant ATG-thymoglobulin) and Interleukin 2 (IL2) receptor blockers (Basiliximab). Either of these agents are combined with 1–2 doses of IV "pulse" methylprednisolone. Although there is no evidence based guidelines for the choice of induction agent, children with low immunological risk and those planned for steroid based immunosuppression protocols typically receive induction with IL-2 receptor blockers. Lymphocyte depleting agents are superior induction agents for pediatric patients who are at high immunological risk (sensitized, greater degree of HLA mismatch, prolonged cold ischemia times) or low immunological risk patients planned for steroid avoidance or steroid minimization maintenance immunosuppression protocols.

Typical Induction Regimen

- IV methylprednisolone 10 mg/kg on day of transplant, prior to surgery.
- AND
- Basiliximab 10 mg (for children <35 kg) prior to surgery on day of transplant and second dose on the fourth day after transplant. For children >35 kg, 20 mg/dose is used.
- OR
- Rabbit antithymocyte globulin (Thymoglobulin): Dose 1.5 mg/kg/dose initiated prior to surgery on the day of transplant, then given daily for a cumulative dose of less than 7.5 mg/kg. One of the authors have used 1 to 1.5 mg/kg successfully as a single dose of thymoglobulin producing significant depletion of lymphocytes there by ameliorating immediate infective episodes and long term PTLD development. In children, it should be ideally administered via central venous access. Premedication (corticosteroids, paracetamol, and diphendramine) to prevent the cytokine release related side effects should be administered, especially for the first dose. Dose reduction is required if leucopenia, neutropenia or thrombocytopenia occur.

The commonly prescribed triple immunosuppressive IS protocol for children is as below:

1. Basiliximab is administered 2 h before transplant surgery (d0) and after 4 days (d4) at the following weight-banded doses: ≥35 kg: 20 mg < 35 kg: 10 mg.
2. Methylprednisolone 600 mg/m^2 (maximum dose 500 mg) is given at induction or reperfusion on the day of transplant (day 0).
3. Initial dosing of Tacrolimus at 0.15–0.2 mg/kg twice daily with a maximum initial dose of 5 mg twice daily.
4. Mycophenolate mofetil (MMF) at as 600 mg/m^2 (maximum 1 g) twice daily on days 0–14, then 300 mg/m^2 twice daily from day 15 onwards.
5. Corticosteroid therapy as below.

Day of transplant (D0) post-transplant days	Steroid dose (mg/m² once daily)	Maximum prednisolone dose (mg)
Day 1–2	Dexa (iv) 0.2 mg/kg/24 h Q12H	60
	(or)	
	Prednisolone 2 mg/kg/24 h	
Day 3–4	Dexa (iv) 0.1 mg/kg/24 h Q12H	50
	(or)	
	Prednisolone 1.5 mg/kg/24 h	
Day 5–7	Prednisolone 1.2 mg/kg/24 h	40
Day 8–14	Prednisolone 1 mg/kg/24 h	30
Day 15–21	Prednisolone 0.67 mg/kg/24 h	20
Day 22–28	Prednisolone 0.3 mg/kg/24 h	10
Day 29–90	Prednisolone 0.25 mg/kg/24 h	7.5
Beyond Day 91	Prednisolone 0.15 mg/kg/24 h	5

Steroid sparing with discontinuation of corticosteroids within the first few weeks (early steroid withdrawal) after transplant has gained widespread acceptance in the United States. Acute rejection rates were higher in well-designed randomized clinical trials of steroid withdrawal. The presence of chronic interstitial fibrosis and tubular atrophy was greater in corticosteroid-free patients. Late steroid therapy withdrawal beyond 1 year is inadvisable. Promoting growth in children is a common argument for steroid withdrawal post-transplant. Recent and safer transplant regimens decrease and keep a maintenance prednisone dosage of 5 mg/day.

Maintenance Immunosuppression

Maintenance immunosuppression in children consists of a combination of drugs to prevent rejection, maximize allograft survival, and to keep patient morbidity from infections, drug side effects, and malignancies as low as possible. Typically, a combination of a calcineurin inhibitor, an anti-proliferative agent, and a glucocorticoid are used. In children, due to the significant side effects of glucocorticoids, there is significant interest in minimizing exposure to steroids. The final choice of immunosuppression protocols is center specific and depends on the child's immunological risk, their native kidney disease, risk of disease recurrence, and practical considerations of drug cost and availability [5].

Steroids

Glucocorticoids have been the mainstay of immunosuppression for decades. After an initial high dose of IV methylprednisolone as induction therapy, maintenance steroids are typically initiated at 1–2 mg/kg/day of oral prednisolone. Over a period of weeks to months, this dosing is reduced until a low dose daily or alternate day dosing regimen is reached. Glucocorticoids are associated with significant side effects in children such as a reduction in linear growth rate and trabecular bone mineral density, obesity, cardiovascular morbidity (hypertension, hyperlipidemia, and glucose intolerance), and cataracts. Short stature is a unique morbidity of CKD in children. Inadequate post-transplant growth results in psychosocial distress, poor self-image and adversely impacts the ability of children to accomplish normal life milestones. Thus, achieving adequate growth in pediatric transplant recipients is an important patient-centered outcome. Other undesirable steroid effects include cushingoid body habitus, rapid weight gain, and mood changes. These are especially distressing for adolescents and contribute to their risk for non-adherence.

Given the expected long-term survival of pediatric transplant recipients, the risks associated from prolonged steroid exposure accumulate over the life course. Therefore, there has been considerable interest in assessing if steroids exposure can be reduced in pediatric transplant patients. The first studies evaluated the effect of late steroid withdrawal (>6 months' post-transplant) in clinically stable patients. In these selected patients, the rates of biopsy proven rejection, patient, and allograft survival were similar in those who continued steroid based protocols and in the group where steroids were withdrawn. Subsequently, early steroid withdrawal (<7 days of transplant) and complete steroid avoidance were studied. In general graft outcomes were similar when compared with patients in whom steroid based immunosuppression was continued. Importantly, patients whose steroids were withdrawn had better growth and lower rates of obesity, hypertension, and cardiovascular morbidity [6].

It is important to note that patients enrolled in steroid minimization studies have been primary graft recipients, of low immunological risk. Sensitized patients and those at risk for recurrent disease not considered ideal candidates for attempting steroid minimization. Children with autoimmune systemic conditions and glomerular diseases at high risk of recurrence should receive steroid based immunosuppression. Also, patients considered for steroid avoidance or early steroid withdrawal received intensive induction and maintenance immunosuppression regimens, exposing children to the toxicity of these agents and risks of over-immunosuppression from maintenance regimen. Overimmunosuppression, especially those who are EBV naïve, increases the risk of PTLD. Ultimately, including steroids in maintenance immunosuppression protocols is a center-specific decision that must take into account balancing the benefits of steroid withdrawal with the individual patient's immunological risk, underlying native kidney disease recurrence risk, and the need for vigilance and frequent monitoring of graft function.

Calcineurin Inhibitors

Similar to adult transplant recipients, CNIs are the cornerstone of immunosuppression in children, responsible for improved graft survival in recent decades. Although pediatric data is limited, tacrolimus has been associated with lower acute rejection rates and is typically the preferred drug of choice for post-transplant immunosuppression. However, in the absence of tacrolimus or when liquid formulations are not available for children who cannot swallow tablets/capsules, cyclosporine is acceptable for use. Toxicities common to both CNIs include reversible and irreversible nephrotoxicity, neurotoxicity, hypertension, hyperlipidemia, and risk of post-transplant diabetes mellitus. However, cyclosporine is associated with significantly greater hirsutism and gum hypertrophy. These cosmetic side effects may be a risk factor for non-adherence if used in adolescents. There is considerable genetic variability in CNI pharmacokinetics, resulting in the need for regular monitoring of trough levels to ensure adequate dosing while avoiding the toxicity.

Therapeutic Drug Monitoring

It is imperative to pay special attention to kidney transplant recipient's prescription list at every visit. Patients need to be educated about possible risk of drug interactions that may decrease or increase ID drug levels which may affect the life of allograft adversely. This will mitigate the risk of prescribing medicines that will interact with the maintenance IS [7].

Target ranges for tacrolimus 12 h trough levels are

- 8–12 ng/mL for months 1 and 2 post-transplant
- 5–8 ng/mL for months 3 to 12 post-transplant
- 4–6 ng/mL beyond the first year post-transplant

Target MMF AUC is

- 40–60 mg h/L in the first 6 months
- 40–50 mg h/L 1–2 years
- 30–40 mg h/L beyond 2 years

Cyclosporine is metabolized more rapidly in children and thus a higher dose and more frequent administration may be required compared to adults. Cyclosporine is initiated at a dose of 8–12 mg/kg/day in 2–3 divided doses. Trough levels may be used for drug level monitoring, aiming for 150–300 mcg/L for the first 3–6 months. Subsequently, over a period of several months, the dose may be reduced to 4–6 mg/kg/day.

Anti-proliferative Agent

Mycophenolate mofetil or its enteric coated equivalent, mycophenolate sodium is the drug of choice in this category. Azathioprine has the advantage of being cheaper but has greater bone marrow toxicity and is associated with a higher risk of rejection. Mycophenolate toxicity includes leucopenia and gastrointestinal side effects such as diarrhea, and reduced appetite. Mycophenolate is dosed at 600–1000 mg/m^2 in 2 divided doses and then reduced to 600 mg/m^2 when adequate CNI levels have been established.

Newer molecules like eculizumab and bortezomib (inhibitor of complement protein C5 and proteasome inhibitor). Four doses of bortezomib 1.3 mg/m^2 at day 1, 4, 8, and 11 was effective and safe for rapid reduction of DSA levels among children. New molecules including belatacept and alemtuzumab carry serious side effects and hence IL2 receptor antagonist and thymoglobulin remain safe and effective agents for pediatric kidney transplantation.

Graft Rejection

According to a recent registry report, the probability for acute rejection in recent pediatric transplant recipients is 13.2% for deceased donor graft recipients and 12.7% in recipients of a living donor kidney [7]. Although the rates of early rejection in children have declined with advances in immunosuppression, the rates of late and chronic rejection are still significant and have not improved over time. Chronic rejection is the most frequent cause of graft failure in children.

The classification of allograft rejection in children is similar to adults. Rejection is suspected when there is a rise in the serum creatinine (decline in the allograft function). However, as is well known, serum creatinine is not a sensitive indicator of a reduction in GFR. This is especially true in children who typically have a large sized graft relative to their body size. A kidney biopsy remains the gold standard for diagnosis of rejection. Rejection is classified into T-cell-mediated rejection (TCMR), antibody-mediated rejection (ABMR), or a mixed rejection on the basis of the Banff classification of Renal Allograft Pathology. Pediatric transplant centers typically monitor serum creatinine frequently, weekly, and then monthly for the first year. Some centers by protocol perform surveillance biopsies, in addition to clinical monitoring, to detect subclinical TCMR rejection. Surveillance biopsies are typically performed between 6 and 12 months after transplant and subclinical rejection is treated similarly to a clinical rejection.

Many risk factors for rejection in children are similar to adults [8]. These include the extent of HLA mismatch, especially Class II HLA mismatches at the DQ and DR loci, and the presence of donor specific antibodies (DSA). Acute/active antibody mediated rejection (AMR) occurs due to the presence of pre-formed donor specific antibodies (DSA) or when de novo DSA develop after transplant. Sensitization to HLA antigens can occur prior to transplant resulting in an accelerated immunological response after exposure to the allograft antigens. In children, a

history of sensitizing events such as blood product transfusions and prior organ transplant should be obtained. After transplant, delayed graft function is an important risk factor for rejection. Any significant infection can trigger an acute rejection. Viral infections such as CMV, EBV, and BK virus also increase rejection risk due to their stimulation of the recipient's immune system or due to the reduction of immunosuppression for control of infection.

Non-adherence

The highest risk of graft loss occurs during adolescence and early adulthood. Non-adherence is a significant, preventable risk factor for graft loss in adolescents. Non-adherence to immunosuppression medications results in an increased risk of late acute and chronic rejection. The reasons for non-adherence are a combination of the complexity of prescribed regimen and pill-burden, adolescent immaturity, and difficulties with management of medication schedules along with their other activities, coinciding with waning parental supervision.

Addressing adherence is crucial to the success of transplant. Assessing readiness prior to transplant requires that the transplant team evaluate the patient's (and family's) knowledge of the transplant process, the need for taking medications for the foreseeable future, and consequences of non-adherence. After transplant, adherence must be monitored at every follow-up visit. Adherence may be assessed via self-reporting of missed doses by the patient or family. In addition, the transplant team can also monitor the frequency with which medication prescriptions are being filled and the variability in immunosuppression drug levels. It is essential to stress the importance of parents continuing to observe the adolescent kidney transplant recipient take his medications [9]. A structured transition program should be followed to ensure that an adolescent transferring to adult healthcare systems does not become non-adherent.

Infections and Malignancy in Pediatric Transplant Recipients

Immunosuppression regimens to prevent and treat rejection come with the associated risk of infections and malignancy. Infections result in frequent hospital admissions after transplant. Urinary tract infections are the most common bacterial infections in children after transplant. This is due to a combination of intense immunosuppression immediately in the peri-transplant period, the increased prevalence of bladder and lower urinary tract abnormalities in children and placement of urinary stents. Opportunistic viral infections are especially important in children because of their lack of exposure or "naivety" to these infections prior to transplant [10]. Cytomegalovirus (CMV) results in significant patient morbidity and mortality. Pediatric recipients are more likely to be seronegative for CMV than their adult recipient counterparts. Their risk for CMV infection and disease is highest when they receive kidneys from seropositive donors. Administration of leucocyte

depleting agents and high dose steroids as induction therapy as well as for treatment of rejection are significant risk factors for CMV infection. Similarly, children who are seronegative for EBV infection are also at risk for EBV viremia and post-transplant lymphoproliferative disease (PTLD). Young age, those receiving intense induction regimens because of sensitization or in preparation for steroid avoidance, or early withdrawal protocols are at increased risk of EBV infection. Although cancer is rare in pediatric kidney transplant recipients, PTLD is commonest malignancy post-transplant. The risk of cancer depends on the intensity and duration of immunosuppression received, and increases with time since transplant [11].

Vaccinations in Pediatric Transplant Recipients

Adequate and complete vaccination of transplant candidates is essential to prevent morbidity in immunosuppressed children from vaccine-preventable infections. Pre-transplant evaluation should confirm that children are up to date on their vaccination schedule as per national guidelines [12]. Serologic evidence of immunity should be documented when possible, especially against Hepatitis B. Inactive vaccines should ideally be completed 2 weeks before transplant but may be administered within 3–6 months after transplant when maximal immunosuppression has reduced to baseline levels.

Live virus vaccines are contraindicated after transplant as the attenuated vaccine virus could cause disease in immunosuppressed patients. Therefore, the measles, mumps, and rubella (MMR) vaccine, varicella, and rotavirus vaccines should be administered a minimum of 4 weeks prior to transplant to ensure that viremia from the vaccine strain has resolved. Completing vaccination against Hepatitis B, Influenza, Pneumococcus, and Varicella is essential. Meningococcal vaccines are indicated in children with complement pathway deficiencies (due to disease or therapy).

Conclusions

The basic immunological aspects of the evaluation and management of pediatric kidney transplant recipients are similar to adults. However, there are important aspects for transplant teams to keep in mind when assessing for immunological risk and when planning induction and maintenance immunosuppression regimens. To ensure optimum graft survival, regular monitoring of graft function is necessary in order to diagnose and treat rejection early.

Aside from the immunological risk factors for rejection, a significant, modifiable risk factor unique to pediatrics in non-adherence, especially in adolescents. Heightened intensity of immunosuppression and an immature immune system results in an increased risk of vaccine preventable diseases, bacterial infections, and opportunistic viral infections. Thus all pediatric transplant recipients must be

evaluated in advance for their risk of developing CMV and EBV disease and vaccination schedules must be completed prior to transplant.

The child described in the case vignette will be assessed as having low immunological risk. The transplant evaluation process should include detailed patient and family counselling to stress the importance of medication adherence and compliance with CIC. After transplant, graft function should be monitored daily until discharge, then twice weekly and slowly spaced out to weekly serum creatinine monitoring. It is necessary to measure her CNI trough levels at regular intervals to ensure adequate drug level, especially after dose adjustments. The child will need to be monitored for side effects of her maintenance immunosuppression. She is at risk for UTI due to dependence on CIC for complete bladder emptying. If a UTI is suspected, early therapy prevents allograft dysfunction. At every clinic visit, the importance of adherence must be stressed with the child and her parents. Lastly, her growth velocity must be monitored to ensure that she shows catch up growth post-transplant. This level of care is best provided by a multidisciplinary team, who will follow the child into adolescence, provide support to the caregivers, and hopefully achieve a successful transition into adult transplant care services when she turns 18 year of age.

References

1. Mishra VC, Chandra D, Raina V. Histocompatibility testing: a fundamental aspect of renal transplant workup. Transplantology. 2024;5(2):85–97. https://doi.org/10.3390/transplantology5020009.
2. Fernandez HE. Application and interpretation of histocompatibility data in pediatric kidney transplantation. Curr Opin Organ Transplant. 2017;22(4):426–32.
3. Xing Y, Guo Q, Wang C, Shi H, Zheng J, Jia Y, Li C, Hao C. Donor-derived cell-free DNA as a diagnostic marker for kidney-allograft rejection: a systematic review and meta-analysis. Biomol Biomed. 2024;24(4):731–40.
4. Laroche C, Engen RM. Immune monitoring in pediatric kidney transplant. Pediatr Transplant. 2024;28(4):e14785. https://doi.org/10.1111/petr.14785.
5. Pratschke J, Dragun D, Hauser IA, Horn S, Mueller TF, Schemmer P, Thaiss F. Immunological risk assessment: the key to individualized immunosuppression after kidney transplantation. Transplant Rev (Orlando). 2016;30(2):77–84. https://doi.org/10.1016/j.trre.2016.02.002.
6. Kizilbash SJ, Jensen CJ, Kouri AM, Balani SS, Chavers B. Steroid avoidance/withdrawal and maintenance immunosuppression in pediatric kidney transplantation. Pediatr Transplant. 2022;26(2):e14189. https://doi.org/10.1111/petr.14189.
7. Chua A, Cramer C, Moudgil A, Martz K, Smith J, Blydt-Hansen T, Neu A, Dharnidharka VR, NAPRTCS Investigators. Kidney transplant practice patterns and outcome benchmarks over 30 years: the 2018 report of the NAPRTCS. Pediatr Transplant. 2019;23(8):e13597. https://doi.org/10.1111/petr.13597.
8. Cippà PE, Schiesser M, Ekberg H, van Gelder T, Mueller NJ, Cao CA, Fehr T, Bernasconi C. Risk stratification for rejection and infection after kidney transplantation. Clin J Am Soc Nephrol. 2015;10(12):2213–20. https://doi.org/10.2215/CJN.01790215.
9. Singer P. Post-transplant education for kidney recipients and their caregivers. Pediatr Nephrol. 2023;38(7):2033–42. https://doi.org/10.1007/s00467-022-05744-6.

10. Epperson K, Crane C, Ingulli E. Prevention, diagnosis, and management of donor derived infections in pediatric kidney transplant recipients. Front Pediatr. 2023;21(11):1167069. https://doi.org/10.3389/fped.2023.1167069.
11. Green M, Squires JE, Chinnock RE, Comoli P, Danziger-Isakov L, Dulek DE, Esquivel CO, Höcker B, L'Huillier AG, Mazariegos GV, Visner GA, Bollard CM, Dipchand AI, Ferry JA, Gross TG, Hayashi R, Maecker-Kolhoff B, Marks S, Martinez OM, Metes DM, Michaels MG, Preiksaitis J, Smets F, Swerdlow SH, Trappe RU, Wilkinson JD, Allen U, Webber SA, Dharnidharka VR. The IPTA Nashville consensus conference on post-transplant lymphoproliferative disorders after solid organ transplantation in children: II-consensus guidelines for prevention. Pediatr Transplant. 2024;28(1):e14350. https://doi.org/10.1111/petr.14350.
12. Fox TG, Nailescu C. Vaccinations in pediatric kidney transplant recipients. Pediatr Nephrol. 2019;34(4):579–91. https://doi.org/10.1007/s00467-018-3953-z.

19 An Approach to Rational Therapeutics by Navigating Pathophysiology and Evidence for Complex Clinical Situations

Vishnu Eriyat, Yuvaram Reddy, and Basu Gopal

Introduction

It is not a rarity in practice to see a prospective transplant recipient who does not conform to the textbook versions we are usually familiarized with. Patients with high degree of sensitization and other immunological barriers pose a challenge to the usual armamentarium of therapies. This calls for an overview of therapeutic agents and regimens which can be tailored to address these clinical scenarios. In this chapter, we will explore a complex clinical situation of intersecting potentially common scenarios in approaching a candidate for kidney transplantation, reviewing the available therapeutic options.

V. Eriyat
Clinical Pharmacology Unit, Department of Pharmacology and Clinical Pharmacology, Christian Medical College Vellore, Vellore, Tamil Nadu, India

Y. Reddy
Renal-Electrolyte and Hypertension Division, Perelman School of Medicine, University of Pennsylvania, Philadelphia, PA, USA
e-mail: yuvaram.reddy@pennmedicine.upenn.edu

B. Gopal (✉)
Monash University (CCS), Melbourne, VIC, Australia

Renal Medicine and Kidney Transplantation, The Alfred, Melbourne, VIC, Australia
e-mail: basu.gopal@monash.edu

G. Abraham et al. (eds.), *Immunological Challenges in Organ Transplantation*, https://doi.org/10.1007/978-981-95-1533-2_19

Case Scenario

A 36-year-old multiparous woman with end-stage kidney disease (ESKD) secondary to nephrotic syndrome due to PLA2R-associated Membranous Nephropathy on haemodialysis for the past 1 year. She has no residual renal function and continues to have a high titer of anti-PLA2R antibodies in the serum at 140 RU/mL.

She was being worked up for a living donor kidney transplantation with her 57-year-old mother as a potential kidney donor. She is blood group B positive, while her mother is blood group A positive.

Her history reveals no prior organ transplants, but she has multiple blood transfusions due to postpartum haemorrhage in her second pregnancy 6 years before the diagnosis of membranous nephropathy.

Immunologic evaluation revealed that she had Anti-A antibody titres by gel column method as 1:16 and 1:32 on two separate occasions.

In addition, she was also HLA haplotype matched with the donor across HLA-A, B, and DR antigens. A CDC crossmatch was positive at 40% against the mother's lymphocytes. Luminex screen was positive for both Class I and Class II anti-HLA Antibodies. Single antigen bead assay (SAB) revealed a current single donor-specific antibody (DSA) against HLA-B*58 01 mismatch at an MFI value of 6500.

On further exploration, there are no other living donors. She has multiple (non-donor specific) Class I and Class II Anti-HLA antibodies in the SAB assay, amounting to a calculated panel reactive antibodies from the Luminex assay (cPRA) of 95%.

The HLA lab director further interrogated her previous serum tested with other potential living and deceased donors in the last 1 year and reported that the Anti-HLA-B*58 01 antibody has not been the dominant Anti-HLA Ab and has been present at an MFI of 6000–8000.

Discussion

This case presents a unique set of complex immunological barriers, which although appear rare, are not unfathomably uncommon to encounter in clinical practice. The complex set of immunological issues could be summarized as:

1. ABO incompatibility with a blood group A donor to blood group B recipient and a peak Antibody titer of 1:32 predisposing to hyperacute rejection or early accelerated—antibody mediated rejection (ABMR).
2. High degree of sensitization with 95% PRA precluding a reasonable potential for finding a compatible deceased donor without prolonged wait times or finding a matched living donor–candidate pair for a paired kidney exchange.
3. HLA incompatible living donor with a positive CDC crossmatch with a single DSA against HLA–B antigen at MFI of 4500.
4. Increased risk of recurrence of Membranous Nephropathy with a High ANTIPLA2R antibody level of 140 RU/mL pre-transplantation.

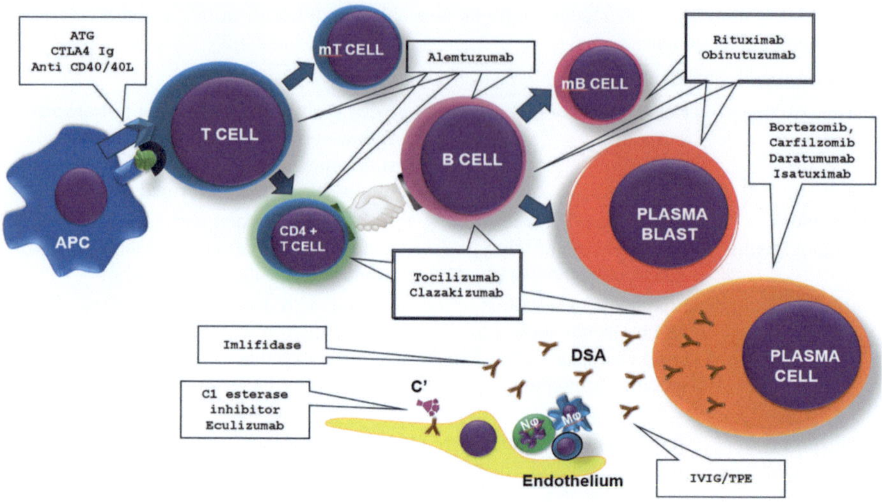

Fig. 19.1 Therapeutic agents and their potential mechanistic site of action in preventing antibody mediated rejection. (Illustration: Basu Gopal)

Such convergence of complex risk factors presents unique challenges in transplantation, requiring a comprehensive approach including devising a desensitization plan drawing from varied literature, and considering the various pathophysiologic processes involved (Fig. 19.1).

Agents/Interventions for Desensitization in Highly Sensitized Patients

Several therapeutic strategies can be utilized in reducing the sensitization and risk of rejection in such patients. Table 19.1 and Fig. 19.1 provide supporting information on several such therapeutic options.

Plasma Exchange

The most critical risk is that of hyperacute rejection, as evidenced by the positive CDC crossmatch and positive DSA in the Luminex single antigen bead assay. Reducing the circulating preformed DSAs is the logical first step in desensitization therapy to prevent hyperacute rejection and reduce the incidence of accelerated acute antibody-mediated rejection in the early post-transplant period. The primary intervention for antibody removal is therapeutic plasma exchange (TPE) or plasmapheresis, where plasma is separated from the patient's blood and replaced by a replacement fluid such as 4% albumin and/or fresh frozen plasma [1]. As depletion of coagulation factors and a consequent increase in bleeding risk is a common

Table 19.1 Summary of therapeutic options available for desensitization in highly sensitized patients and induction agents

Drug/intervention	Mechanism	General remarks on use
Plasma exchange	Removes the donor specific antibodies	Need multiple sessions prior to transplant followed by post-transplant. Need to be supplemented with other agents as it removes only the preformed antibodies
Intravenous immunoglobulin	Saturation of FC-receptors and formation of idiotype-anti idiotype networks	Used in conjunction with plasma exchange and this combination forms the foundation of most desensitization regimens
Imlifidase	Enzymatic degradation of the IgG molecule leading to rapid clearance of the donor specific antibodies	Immediate reduction of antibodies facilitating transplant but needs to be coupled with other therapies which prevent B cell activation or abrogate fresh antibody production from plasma cells
Eculizumab	Terminal complement pathway inhibitor, reducing the downstream inflammatory response to donor specific antibodies	Usually combined with an endopeptidase for effective usage. Lack of availability and high cost are concerns
Rituximab	A chimeric anti-CD20 monoclonal antibody which blunts the B cell response	Used as an adjunct to plasma exchange and IVIG, with reduction in their need. Is particularly useful in the setting of ABO incompatibility
Bortezomib	Proteasome pathway inhibitor which triggers plasma cell apoptosis and thus gets rid of the activated plasma cells	The response on antibody suppression is short and variable with an unacceptable incidence of peripheral neuropathy and hematological toxicity
Carfilzomib	Irreversible inhibitor of proteasome pathway	Can be combined with rituximab for additional antibody suppression. Renal and cardiac toxicity necessitates careful patient selection and pre intervention workup
Daratumumab	CD38 inhibitor which causes depletion of plasma cells	Its effect on Treg cell function and resultant high incidence of cell mediated rejection remains a concern
Isatuximab	CD38 inhibitor which causes depletion of plasma cells	Produced a durable antibody suppression but requires long treatment duration and only modest efficacy as monotherapy

(continued)

Table 19.1 (continued)

Drug/intervention	Mechanism	General remarks on use
Anti IL-6 agents	Tocilizumab (targeting IL-6 receptors) and Clazakizumab (targeting IL-6 itself) can bring down the IgG production by plasma cells	Can be considered as an adjuvanting agent to rituximab for add on effect
Anti Thymocyte globulin	Binds to T cell membrane proteins, resulting in T cell depletion	Induction agent for high immunologic risk transplantation. Treatment of acute rejection. Side effects including cytokine release syndrome, Lymphopenia, post-transplant lymphoma, CMV and BK virus nephropathy
Alemtuzumab (Campath)	Anti-CD52, thereby depleting T cells, B cells, NK cells and monocytes	Induction agent for high immunologic risk transplantation. Treatment of acute rejection. Cytokine release syndrome, Lymphopenia, post-transplant lymphoma, CMV and BK virus nephropathy. Autoimmune phenomena
Basiliximab	Anti IL2 receptor monoclonal ab. Inhibits IL2 induced T cell activation of proliferation	Induction agent for low immunologic risk transplantation. Adverse events are rare—hypersensitivity reactions

complication of TPE, use of double filtration plasmapheresis (DFPP) or specific immunoadsorption (IA) columns provides an equivalent or even superior achievement of antibody reduction without attendant bleeding risk [1–3]. TPE sessions ranging from 1 to 12 reduce the antibody level in mild to moderate MFI anti-HLA antibody levels to achieve a negative CDC crossmatch before the kidney transplant [1–5]. Redistribution of antibodies from tissue to circulation by equilibration and resurgence of antibody production could be rapid, reaching 50% of pre-intervention levels within a week [2]. Monitoring the antibody titers and scheduling the transplant surgery at the time of appropriately controlled antibody levels is critical to avoid hyperacute rejection. In addition, further about 3–4 TPE sessions will be required post-transplantation to avoid early ABMR [5–7].

In addition to reducing anti HLA Antibodies, Anti ABO antibodies also will be reduced by TPE. The required number of TPE sessions pre- and post-transplant for sustained reduction of anti ABO antibodies to a target titer of ≤1:8 depends on the initial antibody titer [8] and concomitant use of anti B cell therapy.

High anti-PLA2R antibodies pre-transplant are likely to cause significant relapse of membranous nephropathy in kidney transplant recipients. Reduction of anti-PLA2R Ab by cytotoxic therapy or Rituximab is the standard of care in managing native kidney MN. However, in cases where pre-transplant antibody levels are high, there are no clear guidelines or therapeutic trial-based regimen for emergent reduction of antibodies pre-transplant in patients with native kidney MN and high antibody titers, it is logical and pathophysiologically reasonable to undertake antibody reduction with a combination of TPE and Rituximab therapy. This is also supported

by the primacy of TPE as a therapeutic regimen in other early recurrent diseases, such as primary FSGS [9]. TPE will achieve a rapid reduction in antibody titers to facilitate transplant without an early relapse of MN.

Intravenous Immunoglobulin

The use of plasmapheresis to remove DSA complemented with other measures such as high-dose intravenous immunoglobulin (IVIG) [2–4]. When pooled together, the immunoglobulin from different donors causes a reduction in antibodies due to multimodal activity. Saturation of Fc receptors by high-dose immunoglobulin [10] and formation of idiotype-anti-idiotypic networks with Anti-HLA Antibodies have been proposed mechanisms in addition to immunosuppressive impacts on various immune cell lines. A common approach is to combine plasmapheresis with IVIG for desensitization protocols with variable additional therapies targeting B cells or T cells, such as Rituximab (Anti CD-20) or anti-thymocyte globulin, respectively [4–7].

When used alone, multiple courses of high-dose IVIG would be required before acceptable desensitization [4, 6, 10], with some effect expected around a median of six courses of high-dose IVIG in 50% of patients [10]. Reports also suggest that a higher dose regimen, i.e., 3 g/Kg, was beneficial in those refractory to the usual dose of 2 g/Kg of IVIG [11]. Apart from the high-dose IVIG monotherapy (2–3 g/kg) [4, 11, 12], it has been more commonly used with rituximab and TPE at lower doses (0.5–1 g/Kg) or as a low-dose adjuvant (100 mg/kg) [6, 10]. The efficacy of desensitization with IVIG alone or with TPE (reduction in PRA) has been demonstrated with >80% success in achieving predefined limits and enabling transplantation without hyperacute rejection [4, 6, 7, 10]. However, both regimens demonstrated a 40% incidence of ABMR [6, 10, 12], with variable impact on renal dysfunction [6]. Both regimens documented an 80% graft survival in long-term studies [6, 7, 12]. One major issue with the use of IVIG and TPE-based desensitization is the long period with multiple cycles of therapy required [4, 6, 10], making it challenging to utilise in an emergent situation such as an incompatible deceased donor transplant. There is a felt need for therapies to deplete DSAs rapidly.

Imlifidase

One such strategy is achieved by pharmacological removal through enzymatic cleavage. IgG endopeptidase (aka IdeS or Imlifidase), an enzyme derived from *Streptococcus pyogenes,* cleaves all circulating immunoglobulin G molecules at the hinge region, decoupling the Fc from the Fab fragments, resulting in rapid deactivation of the antibodies and hence a perceptibly sudden clearance of even high MFI DSAs [13]. The enzyme is given 4–6 h before the transplant (0.25 mg/kg), and by 6 h, it causes a significant reduction in the mean fluorescent intensity and allows the

transplant [14, 15]. This obviates the need for long durations of IVIG/plasmapheresis and increases the probability of getting a deceased donor transplant.

As with TPE, Imlifidase use only temporarily removes the antibodies from circulation. The resurgence of antibodies due to equilibration from tissue, and more importantly, production from active plasma cells, results in severe antibody-mediated rejection and remains a significant impediment to successful outcomes in desensitization. The antibody rebound after Imlifidase treatment is invariably expected from 7 to 28 days post-transplant [14, 15], unless adjunct therapies aimed at controlling antibody production such as IVIG, anti B cell therapies such as rituximab (375 mg/m^2) or anti plasma cell therapies (see below) are co-administered to abrogate this response after 5 days when endopeptidase activity should have been eliminated due to its short half-life [13, 15].

Complement Pathway Inhibitors

The DSA's exert their pathogenic effect on the graft by complement activation. Hence, the terminal complement pathway inhibitors, such as Eculizumab (C5 complement inhibitor), can reduce the downstream inflammatory response even in the presence of antibodies [16] and reduce histological changes in peritubular capillaritis, glomerulitis, and interstitial inflammation [17]. The use of eculizumab with a loading dose of 1200 mg just before transplant, followed by 600 mg doses weekly for 4 weeks, has been found to reduce the incidence of antibody-mediated rejection in the early post-transplant period as well as at 1-year post-transplant compared to standard of care consisting of plasmapheresis and IVIG [18, 19]. By immediately reducing the cytotoxic nature of the DSA, terminal complement inhibitors such as Eculizumab and Ravulizumab could help support HLA-incompatible deceased donor transplantation. The combined use of an endopeptidase before transplant, followed by eculizumab in the post-transplant period, could reduce the need for prolonged therapy with IVIG/plasmapheresis and provide a rapid reduction in the risk of hyperacute rejection at the time of deceased donor transplant. However, the lack of widespread availability of these drugs, the absence of clinical trial data, and the exorbitant price impede our ability to use these agents in resource-constrained and well-resourced healthcare settings.

Anti B Cell Therapies

While removing an antibody provides a window of opportunity to perform a transplant without hyperacute rejection, sustained reduction in antibody levels by curtailing the antibody rebound and further antibody production (B cells/plasma cells) remains a critical need. The antibody rebound is due to an anamnestic response from the memory B cells [20]. This response could be prevented by Anti-B cell therapy Rituximab (A chimeric anti-CD20 monoclonal Ab), resulting in a blunting of the pre-sensitized antibody response [11, 15, 18, 19, 21]. B cells with Rituximab

(anti-CD20 chimeric antibody) administered as two doses, 2 weeks apart, of 1 g each as a 6-h infusion have been found to produce a significant reduction in the DSA titer and result in a 1-year graft survival of 94% [22]. This regimen and high-dose IVIG (2 g/kg, day 1 and day 30) eliminate the need for a higher number of plasmapheresis treatments. A single-dose protocol with Rituximab administered midway in the desensitization treatment (2 g/kg, day 15) has also resulted in 2 a 2-year graft survival of 84% [23]. Rituximab has also been employed successfully as an induction strategy in highly sensitized patients with a prior history of multiple transplants [21].

It must be noted that Rituximab interferes with immunological assessment by causing a B-cell CDC crossmatch positivity. This can be partly offset by using Pronase pre-treatment on donor lymphocytes. Additionally, repeated use of Rituximab could trigger an immune response against the murine Fab segment of the antibody, reducing its efficacy and utility.

Rituximab has also found its use is in reducing rebound of anti-ABO antibodies after TPE in patients undergoing ABO incompatible transplantation and prevented hyper acute rejections [24]. Rituximab has essentially replaced the need for splenectomy in ABO incompatible renal transplants [25], without significantly predisposing to. Unlike splenectomy, Rituximab does not increase the chances of acquiring infections by capsulated organisms, and thus, the need for prior immunization against *pneumococcus* and *meningococcus* at least 2 weeks before transplant and ongoing need for oral antibiotic prophylaxis [26].

Obinutuzumab is a third-generation anti-CD20 fully human monoclonal antibody with a defucosylated Fc region that emphasizes antibody-dependent cellular cytotoxicity and hence effectively depletes not only circulating B cells but also those in secondary lymphoid organs, potentially offering more enduring effects on memory B cells and plasma cells compared to Rituximab. In addition to a lack of immune response against itself, the additional benefit of Obinutuzumab has been that it does not cause a positive B cell CDC crossmatch [27]. Use of Obinutuzumab in desensitization and treatment of ABMR has been studied with limited/no success in some small case series [27, 28]. However, it is notable that Obinutuzumab, while very effective in reducing/eliminating autoantibodies generated by CD20-positive Plasmablasts, fails to significantly impact the alloantibodies that are produced by long-lived CD20-deficient Plasma cells. This explains the critical difference in its outcomes for the glomerular diseases such as lupus nephritis, ANCA vasculitis or Membranous Nephropathy, but not in HLA-desensitization or ABMR.

Rituximab therapy is now well established in the treatment of membranous nephropathy, especially in the anti-PLA2R Ab positive membranous nephropathy [29]. Recent reports support the use of Obinutuzumab, especially in refractory cases of MN not responding well to Rituximab [30]. Both agents have been very effective in reducing the anti-PLA2R antibody titers in those with native kidney membranous nephropathy. Given the risk of recurrence of membranous nephropathy highly correlates with the titer of Anti-PLA2R antibody level pre-transplant [31], despite ESKD, these agents could still be pathophysiologically justified in reducing

pre-transplant anti-PLA2R antibodies to reduce the risk of membranous nephropathy recurrence.

Anti-plasma Cell Therapies

The treatment strategy of targeting mature B cells still does not get rid of the activated long-lived CD20-negative plasma cells or memory B cells [32]. Their survival in immunological niches causes rebound and persistence of antibodies in the post-transplant period following desensitization. This calls for targeting the plasma cells in desensitization therapy. There are two targets explored for this purpose, one is the proteasome pathway, and the other is a surface marker, CD38, on the plasma cells. Reversible proteasome inhibitor Bortezomib, which blocks enzymatic proteolysis and thereby triggers plasma cell apoptosis, has been helpful in plasma cell disorders such as myeloma [33, 34]. While useful in the control of Myeloma, this did not translate into successful application in desensitization, resulting in short duration and variable degree of antibody suppression [35]. In addition, the high doses necessary for such a modest effect bore an unacceptable degree of hematological toxicity and frequent peripheral neuropathy [36, 37].

Carfilzomib is an irreversible inhibitor of proteasome enzymes and has been studied in the suppression of DSAs [38]. While the combination of Carfilzomib with TPE/IVIG demonstrated significant antibody reduction, there was a rapid rebound within a few months [39, 40]. The evidence from disease states such as diffuse large B cell lymphoma, where the NF-KB signalling in B cells is constitutively activated, indicates that a combined inhibition of proteasome and CD20 may produce sufficient inactivation of B cell activation [41–43]. Thereby, a combined desensitization strategy using Rituximab and carfilzomib could be explored. The selection of patients for receiving carfilzomib should consider the possible reported cardiovascular and renal adverse effects [38, 44]. The patients should undergo a cardiac workup to detect preexisting decreased ejection fraction. Though the renal toxicity could preclude its use after transplant, the acute kidney injury seems to be responsive to intravenous fluids before and during the therapy [44].

The monoclonal antibody Daratumumab targeting CD38 showed depletion of plasma cells and a reduction in the donor-specific antibody. However, due to a blockade of the T regulatory cells, the incidence of T cell-mediated rejection after transplant is concerning [45]. A newer congener, Isatuximab, has been tried and has shown a durable response in reducing the antibody levels. However, when used as monotherapy, the long duration of therapy required may be non-viable, and the magnitude of the response was not as good as existing therapies to enable transplantation [46].

Anti IL6 Agents

The cytokine IL-6 has several functions critical for T-follicular/helper cells (Tfh), Th17 cells, Plasmablasts, and adaptive immune responses. Targeting IL6 receptors with Tocilizumab, a humanized monoclonal antibody, has been employed in patients resistant to desensitization using IVIG and Rituximab and in highly sensitized patients naïve to desensitization therapies [47, 48]. While the Tocilizumab alone did not significantly reduce the preformed DSA MFI, when used in conjunction with IVIG, TPE and Rituximab, it did reduce antibody rebound after desensitization [49]. Clazakizumab, a monoclonal antibody against IL6 cytokine, has been studied as an adjuvant to standard desensitization as monthly treatments from 6 months pre-transplant to 12 months post-transplant with some degree of success [50]. Further reports have explored its utility in the treatment of chronic ABMR [51].

There are several other agents that are being explored in various studies for desensitization, such as CTLA4-Ig (co-stimulatory pathway blockade), efgartigimod (anti FcRn monoclonal antibody inhibiting FcRn recycling of IgG), bispecific antibody targeting B cell maturation antigen and CD3, Chimeric antigen receptor T cell therapy, C1 esterase inhibitor, and so on.

Anti T Cell Therapies

Anti-Thymocyte globulin (ATG) is a polyclonal antibody that causes a significant reduction in T cell counts, working through multiple mechanisms, including complement-mediated T cell lysis, antibody-dependent cellular cytotoxicity, and phagocytosis of T cells. It has been used extensively for both the prevention and treatment of acute rejection. It has permissive effect of preserving T regulatory cells resulting in potential benefits in promoting tolerance. Use of ATG as an induction agent forms an integral basis for all high immunologic risk kidney transplantation, is part of most desensitization protocols [52, 53], and provides better graft survival. ATG is also well established in the management of severe acute T cell-mediated and acute antibody-mediated rejection episodes [54]. ATG use has several adverse effects including infusion reactions, cytokine release syndrome, allergic reactions, delayed hypersensitivity reactions (serum sickness) as well as significant increase in long-term risk of CMV disease, BK nephropathy, and malignancies especially when used in higher doses and prolonged duration. ATG use can be titrated using CD2 counts to assess efficacy and duration of impact [55].

Alemtuzumab is pan-lymphocyte depleting monoclonal antibody directed against the CD52 marker that has been used in kidney transplantation. It causes a profound and prolonged reduction in circulating T and B cells, NK cells, dendritic cells, and monocytes, this depletion being mediated by complement-mediated T cell lysis, antibody-dependent cellular cytotoxicity, and lymphocyte apoptosis [56]. It is usually administered as 30 mg dose—once or twice subcutaneously or intravenously at the time of induction [56]. Alemtuzumab has also been used in the prevention of acute rejection, especially in high immunologic risk transplants. While the

efficacy has been studied well in several studies, several significant adverse events limit its use including thrombocytopenia, neuropathy, hepatitis, apart from infusion reactions (IV use), cytokine release syndromes, hypersensitivity reactions, and predisposition to infections (CMV, fungal infections, EBV, JC-PMLE and reactivation of latent TB) and malignancies in the long term [57]. Most transplant centers use alemtuzumab as a second-line induction agent, after ATG, for high immunologic risk transplantation, due to the relatively higher risk of adverse events like infections and malignancies [57].

Devising a Comprehensive Strategy

Several factors need to be considered while devising a therapeutic strategy for desensitization. Firstly, in the case discussed, given the high PRA and single DSA of moderate titer, desensitization is the best option to provide a quick and successful kidney transplant. Secondly, combination of ABO incompatibility and DSA increases risk of rejection. Thirdly, desensitization regimes can work to reduce both DSA and anti ABO antibodies to transplant permissible levels at the same time. Fourthly, the high anti-PLA2R antibody titers portend a significant risk of recurrent membranous nephropathy, which could potentially epitope spread to trigger alloimmune responses. Focusing on controlling both the alloimmune responses, along with the autoimmune response, during desensitization is prudent. The strategies should then be considered to maximize benefits across all these barriers. It should also consider the potential for cumulative toxicity, costs, availability, and clinical ability to administer the treatments, monitor the response, and manage SAE, as well as potential complications effectively.

No clinical trial could provide a direct answer to the presented clinical scenario. Weighing the various therapeutic options is a strenuous exercise due to various factors, such as lack of direct comparisons, observational study designs with minimal control over biases, and heterogeneous and centre-specific standards of care. However, with the current experience in desensitization, it is prudent to build the strategy using the standard desensitization regimen of TPE and IVIG as the base. Addition of widely available agent such as Rituximab is prudent in addressing not only ABO antibodies, but also the PLA2R antibodies as well as the DSA, in addition to redoing the need for multiple sessions of TPE and reducing the potential for antibody rebound. Consideration of either the addition of carfilzomib or Clazakizumab to rituximab can be contemplated; however, given the modest alloimmune activity (MFI 6500 and Anti-A titer of 1:32), this may not be necessary. No availability of eculizumab, Imlifidase, and other experimental therapies in the country will also need to be considered. Careful attention to prophylactic measures to reduce infection risk, including pre-transplant immunization and post-transplant antibiotic prophylaxis, must be incorporated.

Even though the high rates of antibody-mediated rejection remain an unresolved issue, the currently available therapies (Table 19.1) enable transplantation and

reduce the waiting time in highly sensitized patients and provide an opportunity that was not available to them [58].

Agents for Maintenance Immunosuppressive Therapy in Post-transplant Patients

The agents for maintenance immunosuppression in renal transplant recipients (Table 19.2) mainly focus on the reduced activation of T cells and reduced proliferation of immune cells. The agents act on several steps in the three signal pathway to

Table 19.2 Maintenance immunosuppressive agents in kidney transplantation

Drug	Metabolism	Mechanism of action	Selected side effects
Cyclosporine	Liver (CYP3A4)	Binds to cyclophilin and forms a complex that inhibits calcineurin, leading to reduced cytokine production and decreased T cell proliferation	Acute and chronic nephrotoxicity, thrombotic microangiopathy (TMA), hypomagnesemia, hyperkalemia, neurotoxicity, dysglycemia, gingival hyperplasia, hyperlipidemia, hirsutism
Tacrolimus	Liver (CYP3A4)	Binds to FK506-binding protein 12 and forms a complex that inhibits calcineurin, leading to reduced cytokine production and decreased T cell proliferation	Nephrotoxicity, hyperlipidaemia, hypertension, alopecia, TMA
Belatacept	Proteolytic enzymes mediated degradation	Costimulatory pathway blockade by binding to CD80/CD86 receptors	Increased risk of post-transplant lymphoproliferative disease
Sirolimus	Liver (CYP3A4)	Binds and inhibiting mTOR, resulting in decreased cytokine-driven T cell proliferation	Wound healing impairment, cytopenias, hypertension, hyperlipidaemia, oral ulcers, hypokalemia, diarrhea, TMA, pneumonitis/alveolar proteinosis, TMA
Everolimus	Liver (CYP3A4)		
Azathioprine	Liver, erythrocytes	Inhibits purine synthesis, resulting in reduced T cell proliferation	Nausea, vomiting, diarrhea, leukopenia, hepatotoxicity, increased risk of cutaneous malignancies and PTLD
Mycophenolate mofetil	Liver (glucuronyl transferase)	Inhibits inosine monophosphate dehydrogenase, resulting in inhibition of T and B cell proliferation	Diarrhoea, nausea, dyspepsia, Neutropenia, Lymphopenia, Opportunistic infection, teratogenic effects in women
Mycophenolate sodium	Liver (glucuronyl transferase)		
Dosing is ultimately adjusted based on trough levels and local protocols			

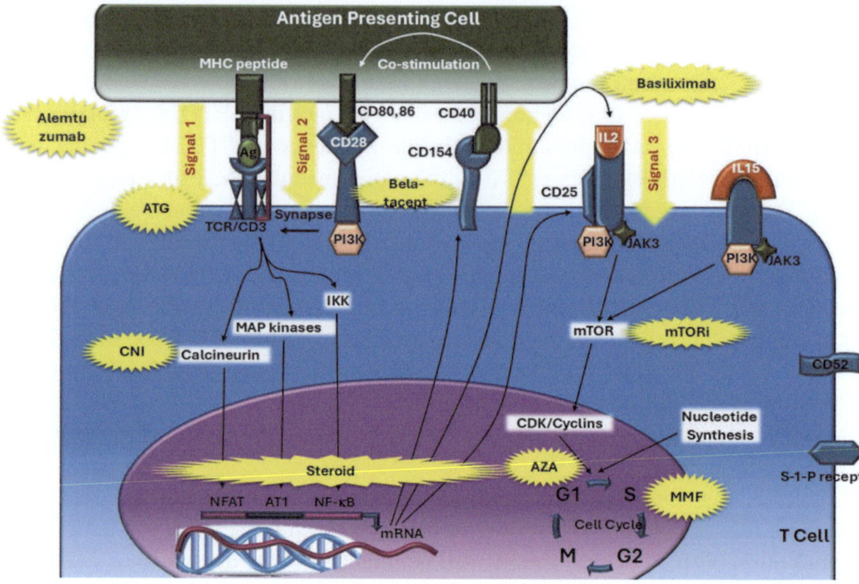

Fig. 19.2 Site of action of induction and maintenance immunosuppressive drugs: (Illustration: Basu Gopal)

antigen recognition and further cellular and humoral pathways of immune activation (Fig. 19.2).

Anti-IL-2 Agents

The monoclonal antibodies which bind the IL-2 receptor and block the action of IL-2 on T lymphocytes and thereby prevents its activation and proliferation has been used successfully in induction of immunosuppression [59]. They are usually started in the immediate preoperative period and continued for two more doses. This has been shown to produce a suppression of IL-2 receptor activity ranging from 4 to 10 weeks. This regimen has shown benefit in low immunological risk renal transplants. In high immunological risk recipients, a more intensive five dose regimen of daclizumab has shown benefit [60].

Calcineurin Inhibitors

The two clinically used calcineurin inhibitors, namely **cyclosporin A** and **tacrolimus,** form the crux of modern immunosuppressive therapy in transplant. They bind to cyclophilin and FK-binding protein respectively and this complex has inhibitory activity on the calcineurin phosphatase. This in turn will reduce the transcription factor activation downstream and reduces the IL-2 gene expression and thereby

reduces the proliferation of T cells [61]. The current evidence clearly supports regular use of tacrolimus in kidney transplant recipients, as it offers better graft survival, lower acute rejection rates as well as lower incidence of steroid resistant rejection [62, 63]. The incidence of post-transplant diabetes mellitus (PTDM) is a considerable comorbidity in this population [64]. The incidence of PTDM is greater with tacrolimus than with cyclosporine [63]. The pharmacokinetics of tacrolimus has been affected by several factors, of these the most important is haematocrit, in the immediate post-transplant period, this is of importance as the recovery in the erythrocyte mass leads to reduced clearance of tacrolimus and improved levels of tacrolimus [65, 66]. Other concomitant illnesses in immunosuppressed individuals such as gastrointestinal infections as well as sepsis can increase the tacrolimus levels owing to its effects on the CYP enzyme system and p-glycoprotein levels [67, 68]. These eliminatory mechanisms are also grounds for several drug-drug interactions with commonly used azole antifungals, macrolides, fluroquinolones, rifamycin, and herbal/dietary agents [65].

Longer acting once daily formulations of tacrolimus are equally efficacious as immediate release Tacrolimus for the prevention of acute rejection and patient and allograft survival [69] but with improved adherence profile due to once daily dosage [70].

Antimetabolites

Mycophenolic acid is the leading antimetabolite in the array of agents used in immunosuppression. The T cells and B cells are dependent on de novo purine synthesis pathway for the synthesis of precursors of various types of RNA molecules. Mycophenolate is a reversible inhibitor of Inosine monophosphate dehydrogenase (IMPDH) which is a rate limiting step in the de novo purine synthesis [71, 72]. It is available in two forms, the esterified mofetil preparation (MMF), which is de-esterified in the acidic environment in stomach and rapidly absorbed, as well as the enteric coated sodium preparation (MPS) which is slowly released in the small intestine and is gradually absorbed [72]. MMF dose reductions are associated with increased risk of acute rejection [73]. MMF is also associated with gastrointestinal side effects such as diarrhea, where conversion to enteric-coated MPS might improve the symptoms [74]. Various comorbidities can affect the pharmacokinetics and thus exposure to mycophenolate. One major factor is those affecting enterohepatic circulation such as concomitant antibiotic therapy, which reduces the bacterial flora available for deconjugating and recirculating the glucuronidated mycophenolate [75]. Hypoalbuminemia is also a factor which can increase the clearance of mycophenolate and can reduce the mycophenolate levels [76]. Apart from these, the immediate post-transplant period with delayed graft function recovery itself has been shown to be associated with higher clearance of mycophenolate and lower mycophenolate levels [77].

Therapeutic drug monitoring with AUC based on several limited sampling strategies plays a significant role in individualized Mycophenolate dosing and reduces risks of rejection as well as adverse events [78].

Azathioprine is another anti-metabolite in use, and this is a precursor of 6-mercaptopurine, which produces thioguanine nucleotides intracellularly and results in inhibition of lymphocyte proliferation [79]. Several investigations have found that mycophenolate offers better transplant outcomes as compared to azathioprine, so as a result azathioprine is a second line agent as of now [80].

Glucocorticoids

Glucocorticoids were the cornerstone of immunosuppression regimens at one time, but currently with combination of calcineurin inhibitors and anti-metabolites being the standard regimen, the glucocorticoids have become an adjuvant agent. The action of glucocorticoids spans both innate and adaptive immune mechanisms [81]. Their effect on adaptive immunity in reducing IL-2 transcription is similar to calcineurin inhibitors, but due to the low association of transcription mechanisms to the glucocorticoid response elements (GRE), this effect is less robust than tacrolimus/cyclosporine A [61]. Due to the various long-term adverse effects such as preponderance to infections, bone resorption, and osteopenia, as well as obesity and metabolic syndrome, there is a general tendency for transplant regimens to avoid or limit the use of glucocorticoids [81]. However, steroid reduction/avoidance protocols are associated with higher risk of rejection needing pulse steroid therapy as well as increased risk of recurrent GN.

mTOR (Mammalian Target of Rapamycin) Inhibitors

As opposed to the reduction of IL-2 synthesis effected by calcineurin inhibitors, mTOR inhibitors cause inhibition of a more downstream step in the activation-proliferation of T cells [61]. The cell cycle progression has been shown to be inhibited by rapamycin (**Sirolimus**) by removal of the effect of IL-2, this is the mechanism by which agents such as everolimus produce their immunosuppressive effect [82]. **Everolimus** can be combined with reduced exposure tacrolimus with its levels maintained at 3–8 ng/mL, this has been shown to provide equivalent graft outcomes, with lesser incidence of CMV and BK-virus infections [83]. Due to its disposition being affected by CYP3A4 and ABCB1, it's prone to drug-drug interactions. Most notable among this is an inhibition of metabolism by calcineurin inhibitor (Cyclosporine A) as much as 50%, which leads it to require a dose reduction. Glucocorticoids are said to have a dual effect, with acute high dosage causing inhibition in metabolism and chronic dosing causing moderate amount of induction and thereby reducing the levels [83]. mTOR inhibitors are associated with several adverse event profiles including delayed wound healing, lymphoedema, cytopenias, oral ulcers, hyperlipidemia [84].

Targeting Costimulatory Molecules

CTLA-4 is a surface molecule on the T cells which recognizes the CD80/86 on the antigen presenting cells and is important in regulating the autoimmune responses. Belatacept is a fusion protein which blocks this interaction and thus prevent the antigen recognition by the T cells [85]. In Phase 3 trials where it has been used as a replacement of calcineurin, it has shown good graft outcomes, but a higher incidence of early acute rejections and a very high risk of Post transplant lymphoma in patients seropositive to EBV. As the trial evaluated only low immunological renal transplants, the use is only limited to them and in select cases where calcineurin inhibitors need to be avoided [86].

Case Progress

The lady successfully undergoes desensitization and reduction in anti PLA2R antibody levels and undergoes the ABO and HLA incompatible living related donor kidney transplantation. There is immediate graft function. She is started on maintenance immunosuppression. She continues to improve her kidney function and reaches a Nadir creatinine of 0.85 mg/dL and has a urine protein creatinine ratio of 0.3 mg/mg.

Choosing a Maintenance Immunosuppression Regimen

In this lady there are several considerations for choosing the maintenance immunosuppression. Initially the best immunosuppression for low-risk immunologic transplant is IL2 blocker induction followed by low-dose tacrolimus, mycophenolate, and prednisolone as per the SYMPHONY trial [87]. This provides the least risk of acute rejection and better 1-year graft survival. For high immunologic risk transplants, most centers use the same combination of drugs but increase the therapeutic immunosuppression exposure targets for each of the individual medication. Considering the wide variation in therapeutic drug exposure despite uniform dosing, and significant intra-patient variability in exposure, such measures help ensure that, the periods of low exposure are limited or avoided and thus potentially reduce the risk of alloimmune responses and rejection. These strategies help reduce the risk of rejection as well as the risk of de novo DSA formation after a high immunologic risk transplantation [88, 89].

Prednisolone may be started at a higher dose (>30 mg/day) or tapered relatively slowly (>3 months), or the lowest dose might be raised from 5 mg daily to 7.5 or 10 mg daily. Steroid withdrawal is completely avoided in such situations due to increased risk of early rejections despite metabolic benefits including reduced post transplant DM and improved cardiovascular and bone mineralization profiles.

Tacrolimus trough level targets might be increased by 2 ng/mL above the standard targets proposed in the SYMPHONY trial for the respective post-transplant

times to cover for the additional immunological risk. In the long run it is expected that the Tacrolimus exposure is kept at 6–8 ng/mL to reduce the risk of rejection and de novo DSA formation [89].

Cyclosporine is usually not a preferred alternative but can be used in patients not tolerating tacrolimus due to serious adverse events. Cyclosporine may be considered in those with tacrolimus related exaggerated neurotoxicity, thrombotic microangiopathy, posterior reversible encephalopathy syndrome (PRESS), and similar such significant adverse events. Cyclosporine however may be chosen later in the period to either benefit the risk of reducing post-transplant diabetes [90] or more commonly to help with managing BK virus nephropathy [91] or using in high doses to treat some recurrent GN (like FSGS or membranous nephropathy) [92], especially when recurrence has progressed while on adequate tacrolimus exposure.

Belatacept replacement of CNI was studied in the BENEFIT trials [86, 93] that showed increased incidence of acute rejection but better renal function. The risk of EBV-associated PTLD and the lack of widespread availability limit the use of Belatacept in many countries. In countries where available, combination therapies with Belatacept and CNI in the early post-transplant period to reduce the initial risk of rejection, followed by tapering of CNI, have been considered in low immunologic risk transplants. Nevertheless, the increased risk of early rejections in the initial trials makes the use of this agent in high immunologic risk kidney transplant an uneasy proposition. Initial phase—10 mg/kg IV on day of transplant (prior to implantation); repeat this dose on Day 5 and at end of Weeks 2, 4, 8, and 12 after transplantation. Maintenance phase: 5 mg/kg IV at end of Week 16 after transplantation and then every 4 weeks (+/− 3 days) thereafter.

On considering the initial dose of mycophenolate, the MMF dose may be increased to 3 g/Day instead of 2 g/day if tolerated [94] and therapy may be closely guided by MMF AUC_{0-12h} targets of 45–60 mcgh/L rather than 30–60 mcg h/L [95]. Mycophenolate sodium may be preferred in patients who have significant diarrhea from MMF use [96] and in those on high doses of proton pump inhibitors due to lack of interactions [97]. Mycophenolate has the additional benefit of reduced risk of skin cancers compared to Azathioprine and no interaction with allopurinol in those on long-term treatment of Gout.

Notably, other antiproliferative agents, such as with Aza or mTOR inhibitors, are generally avoided. Meta-analytic studies have confirmed the superiority of MMF over Aza in reducing the risk of acute rejection [98]. CNI-mTORi combinations such as that used in TRANSFORM study are an alternative option only in the low immunologic risk transplantation [99]. mTOR inhibitors may be preferred in patients with increased risk of skin cancers and viral infections (CMV and BK). However, mTORi combination therapies have been associated with increased risk of either acute rejection or de novo DSA formation which precludes their choice as the first line agent in high immunologic risk transplantation.

Given such high exposures, patients would need adequate and prolonged antibiotic prophylaxis including Valganciclovir prophylaxis for 200 days (or longer—to be guided by CMV CMI testing) and life-long Cotrimoxazole prophylaxis. In

addition, close monitoring of other infectious agents such as BK virus and EBV virus must be undertaken.

In addition, extrapolating from the benefits of steroids and Calcineurin inhibitors in treatment of native kidney membranous nephropathy, they may have a deterrent effect on the recurrence of membranous nephropathy in the transplant, although this is not well studied. Similarly, mTOR inhibitors may be avoided in cases with high risk of recurrence of glomerular disease due to their potential for podocytopathy and triggering proteinuria.

Thus, this patient is best served by remaining on higher exposure triple immunosuppression regimen of Tacrolimus, Mycophenolate, and Prednisolone. The immunosuppressive agents used for maintenance therapy and their adverse effects are given in Table 19.2.

Conclusion

The transplant clinician often faces challenging clinical situations that do not have a straightforward answer from clinical trial literature. In these situations, it is imperative that the clinician consider a multitude of factors such as a clear understanding on the pathophysiology of immune response, the literature on the mechanisms of action of the drugs, the site of action, the pharmacokinetics and pharmacodynamics, the drug interactions, efficacy in certain clinical situations, their safety profile, ease of availability and use in clinical scenario and short and long term adverse events profile is essential to formulate an effective, safe, and viable immunosuppression management strategy for successful kidney transplantation.

References

1. Jambon F, Merville P, Guidicelli G, Taton B, De Précigout V, Couzi L, et al. Efficacy of plasmapheresis and semi-selective immunoadsorption for removal of anti-HLA antibodies. J Clin Apheresis. 2021;36(3):291–8.
2. Hakim RM, Milford E, Himmelfarb J, Wingard R, Lazarus JM, Watt RM. Extracorporeal removal of anti-HLA antibodies in transplant candidates. Am J Kidney Dis. 1990;16(5):423–31.
3. Lorenz M, Regele H, Schillinger M, Kletzmayr J, Haidbauer B, Derfler K, et al. Peritransplant immunoadsorption: a strategy enabling transplantation in highly sensitized crossmatch-positive cadaveric kidney allograft recipients. Transplantation. 2005;79(6):696–701.
4. Glotz D, Antoine C, Julia P, Suberbielle-Boissel C, Boudjeltia S, Fraoui R, et al. Desensitization and subsequent kidney transplantation of patients using intravenous immunoglobulins (IVIg). Am J Transplant. 2002;2(8):758–60.
5. Stegall MD, Gloor J, Winters JL, Moore SB, DeGoey S. A comparison of plasmapheresis versus high-dose IVIG desensitization in renal allograft recipients with high levels of donor specific alloantibody. Am J Transplant. 2006;6(2):346–51.
6. Gloor JM, DeGoey SR, Pineda AA, Moore SB, Prieto M, Nyberg SL, et al. Overcoming a positive crossmatch in living-donor Kidney transplantation. Am J Transplant. 2003;3(8):1017–23.
7. Magee CC, Felgueiras J, Tinckam K, Malek S, Mah H, Tullius S. Renal transplantation in patients with positive lymphocytotoxicity crossmatches: one center's experience. Transplantation. 2008;86(1):96–103.

8. Lawrence C, Galliford JW, Willicombe MK, McLean AG, Lesabe M, Rowan F, et al. Antibody removal before ABO-incompatible renal transplantation: how much plasma exchange is therapeutic? Transplantation. 2011;92(10):1129–33.
9. Kashgary A, Sontrop JM, Li L, Al-Jaishi AA, Habibullah ZN, Alsolaimani R, et al. The role of plasma exchange in treating post-transplant focal segmental glomerulosclerosis: a systematic review and meta-analysis of 77 case-reports and case-series. BMC Nephrol. 2016;17(1):104.
10. De Sousa-Amorim E, Revuelta I, Blasco M, Diekmann F, Cid J, Lozano M, et al. Desensitization before living donor kidney transplantation in highly HLA-sensitized patients: a single-center study. Transplant Proc. 2015;47(8):2332–5.
11. John R, Lietz K, Burke E, Ankersmit J, Mancini D, Suciu-Foca N, et al. Intravenous immunoglobulin reduces anti-HLA alloreactivity and shortens waiting time to cardiac transplantation in highly sensitized left ventricular assist device recipients. Circulation. 1999;100(19 Suppl):II229–35.
12. Ulisses LRDS, Paixão JO, Agena F, Souza PSD, Paula FJ, Bezerra G, et al. Desensitization using IVIG alone for living-donor kidney transplant: impact on donor-specific antibodies. Braz J Nephrol. 2022;44(4):527–32.
13. Winstedt L, Järnum S, Nordahl EA, Olsson A, Runström A, Bockermann R, et al. Complete removal of extracellular IgG antibodies in a randomized dose-escalation Phase I study with the bacterial enzyme IdeS—a novel therapeutic opportunity. PLoS One. 2015;10(7):e0132011.
14. Jordan SC, Lorant T, Choi J, Kjellman C, Winstedt L, Bengtsson M, et al. IgG endopeptidase in highly sensitized patients undergoing transplantation. N Engl J Med. 2017;377(5):442–53.
15. Jordan SC, Legendre C, Desai NM, Lorant T, Bengtsson M, Lonze BE, et al. Imlifidase desensitization in crossmatch-positive, highly sensitized Kidney transplant recipients: results of an international phase 2 trial (Highdes). Transplantation. 2021;105(8):1808–17.
16. Park Y, Ko EJ, Chung BH, Yang CW. Kidney transplantation in highly sensitized recipients. Kidney Res Clin Pract. 2021;40(3):355–70.
17. Lefaucheur C, Viglietti D, Hidalgo LG, Ratner LE, Bagnasco SM, Batal I, et al. Complement-activating anti-HLA antibodies in kidney transplantation: allograft gene expression profiling and response to treatment. J Am Soc Nephrol. 2018;29(2):620–35.
18. Stegall MD, Diwan T, Raghavaiah S, Cornell LD, Burns J, Dean PG, et al. Terminal complement inhibition decreases antibody-mediated rejection in sensitized renal transplant recipients. Am J Transplant. 2011;11(11):2405–13.
19. Marks WH, Mamode N, Montgomery RA, Stegall MD, Ratner LE, Cornell LD, et al. Safety and efficacy of eculizumab in the prevention of antibody-mediated rejection in living-donor kidney transplant recipients requiring desensitization therapy: a randomized trial. Am J Transplant. 2019;19(10):2876–88.
20. Zachary AA, Lucas DP, Montgomery RA, Leffell MS. Rituximab prevents an anamnestic response in patients with cryptic sensitization to HLA. Transplantation. 2013;95(5):701–4.
21. Jackson AM, Kraus ES, Orandi BJ, Segev DL, Montgomery RA, Zachary AA. A closer look at rituximab induction on HLA antibody rebound following HLA-incompatible kidney transplantation. Kidney Int. 2015;87(2):409–16.
22. Vo AA, Wang J, Peng A. Rituximab and intravenous immune globulin for desensitization during renal transplantation. N Engl J Med. 2008;359(3):242–51.
23. Vo AA, Peng A, Toyoda M, Kahwaji J, Cao K, Lai CH, et al. Use of intravenous immune globulin and rituximab for desensitization of highly HLA-sensitized patients awaiting Kidney transplantation. Transplantation. 2010;89(9):1095–102.
24. Tydén G, Kumlien G, Genberg H, Sandberg J, Sedigh A, Lundgren T, et al. The Stockholm experience with ABO-incompatible kidney transplantations without splenectomy. Xenotransplantation. 2006;13(2):105–7.
25. Gloor JM, Lager DJ, Moore SB, Pineda AA, Fidler ME, Larson TS, et al. ABO-incompatible kidney transplantation using both A2 and non-A2 living donors. Transplantation. 2003;75(7):971–7.

26. Viganò M, Beretta M, Lepore M, Abete R, Benatti SV, Grassini MV, et al. Vaccination recommendations in solid organ transplant adult candidates and recipients. Vaccine. 2023;11(10):1611.
27. NasrAllah MM, Elalfy M, El Ansary M, Elmeseery Y, Amer I, Malvezzi P, et al. Obinutuzumab in kidney transplantation: effect on B-cell counts and crossmatch tests. Transplantation. 2022;106(2):369–72.
28. Redfield RR, Jordan SC, Busque S, Vincenti F, Woodle ES, Desai N, et al. Safety, pharmacokinetics, and pharmacodynamic activity of obinutuzumab, a type 2 anti-CD20 monoclonal antibody for the desensitization of candidates for renal transplant. Am J Transplant. 2019;19(11):3035–45.
29. Gauckler P, Shin JI, Alberici F, Audard V, Bruchfeld A, Busch M, et al. Rituximab in membranous nephropathy. Kidney Int Rep. 2021;6(4):881–93.
30. Sridharan K, Gopal B, Wilson S, Pham A, Hutton H. Obinutuzumab in rituximab-resistant and recurrent membranous nephropathy: a case-series. J Nephrol. 2025; https://doi.org/10.1007/s40620-025-02224-6.
31. Hullekes F, Uffing A, Verhoeff R, Seeger H, von Moos S, Mansur J, et al. Recurrence of membranous nephropathy after kidney transplantation: a multicenter retrospective cohort study. Am J Transplant. 2024;24(6):1016–26.
32. Rosado MM, Gesualdo F, Marcellini V, Di Sabatino A, Corazza GR, Smacchia MP, et al. Preserved antibody levels and loss of memory B cells against pneumococcus and tetanus after splenectomy: tailoring better vaccination strategies. Eur J Immunol. 2013;43(10):2659–70.
33. Lemy A, Toungouz M, Abramowicz D. Bortezomib: a new player in pre- and post-transplant desensitization? Nephrol Dial Transplant. 2010;25(11):3480–9.
34. Perry DK, Burns JM, Pollinger HS, Amiot BP, Gloor JM, Gores GJ, et al. Proteasome inhibition causes apoptosis of normal human plasma cells preventing alloantibody production. Am J Transplant. 2009;9(1):201–9.
35. Everly MJ, Terasaki PI, Trivedi HL. Durability of antibody removal following proteasome inhibitor-based therapy. Transplantation. 2012;93(6):572–7.
36. Jeong JC, Jambaldorj E, Kwon HY, Kim MG, Im HJ, Jeon HJ, et al. Desensitization using bortezomib and high-dose immunoglobulin increases rate of deceased donor kidney transplantation. Medicine (Baltimore). 2016;95(5):e2635.
37. Woodle ES, Shields AR, Ejaz NS, Sadaka B, Girnita A, Walsh RC, et al. Prospective iterative trial of proteasome inhibitor-based desensitization. Am J Transplant. 2015;15(1):101–18.
38. Georgoulis V, Haidich AB, Bougioukas KI, Hatzimichael E. Efficacy and safety of carfilzomib for the treatment of multiple myeloma: an overview of systematic reviews. Crit Rev Oncol Hematol. 2022;180:103842.
39. Razia D, Hu C, Cherrier L, Nasar A, Walia R, Tokman S. Carfilzomib versus rituximab for treatment of de novo donor-specific antibodies in lung transplant recipients. Transpl Immunol. 2022;75:101703.
40. Tremblay S, Driscoll JJ, Rike-Shields A, Hildeman DA, Alloway RR, Girnita AL, et al. A prospective, iterative, adaptive trial of carfilzomib-based desensitization. Am J Transplant. 2020;20(2):411–21.
41. Roschewski M, Phelan JD, Wilson WH. Molecular classification and treatment of diffuse large B-cell lymphoma and primary mediastinal B-cell lymphoma. Cancer J. 2020;26(3):195–205.
42. Torka P, Groman A, Wong J, Nichols J, Kader A, Mavis C, et al. Carfilzomib combined with rituximab, ifosfamide, carboplatin, and etoposide for relapsed or refractory DLBCL. Blood Adv. 2023;7(7):1146–55.
43. Dunleavy K, Pittaluga S, Czuczman MS, Dave SS, Wright G, Grant N, et al. Differential efficacy of bortezomib plus chemotherapy within molecular subtypes of diffuse large B-cell lymphoma. Blood. 2009;113(24):6069–76.
44. Rosenthal A, Luthi J, Belohlavek M, Kortüm KM, Mookadam F, Mayo A, et al. Carfilzomib and the cardiorenal system in myeloma: an endothelial effect? Blood Cancer J. 2016;6(1):e384.

45. Kwun J, Matignon M, Manook M, Guendouz S, Audard V, Kheav D, et al. Daratumumab in sensitized kidney transplantation: potentials and limitations of experimental and clinical use. J Am Soc Nephrol. 2019;30(7):1206–19.
46. Vincenti F, Bestard O, Brar A, Cruzado JM, Seron D, Gaber AO, et al. Isatuximab monotherapy for desensitization in highly sensitized patients awaiting Kidney transplant. J Am Soc Nephrol. 2024;35(3):347–60.
47. Vo AA, Choi J, Kim I, Louie S, Cisneros K, Kahwaji J, et al. A phase I/II trial of the Interleukin-6 receptor-specific humanized monoclonal (tocilizumab) + intravenous immunoglobulin in difficult to desensitize patients. Transplantation. 2015;99(11):2356–63.
48. Daligault M, Bardy B, Noble J, Bourdin A, Masson D, Naciri Bennani H, et al. Marginal impact of tocilizumab monotherapy on anti-HLA alloantibodies in highly sensitized Kidney transplant candidates. Transplant Direct. 2021;7(5):e690.
49. Jouve T, Daligault M, Noble J, Terrec F, Imerzoukene F, Dard C, et al. Tocilizumab evaluation in HLA-desensitization before kidney transplantation as an add-on therapy to apheresis: the TETRA study. J Clin Med. 2023;12(2):424.
50. Vo AA, Huang E, Ammerman N, Toyoda M, Ge S, Haas M, et al. Clazakizumab for desensitization in highly sensitized patients awaiting transplantation. Am J Transplant. 2022;22(4):1133–44.
51. Jordan SC, Ammerman N, Choi J, Huang E, Najjar R, Peng A, et al. Evaluation of clazakizumab (anti-Interleukin-6) in patients with treatment-resistant chronic active antibody-mediated rejection of kidney allografts. Kidney Int Rep. 2022;7(4):720–31.
52. Knight RJ, Devos JM, Patel SJ, Land G, Moore LW, Gaber L, et al. Outcomes of living donor renal transplants with a negative cross-match and pretransplant donor-specific antibody. Transplant Proc. 2013;45(4):1399–401.
53. Bächler K, Amico P, Hönger G, Bielmann D, Hopfer H, Mihatsch MJ, et al. Efficacy of induction therapy with ATG and intravenous immunoglobulins in patients with low-level donor-specific HLA-antibodies. Am J Transplant. 2010;10(5):1254–62.
54. Brennan DC, Daller JA, Lake KD, Cibrik D, Del Castillo D, Thymoglobulin Induction Study Group. Rabbit antithymocyte globulin versus basiliximab in renal transplantation. N Engl J Med. 2006;355(19):1967–77.
55. Webster AC, Pankhurst T, Rinaldi F, Chapman JR, Craig JC. Monoclonal and polyclonal antibody therapy for treating acute rejection in kidney transplant recipients: a systematic review of randomized trial data. Transplantation. 2006;81(7):953–65.
56. van der Zwan M, Baan CC, van Gelder T, Hesselink DA. Review of the clinical pharmacokinetics and pharmacodynamics of alemtuzumab and its use in kidney transplantation. Clin Pharmacokinet. 2018;57(2):191–207.
57. Chukwu CA, Spiers HVM, Middleton R, Kalra PA, Asderakis A, Rao A, et al. Alemtuzumab in renal transplantation. Reviews of literature and usage in the United Kingdom. Transplant Rev Orlando Fla. 2022;36(2):100686.
58. Jordan SC, Tyan D, Stablein D, McIntosh M, Rose S, Vo A, et al. Evaluation of intravenous immunoglobulin as an agent to lower allosensitization and improve transplantation in highly sensitized adult patients with end-stage renal disease: report of the NIH IG02 trial. J Am Soc Nephrol. 2004;15(12):3256–62.
59. Swiatecka-Urban A. Anti-Interleukin-2 receptor antibodies for the prevention of rejection in Pediatric renal transplant patients: current status. Pediatr Drugs. 2003;5(10):699–716.
60. Sageshima J, Ciancio G, Chen L, Burke GW. Anti-interleukin-2 receptor antibodies—basiliximab and daclizumab—for the prevention of acute rejection in renal transplantation. Biol Targets Ther. 2009;3:319–36.
61. Ho S, Clipstone N, Timmermann L, Northrop J, Graef I, Fiorentino D, et al. The mechanism of action of cyclosporin a and FK506. Clin Immunol Immunopathol. 1996;80(3):S40–5.
62. Webster A, Woodroffe RC, Taylor RS, Chapman JR, Craig JC. Tacrolimus versus cyclosporin as primary immunosuppression for kidney transplant recipients. Cochrane Database Syst Rev. 2005;4:CD003961.

63. Haddad E, McAlister V, Renouf E, Malthaner R, Kjaer MS, Gluud LL. Cyclosporin versus tacrolimus for liver transplanted patients. Cochrane Database Syst Rev. 2006;2006(4):CD005161.
64. Meera M, Manikandan S, Parameswaran S. Adverse effects of tacrolimus and its associated risk factors in renal transplant recipients. Exp Clin Transplant. 2023;21(1):22–7.
65. Brunet M, van Gelder T, Åsberg A, Haufroid V, Hesselink DA, Langman L, et al. Therapeutic drug monitoring of tacrolimus-personalized therapy: second consensus report. Ther Drug Monit. 2019;41(3):261–307.
66. Størset E, Holford N, Midtvedt K, Bremer S, Bergan S, Åsberg A. Importance of hematocrit for a tacrolimus target concentration strategy. Eur J Clin Pharmacol. 2014;70(1):65–77.
67. Percy C, Hassoun Z, Mourad M, De Meyer M, Beguin C, Jadoul M, et al. Impact of acute infection requiring hospitalization on tacrolimus blood levels in Kidney transplant recipients. Transplant Proc. 2017;49(9):2065–9.
68. Asano T, Nishimoto K, Hayakawa M. Increased tacrolimus trough levels in association with severe diarrhea, a case report. Transplant Proc. 2004;36(7):2096–7.
69. Silva HT Jr, Yang HC, Abouljoud M, Kuo PC, Wisemandle K, Bhattacharya P, Dhadda S, Holman J, Fitzsimmons W, First MR, Kidney TE. One-year results with extended-release tacrolimus/MMF, tacrolimus/MMF and cyclosporine/MMF in de novo kidney transplant recipients. Am J Transplant. 2007;7(3):595–608.
70. Kuypers DR, Peeters PC, Sennesael JJ, Kianda MN, Vrijens B, Kristanto P, Dobbels F, Vanrenterghem Y, Kanaan N, ADMIRAD Study Team. Improved adherence to tacrolimus once-daily formulation in renal recipients: a randomized controlled trial using electronic monitoring. Transplantation. 2013;95(2):333–40.
71. Hackl A, Ehren R, Weber LT. Effect of mycophenolic acid in experimental, nontransplant glomerular diseases: new mechanisms beyond immune cells. Pediatr Nephrol Berl Ger. 2017;32(8):1315–22.
72. Bergan S, Brunet M, Hesselink DA, Johnson-Davis KL, Kunicki PK, Lemaitre F, et al. Personalized therapy for mycophenolate: consensus report by the International Association of Therapeutic Drug Monitoring and Clinical Toxicology. Ther Drug Monit. 2021;43(2):150–200.
73. Sollinger HW, Sundberg AK, Leverson G, Voss BJ, Pirsch JD. Mycophenolate mofetil versus enteric-coated mycophenolate sodium: a large, single-center comparison of dose adjustments and outcomes in kidney transplant recipients. Transplantation. 2010;89(4):446–51.
74. Ross DJ, Waters PF, Levine M, Kramer M, Ruzevich S, Kass RM. Mycophenolate mofetil versus azathioprine immunosuppressive regimens after lung transplantation: preliminary experience. J Heart Lung Transplant. 1998;17(8):768–74.
75. Simoons M, Naipal KAT, de Jong H, den Hoed CM, de Winter BCM, Mulder MB. Oral antibiotics lower mycophenolate mofetil drug exposure, possibly by interfering with the enterohepatic recirculation: a case series. Pharmacol Res Perspect. 2023;11(3):e01103.
76. Jain A, Venkataramanan R, Hamad IS, Zuckerman S, Zhang S, Lever J, et al. Pharmacokinetics of mycophenolic acid after mycophenolate mofetil administration in liver transplant patients treated with tacrolimus. J Clin Pharmacol. 2001;41(3):268–76.
77. van Gelder T, Silva HT, de Fijter H, Budde K, Kuypers D, Mamelok RD, et al. How delayed graft function impacts exposure to mycophenolic acid in patients after renal transplantation. Ther Drug Monit. 2011;33(2):155–64.
78. Le Meur Y, Büchler M, Thierry A, Caillard S, Villemain F, Lavaud S, Etienne I, Westeel PF, De Ligny BH, Rostaing L, Thervet E. Individualized mycophenolate mofetil dosing based on drug exposure significantly improves patient outcomes after renal transplantation. Am J Transplant. 2007;7(11):2496–503.
79. Chan GL, Canafax DM, Johnson CA. The therapeutic use of azathioprine in renal transplantation. Pharmacotherapy. 1987;7(5):165–77. https://doi.org/10.1002/j.1875-9114.1987.tb04046.x.
80. Kwon O, Cho JH, Choi JY, Park SH, Kim YL, Kim HK, et al. Long-term outcome of azathioprine versus mycophenolate mofetil in cyclosporine-based immunosuppression in Kidney transplantation: 10 years of experience at a single center. Transplant Proc. 2013;45(4):1487–90.

81. Dashti-Khavidaki S, Saidi R, Lu H. Current status of glucocorticoid usage in solid organ transplantation. World J Transplant. 2021;11(11):443–65.
82. Nourse J, Firpo E, Flanagan WM, Coats S, Polyak K, Lee MH, et al. Lnterleukin-2-mediated elimination of the p27Kipl cyclin-dependent kinase inhibitor prevented by rapamycin. Nature. 1994;372(6506):570–3.
83. Masuda S, Lemaitre F, Barten MJ, Bergan S, Shipkova M, Van Gelder T, et al. Everolimus personalized therapy: second consensus report by the International Association of Therapeutic Drug Monitoring and Clinical Toxicology. Ther Drug Monit. 2025;47(1):4–31.
84. Pascual J, Diekmann F, Fernández-Rivera C, Gómez-Marqués G, Gutiérrez-Dalmau A, Pérez-Sáez MJ, Sancho-Calabuig A, Oppenheimer F. Recommendations for the use of everolimus in de novo kidney transplantation: False beliefs, myths and realities. Nefrologia. 2017;37(3):253–66. . English, Spanish. https://doi.org/10.1016/j.nefro.2016.11.007.
85. Schwarz C, Mahr B, Muckenhuber M, Wekerle T. Belatacept/CTLA4Ig: an update and critical appraisal of preclinical and clinical results. Expert Rev Clin Immunol. 2018;14(7):583–92.
86. Vincenti F, Charpentier B, Vanrenterghem Y, Rostaing L, Bresnahan B, Darji P, et al. A phase III study of belatacept-based immunosuppression regimens versus cyclosporine in renal transplant recipients (BENEFIT study). Am J Transplant. 2010;10(3):535–46.
87. Ekberg H, Tedesco-Silva H, Demirbas A, Vítko S, Nashan B, Gürkan A, et al. Reduced exposure to calcineurin inhibitors in renal transplantation. N Engl J Med. 2007;357(25):2562–75.
88. Davis S, Gralla J, Klem P, Tong S, Wedermyer G, Freed B, et al. Lower tacrolimus exposure and time in therapeutic range increase the risk of de novo donor-specific antibodies in the first year of kidney transplantation. Am J Transplant. 2018;18(4):907–15.
89. Wojciechowski D, Wiseman A. Long-term immunosuppression management: opportunities and uncertainties. Clin J Am Soc Nephrol. 2021;16(8):1264–71.
90. Wissing KM, Abramowicz D, Weekers L, Budde K, Rath T, Witzke O, et al. Prospective randomized study of conversion from tacrolimus to cyclosporine a to improve glucose metabolism in patients with posttransplant diabetes mellitus after renal transplantation. Am J Transplant. 2018;18(7):1726–34.
91. Brennan DC, Agha I, Bohl DL, Schnitzler MA, Hardinger KL, Lockwood M, et al. Incidence of BK with tacrolimus versus cyclosporine and impact of preemptive immunosuppression reduction. Am J Transplant. 2005;5(3):582–94.
92. Cravedi P, Kopp JB, Remuzzi G. Recent progress in the pathophysiology and treatment of FSGS recurrence. Am J Transplant. 2013;13(2):266–74.
93. Pestana JOM, Grinyo JM, Vanrenterghem Y, Becker T, Campistol JM, Florman S, et al. Three-year outcomes from BENEFIT-EXT: a phase III study of belatacept versus cyclosporine in recipients of extended criteria donor kidneys. Am J Transplant. 2012;12(3):630–9.
94. Gourishankar S, Houde I, Keown PA, Landsberg D, Cardella CJ, Barama AA, et al. The CLEAR study: a 5-day, 3-g loading dose of mycophenolate mofetil versus standard 2-g dosing in renal transplantation. Clin J Am Soc Nephrol. 2010;5(7):1282–9.
95. Metz DK, Holford N, Kausman JY, Walker A, Cranswick N, Staatz CE, et al. Optimizing mycophenolic acid exposure in kidney transplant recipients: time for target concentration intervention. Transplantation. 2019;103(10):2012–30.
96. Chan L, Mulgaonkar S, Walker R, Arns W, Ambühl P, Schiavelli R. Patient-reported gastrointestinal symptom burden and health-related quality of life following conversion from mycophenolate mofetil to enteric-coated mycophenolate sodium. Transplantation. 2006;81(9):1290–7.
97. Xu L, Cai M, Shi BY, Li ZL, Li X, Jin HL. A prospective analysis of the effects of enteric-coated mycophenolate sodium and mycophenolate mofetil co-medicated with a proton pump inhibitor in kidney transplant recipients at a single institute in China. Transplant Proc. 2014;46(5):1362–5.
98. Knight SR, Russell NK, Barcena L, Morris PJ. Mycophenolate mofetil decreases acute rejection and may improve graft survival in renal transplant recipients when compared with azathioprine: a systematic review. Transplantation. 2009;87(6):785–94.
99. Pascual J, Berger SP, Witzke O, Tedesco H, Mulgaonkar S, Qazi Y, et al. Everolimus with reduced calcineurin inhibitor exposure in renal transplantation. J Am Soc Nephrol. 2018;29(7):1979–91.

Immunological Challenges in Xenotransplantation

20

Georgi Abraham, Ravi Raju, and Sree Bhushan Raju

There is an unmet need for organ transplantation because of the non-availability of appropriate organs. The individuals who require organ transplantation for end stage organ failure (ESOF) are rapidly increasing in spite of both live and deceased donor organs being transplanted. The waiting list is increasing by leaps and bounds in both developing and developed countries. Xenografts can include organs like hearts, kidneys, and livers, as well as tissues like pancreatic islets and skin to be transplanted from animals to humans.

The report of short-term success of liver transplantation from baboon to human being (70 days survival) was reported by late Prof. Thomas Starzl in the late 90s at the WCT in Barcelona, Spain. This experience gave new life to xenotransplantation science. Many research groups have been working tirelessly in this field of transplantation [1].

The obstacles faced were many such as the size of the organ, continuous monitoring and prevention of infection, blood group incompatibilities, immunological challenges, immunosuppressive medications, and surgical skills.

The primary immunological challenges in xenotransplantation are hyperacute rejection (HAR), cellular rejection, and the risk of xenozoonosis (transmission of infectious agents from the donor animal). Hyperacute rejection is triggered by preexisting antibodies against antigens on the donor organ, which occurs immediately after engraftment leading to graft loss. In spite of overcoming HAR, cellular rejection the development of a xenogeneic immune response is faced, challenging the

---

G. Abraham (✉)
The Madras Medical Mission Hospital, Chennai, Tamil Nadu, India

R. Raju
Department of Nephrology, Appollo Hospital, Vishakhapatnam, Andhra Pradesh, India

S. B. Raju
Department of Nephrology, Nizam's Institute of Medical Sciences, Hyderabad, Telangana, India

G. Abraham et al. (eds.), *Immunological Challenges in Organ Transplantation*, https://doi.org/10.1007/978-981-95-1533-2_20

long-term success. Other immunological barriers exist, including the activation of the recipient's own immune system by the graft leading to cross-species incompatibilities in complement systems.

As pigs are more widely available as organ donors, they present greater immunologic barriers. Development of rapid genome editing tools, particularly the advent of CRISPR-Cas9 gene editing technologies, has enabled kidney and heart xenotransplantation in pig-to-nonhuman primates. Three different groups of scientists made a breakthrough in laying the foundation for three recent pig-to-human transplants. Two kidney xenografts in brain dead recipients, ineligible for transplant, and one heart xenograft in the first clinical grade study of pig-to-human transplantation [2].

Recently genetically modified pig kidney and heart xenotransplants in human living recipients have provided insights that adaptive as well as innate immune mediated xenograft injury occurs. We require more information at the cellular and molecular levels to the development of targeted genetic modifications and strategies aimed at preventing rejection and their by inducing tolerance. We also need to understand the bidirectional cross talk between the human recipient and pig xenograft [3].

One of the major barriers to xenotransplantation is α-Gal Epitope which is the principle antigenic target in xenotransplantation. Naturally occurring anti-Gal antibodies from infancy that can account for 1% of circulating immunoglobulins due to exposure to α-gal express by the gut microbiota. This can lead to hyperacute rejection (HAR) [4, 5].

The first homozygous Ggta1 knockout (GTKO) pigs were created in the last decade using homologous recombination to disrupt Ggta1 in porcine fetal fibroblast. These GTKO kidneys are not prone to HAR. Transgenic pigs were created expressing the human compliment regulatory protein CD55, CD46, and CD59 which may provide xenografts with protection from complement mediated damage. Cross species mismatches can lead to inability of pig MHC class I to inhibit human NK cells and failure of porcine CD47 signal to human macrophages. This impaired interaction may contribute to NK and macrophage induced cell mediated injury to xenografts.

Xenotransplantation is going through a breakthrough in the field of organ donation. We do not know what are the ideal immunosuppressive regimens and the genetic edits in the donor required for optimal graft survival and function. We need to have newer crossmatch techniques which will identify humans with high and low risk for rejection.

References

1. Cooper DKC. Early clinical xenotransplantation experiences—an interview with Thomas E. Starzl, MD, PhD. Xenotransplantation. 2017;24(2):e12306. https://doi.org/10.1111/xen.12306.

2. Eisenson DL, Yu H, Kazuhiko Y. Progress in xenotransplantation: immunologic barriers, advances in gene editing, and successful tolerance induction strategies in pig-to-primate transplantation. Front Immunol. 2022;13:899657. https://doi.org/10.3389/fimmu.2022.899657.
3. Tatapudi VS, Mattoo A, Schiff T, Mehta SA, Skolnik EY, Montgomery RA. Xenotransplantation: current understanding of the mechanism of immune-mediated injury. J Am Soc Nephrol. 2025; https://doi.org/10.1681/ASN.0000000745.
4. Galili U. Interaction of the natural anti-gal antibody with alpha-galactosyl epitopes: a major obstacle for xenotransplantation in humans. Immunol Today. 1993;14(10):480–2. https://doi.org/10.1016/0167-5699(93)90261-i.
5. Hirose T, Hotta K, Otsuka R, Seino K-I. Mechanism and regulation of the complement activity in kidney xenotransplantation. Transplant Rev. 2025;39(3):100931.

Conclusions, Recommendations, and Guidelines for Immunological Evaluation of a Transplant Recipient

21

Georgi Abraham, Dolly Daniel, Shruti Tapiawala, Vivek Kute, Narayan Prasad, K. Sampath Kumar, M. Sahay, S. Guditi, S. Mahajan, and A. P. Das

Graft longevity is key to a successful transplant program. This is a multistep ongoing process from the time of donor choice. Of the variables that affect longevity, one stands out—efficient, timely, and high quality immune monitoring. In all forms of organ and cellular transplants, the role of immunological testing in making donor

G. Abraham (✉)
The Madras Medical Mission Hospital, Chennai, Tamil Nadu, India

D. Daniel
Transfusion Medicine and Immunohaematology, Christian Medical College, Vellore, Tamil Nadu, India
e-mail: dollyd@cmcvellore.ac.in

S. Tapiawala
Institute of Renal Sciences, Gleneagles Hospitals, Mumbai, Maharashtra, India

V. Kute
Institute of Kidney Diseases, IKDRC-ITS, Ahmedabad, Gujarat, India

N. Prasad
Department of Nephrology, Sanjay Gandhi Postgraduate Institute of Medical Sciences, Lucknow, Uttar Pradesh, India

K. Sampath Kumar
Meenakshi Mission Hospital, Madurai, Tamil Nadu, India

M. Sahay
Department of Nephrology, Osmania Medical College & Osmania General Hospital, Hyderabad, Telangana, India

S. Guditi
Department of Nephrology, NIMS, Hyderabad, Telangana, India

S. Mahajan
Department of Nephrology, All India Institute of Medical Sciences, New Delhi, Delhi, India

A. P. Das
Sir Gangaram Hospital, New Delhi, Delhi, India

G. Abraham et al. (eds.), *Immunological Challenges in Organ Transplantation*,
https://doi.org/10.1007/978-981-95-1533-2_21

choices, assessing suitability for transplants, and subsequent post-transplant monitoring cannot be underestimated.

The preceding chapters have dealt in depth, various technologies to immune-stratify a patient, clinical scenarios—from the garden variety transplants, to complex one of a kind unusual clinical situations. Refer to Tables 21.1, 21.2, and 21.3 for requirement of standard tests employed towards immune-stratification and their applications. Refer to Figs. 21.1 and 21.2 depicting the relationships of one test with the other their inherent fallacies. From single solid organ to stem cell and composite transplants, the theme remains central—that immune monitoring is center stage to a successful transplant.

The variability in immune monitoring required across different organ transplants, and the complexity of donor choice and maximizing graft longevity in a sensitized recipient highlights the diversity of clinical situations we face. The chapters on the varied organ transplants depict these differences in a lucid manner.

Algorithms will vary depending on some key considerations. These include the sensitization status of the patient, whether it's a first or a second transplant irrespective of alloimmunisation, ABO compatibility or the lack of it, the organ being transplanted, and the evidence available to drive an algorithmic approach. Central to immune evaluation is our added understanding that while algorithms are ideal, there will be situations where a more personalised approach is required and therefore like in most other clinical fields, individualized precision medicine will have to form part of the decision making process.

Refer to Figs. 21.3 and 21.4 for a practical algorithm for prospective living and deceased donor recipient proceeding for transplantation.

However, there are some must do's within most guidelines. These include for example in a renal transplant setting:

(a) Molecular class I and II typing ideally 11 loci for patient and donor.
(b) High resolution typing if possible—this offers significant advantages in terms of accuracy of virtual crossmatch interpretation, epitope mismatch loads etc.
(c) Antibody testing that detects IgG anti HLA antibodies.

Table 21.1 Components of comprehensive CDC crossmatch (CDCXM) [1]

Components	Methods
T and B cell separation from peripheral blood	• Using Nylon wool column method
	• Using Magnetic beads
	• Using The RosetteSep™ Human T Cell and B cell Enrichment Cocktail
Allo and auto crossmatch with proper tray format	• With Neat serum
	• With dilution up to N:16
	• Using DTT treated serum
Anti-human globulin (AHG) enhancement	Crossmatch should enhanced by adding anti-human globulin (AHG) to the reaction wells with serum and Cells
Reading of CDC crossmatch tray with	• NIH scoring system
	• ASHI scoring system

Table 21.2 Components of comprehensive flow crossmatch (FCXM) [1]

Components	Methods
Lymphocyte separation from peripheral blood	• Using density gradient solution
Preparing negative control and deriving cut-off for positive crossmatch	• Pooling of serum and testing against panel of cells
Allo and auto crossmatch with	• Neat serum
	• Dilution up to N:16
When patient has been treated with Monoclonal antibodies like Rituximab	• Pronase treatment should be done
Interpretation	• Proper analysis and gating of T and B-cells
Quality control	• Inter lab comparison/proficiency testing

Table 21.3 Components of solid phase assays (SPA) [2, 3]

Components	Methods
Preparation of serum	• DTT treatment
	• Heat inactivation
	• EDTA treatment
Non-specific positivity	• Denatured antigens on beads
	• High background values and needs cleaning of serum
Interpretation	• Proper analysis
Quality control	• Inter lab comparison/proficiency testing

(d) At least one solid phase assay should be used to detect and characterize HLA class I and II specific antibodies.
(e) An assay to define HLA-A, B, C, DRB1/3/4/5, DQA1, DQB1, DPA1, and DPB1 antibody specificities.
(f) HLA antibody detection and identification techniques should be able to exclude reactivity attributable to IgM antibodies.
(g) HLA-specific antibodies must be characterized at regular agreed intervals prior to transplantation in sensitized patients and whenever a change in HLA antibody profile is suspected.
(h) For patients on the transplant list, regular samples must be sent to the histocompatibility laboratory for antibody testing. (3 monthly is suggested).
(i) A patient's HLA antibody profile must be assessed to determine the acceptable risk, and delineate the antigens regarded as unacceptable [4].

This is in contrast to the requirements for liver transplants where the guidelines are as below:

(a) Prospective HLA antibody definition is not indicated prior to primary deceased donor liver transplantation.

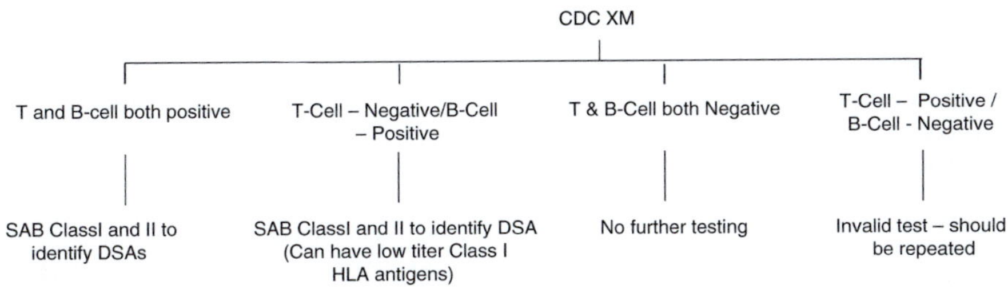

Fig. 21.1 Use of CDCXM along with other platforms. *CDC* complement dependent cytotoxicity, *SAB* single antigen bead assay, *XM* crossmatch, *DSA* donor specific antibodies

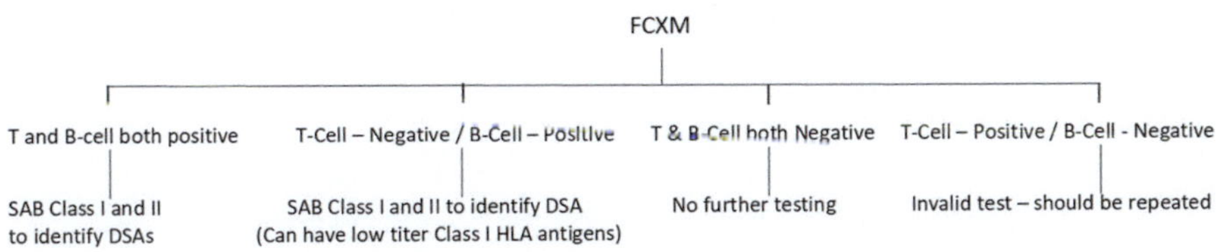

Fig. 21.2 Use of FCXM along with other platforms. *CDC* complement dependent cytotoxicity, *SAB* single antigen bead assay, *XM* crossmatch, *DSA* donor specific antibodies

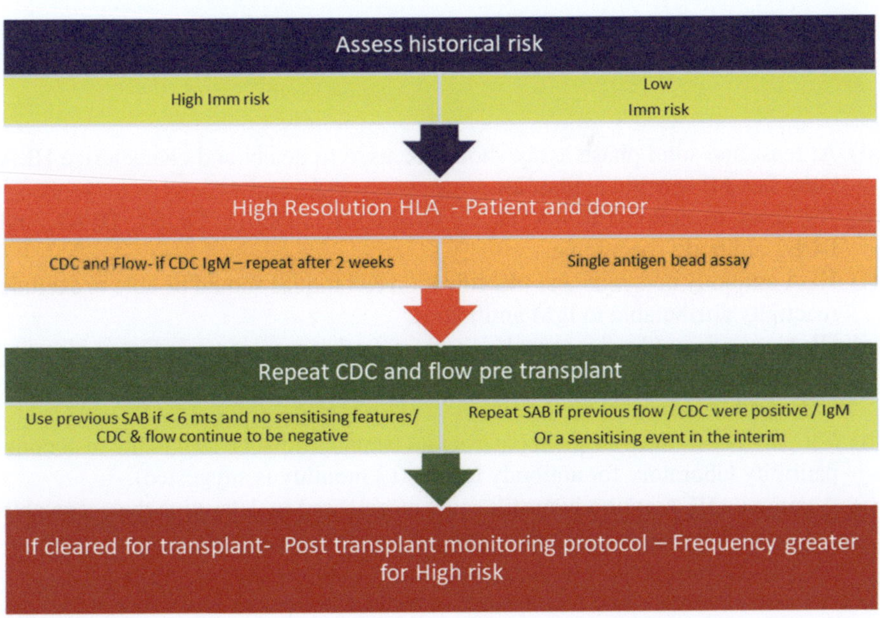

Fig. 21.3 Sample algorithm for a live related transplant

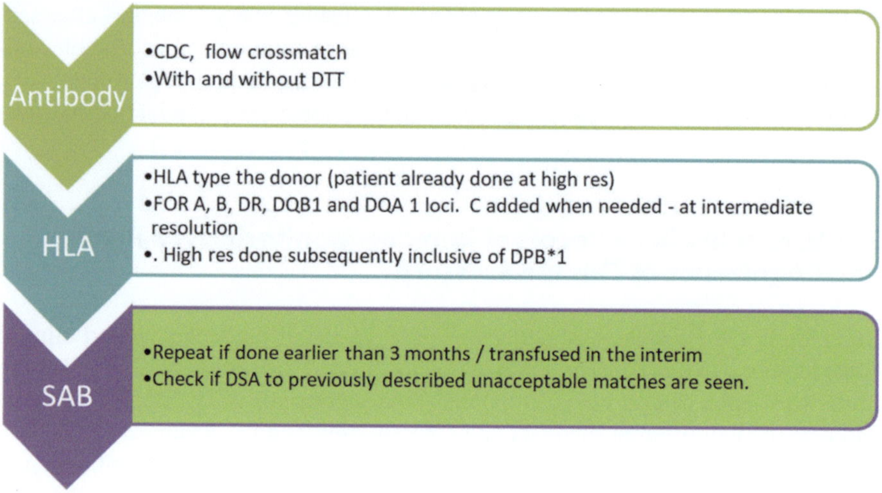

Fig. 21.4 Sample algorithm for patients on a deceased donor list

(b) However, laboratories must store time of transplant samples for liver and SLK transplants (serum and DNA from patient, ideally DNA from donor) to have material to assess baseline HLA antibody levels in the event of development of potential de novo HLA-DSA or declining graft function.

(c) For living donor liver transplantation, especially in paediatric cases, full compatibility testing (HLA antibody screening and typing) of the patient and donor should be undertaken to aid patient management [4].

The American Society for Histocompatibility and Immunogenetics (ASHI) and the European Federation for Immunogenetics (EFI) provide guidelines to standardize and ensure the quality of histocompatibility testing laboratories. These guidelines outline the required testing and assessment protocols for evaluating recipients for transplantation, with a focus on immunogenetics, including HLA typing, antibody screening, and crossmatching [5, 6].

ASHI and EFI Guidelines

- Minimum of low-resolution (antigen-level) typing for HLA-A, -B, -C, -DRB1, and -DQB1 loci for solid organ transplantation. High-resolution (allele-level) typing may be recommended for certain scenarios (e.g., highly sensitized patients, DSA identification, or living donor evaluation).
- Use of SPAs such as Luminex-based bead arrays is recommended over CDC-based methods for sensitivity and specificity for antibody profiling.

- PRA Screening of HLA Class I and Class II at listing, every 3 months, after sensitizing events (e.g., transfusion, pregnancy, transplantation). Unacceptable antigens must be determined based on historical and current antibody profiles.
- Identification of DSA using single antigen bead assays is essential before transplantation. DSA monitoring is recommended post-transplantation.

The Variability in Protocols of Immune Monitoring Is a Reality and Cognizance of This Is Necessary

In addition, the consensus guidelines of the TTS detail the technical and clinical considerations to be taken on board and helps guide broad principles to be kept in mind while each center/region formulates their algorithms of testing [7]. The pros and cons of the different methods and formats used and the technical nitty gritty discussed provide insights on how the combination of an in depth understanding of each assay alongside a comprehensive awareness of the clinical setting is what will enable the development of robust immune monitoring algorithms [7].

Despite good HLA typing methodologies and platforms for Anti HLA antibody testing, rejections have continued to be a challenge. Traditionally the gold standard for diagnosing rejection has been a renal biopsy. The disadvantage of having to repeat invasive procedures alongside the fact that the damage to the graft has already occurred has initiated many to pursue the search for non-invasive markers which will predict rejection before actual damage has occurred [8]. It is in this context that a large array of tests—novel and less invasive tests have emerged. These include blood and urine biomarkers, gene expression, cell-free DNA, molecular imaging, and the use of artificial intelligence [8].

Alongside the tightrope walk that transplant recipients tread with management of their immunosuppression, with graft rejection on one side warring with the risk of infections and malignancy on the other hand is extremely challenging [8].

Combatting this challenge has been a priority for many groups and it has led to the description of many markers that could be used. Though most have not made it into any definitive algorithms, they offer important pathways for both personalized medicine and future research. Some samples of these biomarkers are described below.

The MICA and MICB loci that serve as ligands for the NKG2D receptor and activate natural killer (NK) cells have in this setting gained relevance. Coded for within the MHC and linked closely to the class I gene though they have no role in antigen presentation, anti-MICA antibodies have been shown to be associated with graft rejection [9]. The potential mechanism is the initiation of independent pathways of complement activation. This then leads to vascular thrombosis and graft dysfunction/loss [4]. MICA antigens, capable of activating the humoral and cellular arms, could be a biomarkers in solid organ transplants [10].

Donor derived cell-free DNAs (dd-cfDNA) testing has come into clinical use in many countries. Assessment of this non-encapsulated DNA that may be released following graft injury, this has come into the fray as a biomarker for allograft rejection, the detection being done using single-nucleotide polymorphisms (SNPs) that differentiate donor and recipient cfDNA [11]. Cell-free DNA (cfDNA) has shown to be of use in both predicting rejection and in the diagnosis of both acute cellular and antibody mediated very early in its course [12].

Another minimally invasive method is that of gene expression testing that has been used in the setting of kidney transplant patients. Utilizing the DNA microarray technology, it compares whether a patient's gene expression is similar to a reference population. The reference population consists of a group with adequate immunosuppression and another with inadequate immunosuppression [13].

The review by Song et al. [13] showcases non-invasive biomarkers and their relevance in predicting diagnosing and assessing graft outcomes and injury. Categorizing them as predictive, diagnostic, or prognostic, the role of these biomarkers in enhancing care of transplant patients is highlighted. The value of microRNA (miRNA) has been seen in varied aspects—predicting and diagnosing transplant rejection, as well as in the assessment of graft prognosis.

Biomarkers in a nutshell are evolving and are likely to be part of most future algorithms. They can be of significance either pre or post-transplant or sometime both.

The most significant pre transplant biomarkers are HLA, Preformed donor specific antibodies (DSA) and IL6.

Post-transplant markers include de novo DSA, miRNA, ddcfDNA, neutrophil gelatinase-associated lipocalin (NGAL), chemokine estimation, gene expression profiling (GEP), circulating miRNAs, and Fas ligand (FasL) [13].

Taking the science to a greater height is the use of advanced molecular imaging techniques in the context of monitoring of rejection. Techniques used include fluorescence imaging nuclear imaging technologies including single photon emission computed tomography and positron emission tomography. These allow a dynamic visualization of immune responses within the graft [13].

Critical to note is the complimentary role these markers can play. It is the combined framework of these that will enable tools with a high level of predictive and diagnostic accuracy to be developed.

Ultimately graft longevity is the priority that we look for. Loupy et al. describe an integrative, accurate, and readily implementable risk prediction score for kidney allograft failure. They describe its generalizability across centers and clinical scenarios. Referred to as the iBox risk prediction score, it could be a guide for monitoring of patients and early intervention [14].

Combined with tools and scoring systems such as the iBox risk prediction score biomarkers and molecular markers may add great incremental value in offering personalized care.

Conclusion

The constantly evolving combination of better therapeutics, high end technology, and assays of immune monitoring has impacted positively in the arena of solid organ transplantation both in the short and long term. But as in any space, it is critical that clinical applicability of the varied markers/tests are carefully studied and algorithms made, to ensure that they don't negate a relatively safe transplant.

In addition, management of the "highly sensitized" should take priority and strategies drawn out for this group who otherwise could get marginalized. In developing countries such as ours, sustainability of testing and follow-up strategy and feasibility of implementation should be taken into consideration. Alongside, unique problems that might arise in our population because of the diversity of HLA and local ethnicity unrepresented in assays such as the SAB have to be kept in mind while utilising and interpreting these assays.

Central to ensuring high end care is the multidisciplinary approach, with the clinicians, surgeon, allied health specialists, and the transplant immunologist working alongside each other. The importance of regular discussions and feedback is critical if best algorithms are to be offered to our patients in this constantly evolving field (Fig. 21.5).

A multidisciplinary transplant team, which puts evidence, feasibility, and local availability of resources together, will offer the most practical and sustainable solutions for pre- and post-transplant monitoring. This is crucial as we work towards enhancing graft longevity.

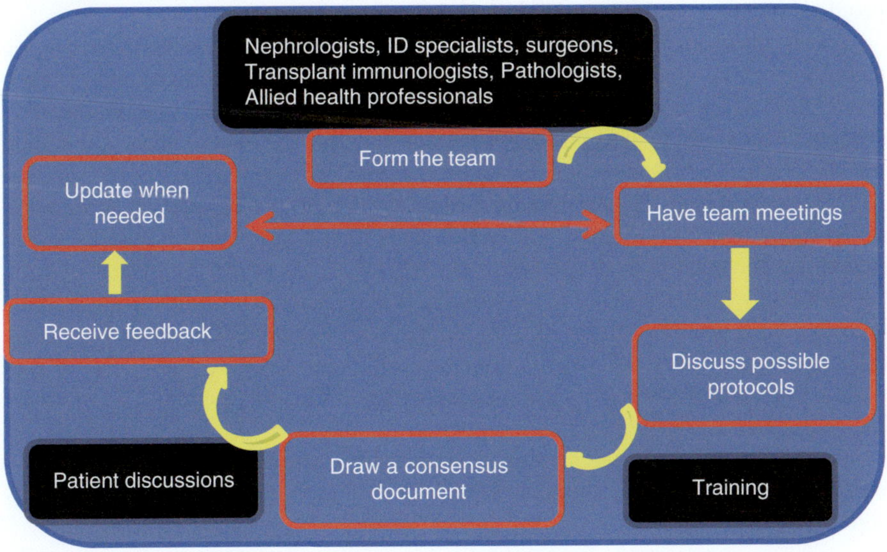

Fig. 21.5 Depicts the working of a multidisciplinary team

References

1. Hahn AB, Land GA, Strothman RM. ASHI laboratory manual. 4th ed. American Society for Histocompatibility and Immunogenetics; 2000.
2. Colombo MB, Haworth SE, Poli F, Nocco A, Puglisi G, Innocente A, Serafini M, Messa P, Scalamogna M. Luminex technology for anti-HLA antibody screening: evaluation of performance and of impact on laboratory routine. Cytometry B Clin Cytom. 2007;72(6):465–71.
3. Tait BD. Detection of HLA antibodies in organ transplant recipients—triumphs and challenges of the solid phase bead assay. Front Immunol. 2016;7:570.
4. Tambur AR, Campbell P, Chong AS, Feng S, Ford ML, Gebel H, Gill RG, Kelsoe G, Kosmoliaptsis V, Mannon RB, Mengel M, Reed EF, Valenzuela NM, Wiebe C, Dijke IE, Sullivan HC, Nickerson P. Sensitization in transplantation: assessment of risk (STAR) 2019 Working Group Meeting Report. Am J Transplant. 2020;20(10):2652–68.
5. Mamode N, Bestard O, Claas F, Furian L, Griffin S, Legendre C, Pengel L, Naesens M. European guideline for the management of kidney transplant patients with HLA antibodies: by the European Society for Organ Transplantation Working Group. Transpl Int. 2022;35:10511.
6. Battle R, Pritchard D, Peacock S, Hastie C, Worthington J, Jordan S, McCaughlan JA, Barnardo M, Cope R, Collins C, Diaz-Burlinson N, Rosser, Foster L, Kallon D, Shaw O, Briggs D, Turner D, Anand A, Akbarzad-Yousefi A, Sage D. BSHI and BTS UK guideline on the detection of alloantibodies in solid organ (and islet) transplantation. Int J Immunogenet. 2023;50(Suppl. 2):3–63. https://doi.org/10.1111/iji.12641.
7. Tait BD, Süsal C, Gebel HM, Nickerson PW, Zachary AA, Claas FHJ, et al. Consensus guidelines on the testing and clinical management issues associated with HLA and non-HLA antibodies in transplantation. Transplantation. 2013;95:19–47. https://doi.org/10.1097/TP.0b013e31827a19cc.
8. Han HS, Lubetzky ML. Immune monitoring of allograft status in kidney transplant recipients. Front Nephrol. 2023;3:1293907. https://doi.org/10.3389/fneph.2023.129390.
9. Fernando JJ, Biswas R, Biswas L. Non-invasive molecular biomarkers for monitoring solid organ transplantation: a comprehensive overview. Int J Immunogenet. 2024;51:47–62. https://doi.org/10.1111/iji.12654.
10. Sapak M, Chrenova S, Tirpakova J, Zilinska Z, Durmanova V, Shawkatova I, Jakus V, Kuba D, Buc M. Donor nonspecific MICA antibodies in renal transplant recipients. Immunobiology. 2014;219(2):109–12. https://doi.org/10.1016/j.imbio.2013.08.006.
11. Gielis EM, Ledeganck KJ, De Winter BY, Del Favero J, Bosmans J-L, Claas FHJ, et al. Cell-free DNA: an upcoming biomarker in transplantation. Am J Transplant. 2015;15:2541–51. https://doi.org/10.1111/ajt.13387.
12. Khachatoorian Y, Khachadourian V, Chang E, et al. Noninvasive biomarkers for prediction and diagnosis of heart transplantation rejection. Transplant Rev-Orlan. 2021;35(1):100590.
13. Friedewald JJ, Kurian SM, Heilman RL, Whisenant TC, Poggio ED, Marsh C, et al. Development and clinical validity of a novel blood-based molecular biomarker for subclinical acute rejection following kidney transplant. Am J Transplant. 2019;19:98–109. https://doi.org/10.1111/ajt.15011.
14. Loupy A, Aubert O, Orandi BJ, Naesens M, Bouatou Y, Raynaud M, et al. Prediction system for risk of allograft loss in patients receiving kidney transplants: international derivation and validation study. BMJ. 2019;366:l4923. https://doi.org/10.1136/bmj.l4923.